Environmental Technologies to Treat Platinum Group Element Pollution: Principles and Engineering

Edited by

Piet N. L. Lens and
Mohanakrishnan Logan

IWA PUBLISHING

Published by **IWA Publishing**
Unit 104–105, Export Building
1 Clove Crescent
London E14 2BA, UK
Telephone: +44 (0)20 7654 5500
Fax: +44 (0)20 7654 5555
Email: publications@iwap.co.uk
Web: www.iwaponline.com

First published 2025

British Library Cataloguing in Publication Data
A CIP catalogue record for this book is available from the British Library

ISBN: 9781789065077 (paperback)
ISBN: 9781789065084 (eBook)
ISBN: 9781789065091 (ePub)

Doi: 10.2166/9781789065084

This eBook was made Open Access in September 2025.

Environmental Technologies to Treat Platinum Group Element Pollution: Principles and Engineering

Contents

Cover images

Centre: Bakubung Platinum Mine, South Africa, iStock

Top left: Microbiologically produced palladium nanoparticles by the bacterium *Shewanella oneidensis*. Photo courtesy of Christopher Egan Morriss, Jinxin Xie and Jon Lloyd, Manchester University, UK

Top centre: Iridium pen nib, © Francis Flinch, https://creativecommons.org/licenses/by-sa/4.0/

Bottom: Osmium crystals, © Heinrich Pniok, http://artlibre.org/licence/lal/en

List of contributors

Aminu Abdullahi
Centre for Biomaterials & Environmental Biotechnology, Department of Biotechnology, Vel Tech Rangarajan Dr. Sagunthala R&D Institute of Science and Technology, Chennai 600062, India

and

Department of Biotechnology, Modibbo Adama University, Yola-PMB 2078, Adamawa State, Nigeria

John Kwame Bediako
Department of Separation Science, School of Engineering Science, Lappeenranta-Lahti University of Technology (LUT), FI-53850 Lappeenranta, Finland

Michael Capeness
School of Biological Sciences, University of Edinburgh, Alexander Crum Brown Road, Edinburgh EH9 3FF, United Kingdom

Rajkuberan Chandrasekaran
Department of Biotechnology, Karpagam Academy of Higher Education, Coimbatore 641021, Tamil Nadu, India

Mariacristina Compagnone
Department of Chemical Engineering and Food Technologies, Wine and Agrifood Research Institute (IVAGRO), Faculty of Sciences, University of Cadiz, Puerto Real, Cadiz, 11510, Spain

Jeba Sweetly Dharmadhas
Department of Biochemistry and Biotechnology, Avinashilingam Institute for Home Science and Higher Education for Women, Coimbatore, Tamil Nadu, 641108, India

Christopher Egan-Morriss
Department of Earth and Environmental Sciences, University of Manchester, Oxford Road, Manchester M13 9PL, United Kingdom

José Joaquín González-Cortés
Department of Chemical Engineering and Food Technologies, Wine and Agrifood Research Institute (IVAGRO). Faculty of Sciences, University of Cadiz, Puerto Real, Cadiz, 11510, Spain

Louise Horsfall
School of Biological Sciences, University of Edinburgh, Alexander Crum Brown Road, Edinburgh EH9 3FF, United Kingdom

Javier Jiménez-Lamana
Institut des sciences analytiques et de physico-chimie pour l'environnement et les matériaux (IPREM), Université de Pau et des Pays de L'Adour, Technopole Hélioparc, 2 avenue du Président Pierre Angot, 64053 Pau, France

Tarangini Korumilli
Centre for Biomaterials & Environmental Biotechnology, Department of Biotechnology, Vel Tech Rangarajan Dr. Sagunthala R&D Institute of Science and Technology, Chennai 600062, India

Selva Kumar T
Centre for Interface & Nanomaterials, Department of Biotechnology, Vel Tech Rangarajan Dr. Sagunthala R&D Institute of Science and Technology, Chennai 600062, India

Piet N. L. Lens
University of Galway, University Road, Galway H91 TK33, Ireland

Mohanakrishnan Logan
Environmental Engineering Division, Department of Civil Engineering, Indian Institute of Technology, Madras, Chennai 600036, India

Talari Madhumanohar
Environmental Engineering Division, Department of Civil Engineering, Indian Institute of Technology, Madras, Chennai 600036, India

Arunabh Meshram
Department of Materials Science and Engineering, Indian Institute of Technology Kanpur, Kanpur 208016, Uttar Pradesh, India

Taslin Mou
School of Biological Sciences, University of Edinburgh, Alexander Crum Brown Road, Edinburgh EH9 3FF, United Kingdom

Rajiv Periakaruppan
Department of Biotechnology, PSG College of Arts & Science, Coimbatore 641014, Tamilnadu, India

Hugues Preud'homme
Institut des sciences analytiques et de physico-chimie pour l'environnement et les matériaux (IPREM), Université de Pau et des Pays de L'Adour, Technopole Hélioparc, 2 avenue du Président Pierre Angot, 64053 Pau, France

Martín Ramírez
Department of Chemical Engineering and Food Technologies, Wine and Agrifood Research Institute (IVAGRO), Faculty of Sciences, University of Cadiz, Puerto Real, Cadiz, 11510, Spain

K. Jagajjanani Rao
Centre for Interface & Nanomaterials, Department of Biotechnology, Vel Tech Rangarajan Dr. Sagunthala R&D Institute of Science and Technology, Chennai 600062, India

Eveliina Repo
Department of Separation Science, School of Engineering Science, Lappeenranta-Lahti University of Technology (LUT), FI-53850 Lappeenranta, Finland

Shinjini Sen
Environmental Engineering Division, Department of Civil Engineering, Indian Institute of Technology, Madras, Chennai 600036, India

Ankur Srivastava
Department of Materials Science and Engineering, Indian Institute of Technology Kanpur, Kanpur 208016, Uttar Pradesh, India

Harish-Kumar Trivedi
Materials Science Programme, Indian Institute of Technology Kanpur, Kanpur 208016, Uttar Pradesh, India

Jinxin Xie
Department of Earth and Environmental Sciences, University of Manchester, Oxford Road, Manchester M13 9PL, United Kingdom

Nang-Htay Yin
Institut des sciences analytiques et de physico-chimie pour l'environnement et les matériaux (IPREM), Université de Pau et des Pays de L'Adour, Technopole Hélioparc, 2 avenue du Président Pierre Angot, 64053 Pau, France

doi: 10.2166/9781789065084_xvii

Preface

Platinum Group Elements (PGEs) are intriguing valuable elements. Their unique properties, which combine resilience and catalytic activity, make them indispensable across a range of industries, from automotive to electronics, jewellery to biomedical implants, and beyond. PGEs are 'critical raw materials' due to their strategic significance in our economies and the high risks associated with their supply chain. Recovery and refinement techniques aimed at enhancing the circularity of the PGEs are essential for their sustainable resource management and the transition towards a circular economy. Next to the circular economy framework, society also requires us to embrace the concepts of enoughness and sufficiency to balance the needs with the limited availability of PGE.

Based on theoretical insights, practical applications, and real-world examples, *Environmental Technologies to Treat Platinum Group Element Pollution: principles and engineering* aims to inspire a paradigm shift toward the sustainable use of these critical elements. The first part of this book covers the biogeochemistry of PGEs. The sources, occurrence, and wide range of applications have been highlighted, showcasing their critical role in modern industries. The state-of-the-art advancements in analytical methodologies aimed at precisely determining PGEs are also presented, including sample pretreatment, chromatographic separation, purification, and innovations in mass spectrometry.

The second part of the book provides a comprehensive discussion of technologies that enable the recovery of PGE from wastewaters and leachates. The integration of cutting-edge techniques, such as adsorption, ion exchange, membrane processes and solvent extraction into existing processes allow their recovery.

Microorganism and plants can also be used in PGE recovery, for example, via biosorption and bioreduction. Alternatively, PGE hyperaccumulators demonstrate their potential application in phytoremediation and phytomining.

The performance, energy efficiency, cost effectiveness and scalability of different processes for PGE recovery are summarised in the third part of the book. These recoveries are demonstrated from synthetic monometallic and mixed metal systems as well as real industrial effluents. Dedicated chapters examine PGE recovery from spent catalysts and electronic waste using state-of-the-art technologies, including mechanical, pyrometallurgical, hydrometallurgical, and biometallurgical processes.

The last part of the book considers the (bio)manufacturing of novel materials from PGE present in waste and wastewater. The chapters show that recovery of PGE from secondary sources is becoming a reality. Equally emerging are PGE nanoparticles due to their widespread applications in the medical and energy sectors. Thus, the present book introduces readers to the sources of PGE emissions up to the recovery of PGE-based value-added products from solid waste and wastewate.

Authored by experts and researchers at the forefront of chemical engineering, analytical chemistry and environmental biotechnology, this comprehensive volume aims to provide an authoritative resource for academics, researchers, industry professionals, and policymakers that will empower readers with knowledge about the latest advancements, challenges, and breakthroughs in the realm of the use and reuse of PGEs. Each contributed chapter is presented on a stand-alone basis, so that the reader will find it helpful to consider the theme of each chapter. There are nevertheless many connections between what may at first seem to be quite different topics. As in all titles in the *Integrated Environmental Technology Series*, one of our purposes was to draw out and emphasise these interdisciplinary linkages. For this reason, a comprehensive index is included to facilitate cross-referencing. We hope that the work described in this book will inspire those working in the field and will encourage those who are beginning to investigate it.

We wish to thank all contributors to the book for their valuable contributions, sharing their expertise in the form of the various chapters included. We also thank all past and present co-workers as well as the collaborators who joined in unravelling different areas of the application of platinum group elements and their application in environmental technology, especially those at National University Ireland Galway and IHE Delft. We would also like to thank all the reviewers who put a lot of effort into improving the quality of this book. In addition, the national and international granting agencies who supported our work on various aspects of metal recovery and green mining over the years are gratefully acknowledged, in particular the research funding provided by Science Foundation Ireland (SFI), who financially supported the open access publication of this book through the SFI Research Professorship Programme

Innovative Energy Technologies for Biofuels, Bioenergy and a Sustainable Irish Bioeconomy (IETSBIO³; grant number 15/RP/2763). We also express our appreciation to the editorial team of IWA Publishing, in particular Katharine Allenby and Andrew Peart, for their help and editorial support in realising this book.

Mohanakrishnan Logan
Indian Institute of Technology Madras, India

Piet N. L. Lens
University of Galway, Ireland

Part I

Environmental technologies for treating pollution by platinum group elements

doi: 10.2166/9781789065084_0003

Chapter 1

Introduction

Mohanakrishnan Logan[1] and Piet N. L. Lens[2]*

[1]Environmental Engineering Division, Department of Civil Engineering, Indian Institute of Technology, Madras, Chennai 600036, India
[2]University of Galway, University Road, H91 TK33, Galway, Ireland
*Corresponding author: mohanlogan@iitm.ac.in

ABSTRACT

The source, occurrence and biogeochemistry of platinum group elements (PGEs) are discussed and the modern applications of these elements in industry are presented. The removal and recovery of PGEs using conventional hydro- and pyro-metallurgy as well as unconventional phytoremediation and bioremediation are overviewed. The 6Rs – reduction, reclamation (removal), reuse, recycling, recovery and rethink – is a way forward to implement the circular economy. A shared vision for PGEs use, guided by the principles of circularity complemented by sufficiency, offers a promising path forward to sustainable use of PGEs.

Keywords: PGMs, pollution, recovery, circular economy, sufficiency

1.1 PLATINUM GROUP ELEMENTS

Platinum group elements (PGEs) or platinum group metals (PGMs) are six rare, noble and precious metallic elements, viz. platinum (Pt), palladium (Pd), rhodium (Rh), iridium (Ir), osmium (Os), and ruthenium (Ru). PGEs find their application in wide range of industrial applications, from biomedical implants (pacemakers and stents) to curing cancer as chemotherapeutic agents, and petroleum refining to electrical devices (Table 1.1). Due to the market risk in their supply chain, PGMs are considered as critical raw materials (CRMs) by many nations including the European Union. Though PGMs tend to be relatively less researched compared to other more abundant metals, they have been receiving traction in recent times and many research and review articles are being published (Figure 1.1). This book volume synthesizes the available

Table 1.1 Primary applications of PGEs.

Element	Chemical Symbol	Primary Applications
Platinum	Pt	Catalytic converters, jewellery, electronics, hydrogen production, fuel cells, petroleum refining, production of pharmaceuticals and bulk chemicals (ammonia, alkylation and chlorination processes), biomedical implants, electrolysis and electroplating
Palladium	Pd	Catalysts (mainly in automotive catalytic converters), electronics (surface-mount devices and circuit boards), hydrogen storage and fuel cells, jewellery, pharmaceuticals, specialty glasses, dental alloys, electrical contacts, switches and connectors
Rhodium	Rh	Catalysts (used in automotive catalytic converters), production of aldehydes in petrochemical industries, jewellery, electrical contacts, and hydrogenation
Ruthenium	Ru	Electronics (electrodes, resistors), catalysts, chemical industry, solar cells and dental alloy
Iridium	Ir	Spark plugs, aerospace components, catalysts, electrical contacts and hardening agents
Osmium	Os	Alloys (used in electrical contacts and in fountain pen tips), chemical catalysts, and specialized scientific (nuclear, radiation, mass spec and electron microscopy) instruments

knowledge on relevant topics in the area of source, extraction, analytics, removal and recovery of PGEs (Figure 1.2).

PGEs emission to air occurs via exhaust fumes, into the soil through PGE-based fertilizers, and into hospital wastewater from PGE-based pharmaceuticals. The deposition of these elements resulting in progressive accumulation of PGMs in the surrounding ecosystem is a concern. For example, PGMs presence has been confirmed in the environmental samples such as drainage from roadways. PGMs as pollutants are a threat to public health since they are reported to be highly toxic, allergic and can even damage DNA. Owing to their bioavailability, PGMs can enter the food chain and cause respiratory sensitization, allergic reactions, dermatitis and asthma (Kalavrouziotis & Koukoulakis, 2009). A significant bioaccumulation of PGEs in human tissues/hair indicates under-studied biogeochemical transformations and posing potential human risks.

1.1.1 Platinum, palladium and rhodium

The book presents in greater detail about the abundant Pt, Pd and Rh, which are significantly produced in the market and treated/recovered. Detection of trace concentrations of these three PGMs is challenging in analytical equipment (Chapter 3). The predominant applications of PGEs (especially Pt, Pd and Rh) are in the production of automobile catalyst to reduce harmful emissions (such as HC, CO and NO_x) from vehicle exhaust systems. Therefore, a significant share of residual PGEs comes from the resulting spent catalysts (Chapter 8).

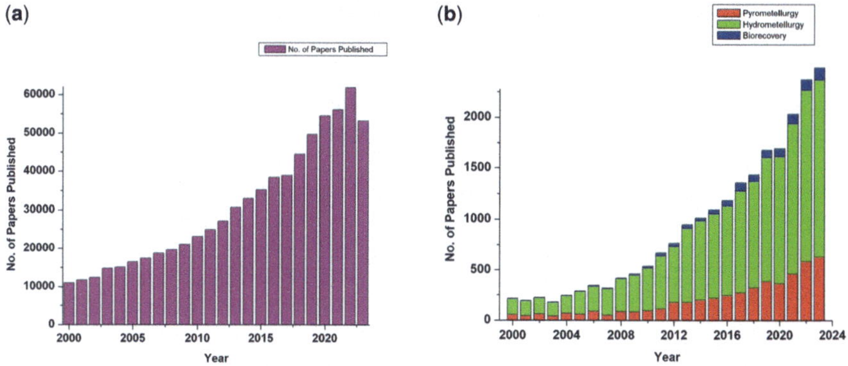

Figure 1.1 Increase in number of publications over the last two decades on (a) PGEs and (b) PGE recovery from waste and secondary resources using different methods (Source: Google Scholar; accessed 27 March 2025. Data obtained by searching using keywords (a) 'platinum group elements' and (b) 'platinum group elements, and the respective figure legend').

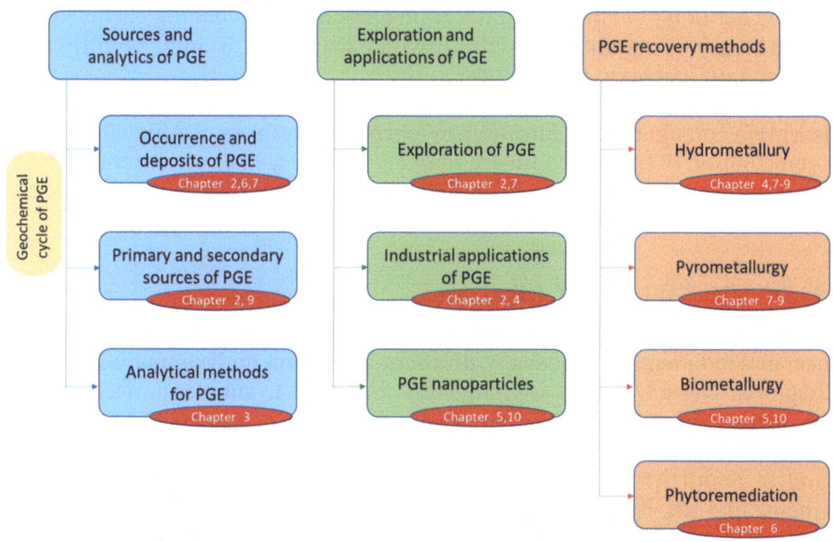

Figure 1.2 Schematic presentation of the topics covered in this book volume.

1.1.2 Iridium, osmium and ruthenium

Iridium (Ir) is a rare, dense element that is less reactive and less toxic than the other PGEs. Ir is used as a catalyst, including in the production of chemicals like acetic acid, hydrogenation processes and the automotive industry for catalytic converters. Ir is used in the manufacturing of electrical and aerospace components. Ir is typically obtained as a byproduct of platinum and nickel mining. Ir compounds, such as iridium chloride or iridium oxide, can be more reactive and toxic. These compounds can cause skin irritation, respiratory issues, and environmental damage if they are released into ecosystems. Iridium-192 needs specialized radioactive handling procedures (Iavicoli & Leso, 2021).

Osmium (Os) is used in various industrial processes, such as in the production of electrical contacts, catalysts, and in the aerospace and automotive industries. Osmium tetroxide (OsO_4) is a powerful oxidizing agent and can be highly toxic, even in small quantities. Similar to the compounds of Ir (Iridium tetrachloride/trichloride) and Ru (Ruthenium tetroxide), it is highly corrosive and can cause severe damage to the eyes, skin and respiratory system. Osmium tetroxide spills can be neutralized with a reducing agent, such as sodium thiosulfate, to minimize its toxic effects.

Ruthenium tetroxide (RuO_4) can be highly toxic and pose significant environmental and health risks. It is a volatile and highly reactive compound, and if inhaled can cause damage to the lungs and mucous membrane. Ruthenium is used as a catalyst in chemical processes, electrical and electronic components and as automotive catalytic converters. Ruthenium is obtained as a byproduct of platinum and nickel mining. Ruthenium tetroxide is highly reactive and a common method of neutralizing RuO_4 involves using a reducing agent, such as sodium bisulfite or sodium thiosulfate, which can safely decompose ruthenium tetroxide into less toxic compounds.

Treatment of these three metals from aqueous solution generally includes adsorption, precipitation, filtration or chemical methods. Unlike Pt and Pd, bioremediation of Ir, Os and Ru are scarcely reported. The role, interaction and underpinning mechanisms pertaining to Ir, Os and Ru are largely untapped, implying the scope for their research and development.

1.2 BIOGEOCHEMISTRY OF PGMs

The presence of PGM in nature is linked to geological processes, which includes placer and sedimentary deposits, hydrothermal systems, and magmatic intrusions (Chapter 2). Notably, PGMs are frequently connected to sulfides and tellurides of base metals. The diverse PGM abundances and patterns of dispersion in various geological settings reflect the variety of PGM in nature. PGMs are extensively found in South Africa, Russia, United States, Canada and Zimbabwe. While elements like Rh, Ru, Ir and Os are uncommon and frequently found in lesser quantities, Pt and Pd are comparatively more plentiful and broadly dispersed. PGEs pose characteristics such as durability, resistance to corrosion, high melting point and catalytic properties (Hughes *et al.*, 2021). Automotive, jewellery and electronics serve as the major market for recovering PGMs through recycling (Chapter 2). Despite the benefits accrued from such open loop recycling, due to the higher cost and complex recycling technique, the majority of PGMs are currently realized from primary sources only (Brown *et al.*, 2024).

The inherent potential of PGEs to undergo intricate transformations in environmental matrices due to their diverse oxidation states and having multiple stable/radiogenic isotopes mandates accurate measurements (Chapter 3). PGEs analyses are crucial to not just monitor environmental/anthropogenic changes and assess health risks, but also to have a comprehensive understanding of geological processes and exploration of valuable mineral deposits. Extremely low concentrations of PGEs pose challenges for the detection limit of even advanced analytical techniques. Sample pretreatment techniques such as fused sample pellet, closed or fusion digestion is a requisite for PGEs analyses to ensure accurate, precise and replicable quantification. In chromatographic column separation, the unique distinct feature of PGEs to form strong anionic chloro complexes in mineral acids, unlike transition metals (like alkali alkaline earths) that form weaker anionic or stable cationic complexes, allows for effective separation of PGEs from matrix elements. Single spiking or isotopic dilution serves as a powerful method for the concentration measurement of PGEs, meanwhile double spiking is employed for the precise measurement of stable isotopic ratios of PGEs. Recent advanced techniques, such as inductively coupled plasma mass spectrometry (ICP-MS) in MS/MS mode and neutron activation analysis, excel in detecting PGEs at ultra-low concentrations (μg/L, ng/L, even pg/L) with high selectivity and sensitivity (Chapter 3).

1.3 CIRCULAR ECONOMY

The conventional 'linear economy' (extract-use-dispose) has resulted in anthropogenic urban waste (such as waste electrical and electronic equipment and spent automotive catalysts) with higher PGEs content than many primary ore deposits. The concept 'urban mining' emphasizes the recovery and recycling of PGEs, and the need for a shift towards a circular economy (CE) focused on the sustainable management of PGEs over their entire lifecycle (Cossu, 2013). In this regard, PGMs recovery is crucial from mine tailings to PGE-laden wastewaters. The challenges of conventional mining as well as challenges (such as economic factors, market volatility, technological advancement and process complexity, energy intensity, carbon footprint, environmental and societal impacts and geopolitical factors) in the supply chain of PGMs, has necessitated the need for alternative supply sources.

The European Commission (EC) defines that CE is a system where the value of products, materials and resources is maintained in the economy for as long as possible, and the generation of waste is minimized. CE can be achieved through sustainable management of PGE resources and more rational disposal and processing of PGE-laden waste/wastewater. A possible CE model framework includes the 6Rs (Figure 1.3): Reduction (of PGE usage and pollution reduction at source), Reclamation/removal (through application of effective technologies for the removal of PGE pollutants), Reuse (of reclaimed PGE), Recycling (recovery of PGE from waste/wastewater), Recovery (of resources from PGE-laden waste/wastewater) and Rethinking (how to use resources to create a sustainable economy free of waste and emissions). This CE model framework integrates all possible ways of implementation of CE principles and by including technological, organization and societal changes (Smol *et al.*, 2020).

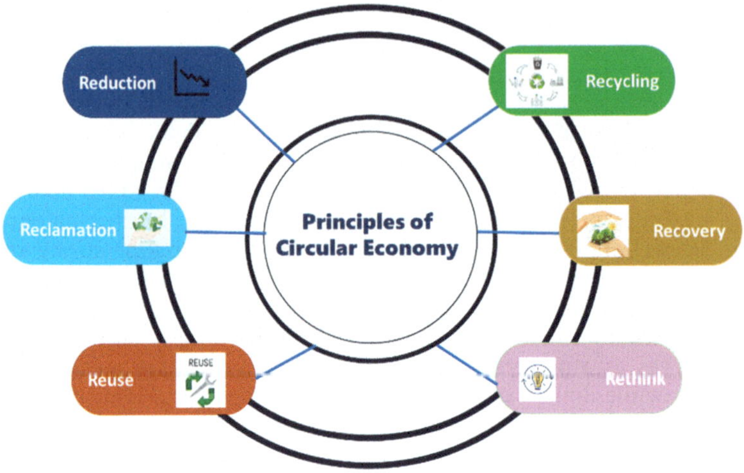

Figure 1.3 6R principles of CE model framework of PGEs.

Sufficiency – securing enough for a good life whilst reducing total resource consumption – is essential to complement circularity and create sustainable and equitable PGM value chains (Pereira *et al.*, 2025). Sufficiency involves balancing extraction to meet basic needs and wellbeing, while aligning with the ecological ceiling of non-renewable resources and upholding intergenerational equity (Figure 1.4). Sufficiency measures can bridge the inequality gap (for instance, between global north and global south) by setting clear consumption limits based on the principle of common, but differentiated responsibilities and capabilities. For example, PGM sufficiency strategies for cars might be modelled after suggestions to curb carbon emissions and include using PGM per capita metrics instead of PGM per geographical area. This could result in reducing the number of cars per capita and promote affordable public transportation. Another way to support systemic demand reduction for PGE jewellery would be use of low-cost alternate materials. Policymakers should also prioritize better stewardship of finite PGE resources over downcycling practices.

1.4 TECHNOLOGIES FOR PGE RECOVERY

The popular approaches for PGMs recovery are the traditional hydrometallurgical and pyrometallurgical approaches, as well as the emerging biotechnological approaches (Figure 1.5). Hydrometallurgical processes rely on chemical reactions in aqueous solutions. Hydrometallurgy offers advantages such as selective extraction and the ability to process complex waste materials; however, corrosive liquid waste is often an issue. A typical hydrometallurgical process generally involves pretreatment and leaching stages where the PGMs are simultaneously dissolved from their matrices and complexes are formed, followed by separation of the dissolved PGMs from the leach solution and

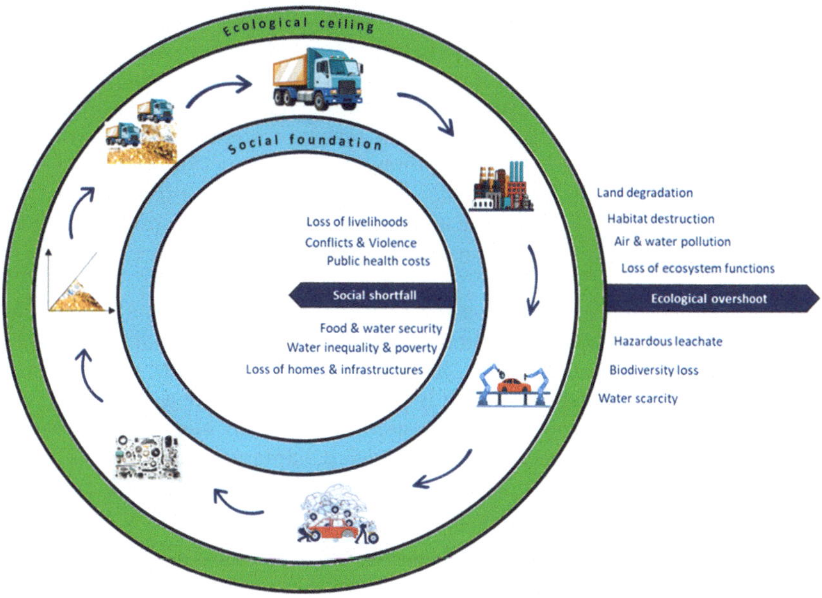

Figure 1.4 PGM sufficiency – the figure illustrates how PGM value chains must operate between an ecological ceiling and a social foundation (modified from Pereira *et al.*, 2025).

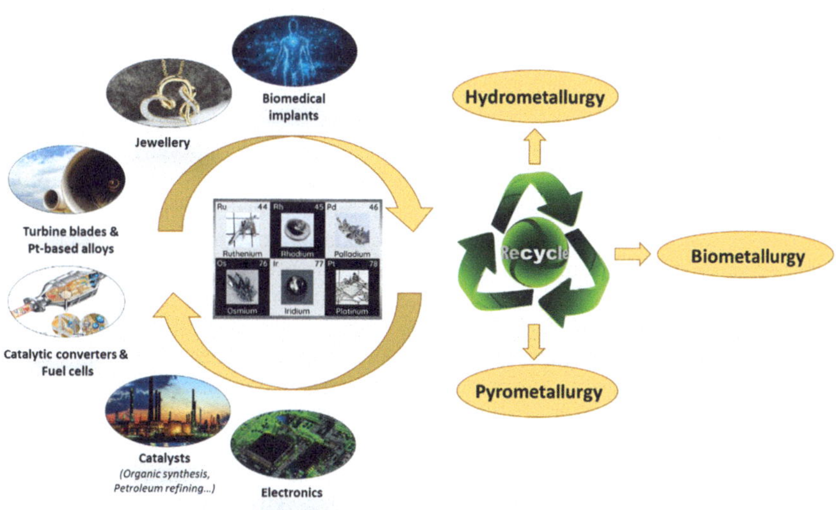

Figure 1.5 Recycling approaches for PGMs.

subsequent purification of PGMs (Chapter 7). Examples of hydrometallurgical processes are adsorption, solvent extraction, solid-phase extraction, chemical precipitation and ion exchange. Adsorption is considered a primary means of recovering PGEs through the interaction of these element ions with the surface of adsorbent materials (Chapter 4). Recent research has tried to explain how different adsorbents, metal-organic frameworks and functionalized polymers interact with PGEs, and the basic principles behind those interactions (MacDonald *et al.*, 2024). The environmental factors, mechanisms, case studies as well as kinetics and thermodynamics of PGE adsorption are reviewed in Chapter 4.

Pyrometallurgical processes are the techniques used to concentrate PGMs by physical and chemical conversion at high temperatures (Chapter 7). These processes involve the application of heat to transform raw PGM materials into metal-rich products. Pyrometallurgy is energy-intensive, yet a traditional and widely employed primary method for metal extraction due to its efficiency and applicability to various types of ores (Chapter 9). Occasionally, the pyrometallurgical process is complemented with hydrometallurgical processes to obtain high-purity metals (Chapter 7). The recent electrometallurgical techniques involve the use of electrical energy to extract and purify PGMs from leachates or other solutions. This process can help separate PGMs from base metals and other impurities, resulting in higher purity metals suitable for industrial use (Chapter 8).

Biorecovery can recycle PGEs without the use of high temperature or harmful chemicals by leveraging the capabilities of microorganisms to bind, accumulate and transform PGE ions from solution into recoverable forms. Biotechniques can be an alternative or complementary process to conventional recovery methods. Chapter 5 covers a detailed review on PGE biorecovery from synthetic monometallic systems, mixed metal solutions, and real industrial effluents. Biosorption involves the adsorption of PGE ions onto biomaterials, whereas bioreduction converts soluble metal ions as insoluble and recoverable metal nanoparticles or metal alloys. Very few studies have extensively investigated microorganisms capable for PGM recovery. *Saccharomyces cerevisiae*, *Escherichia coli*, *Aspergillus*, *Galdieria sulphararia*, *Rhodopseudomonas palustris*, *Pichia pastoris* are reported for biorecovery of PGMs from real mixed metal industrial effluents (Chapter 5). Challenges such as low specificity, competition from contaminant metals in biosorption and complexity and variability of industrial effluents still hinder industrial application.

Phytoremediation and phytomining have emerged as cost-effective and environmental-friendly approach. Phytoremediation involves hyper-accumulating plants and the microorganisms in the rhizosphere to immobilize, translocate or breakdown contaminants found in soil, water, and the surrounding environment. On the other hand, phytomining refers to a process where bioavailable metal species are absorbed and concentrated from the surrounding environment into the plant biomass. After harvesting, PGE can be recovered and the plant biomass can generate biogas through anaerobic digestion or electricity through incineration. Chapter 6 comprehensively reviews the plant species that are capable of efficiently remediating Pd and other PGEs. Successful case studies and best practices are presented in Chapter 6 as well.

Pretreatment of the identified raw feed to increase the PGE concentrations is crucial to recycling PGE. Due to the complex nature of the recycling processes, the separation of valuable resources is a challenge and there is a wide scope of investigating alternative methods for PGE recycling. The scaling up of the existing and newer processes requires extensive technological and economic analysis to ensure their feasibility (Chapter 9).

1.5 MANUFACTURING OF PGM-BASED NANO MATERIALS

Metallic nanoparticles have unique optical, electrical, magnetic and catalytic properties, and find their application in advanced modern industries. Utilization of PGM can be maximized by applying nanoparticles, as lower amounts are required for achieving desired outcomes due to their increased surface-to-volume ratio. PGM nanoparticles can be synthesized using flame pyrolysis, vapour deposition and mechanical milling. These energy-intensive conventional processes can be replaced by green and cost-effective biological approach. Chapter 10 reviews in detail the different types of bacteria that can synthesize nanoparticles of PGMs, especially platinum and palladium.

REFERENCES

Brown S., Cole L., Cowley A., Dickson J., Fujita M., Girardot N., Grant S., Jiang J., Raithatha R., Ryan M., Underwood M. and Wang A. (2024). *PGM market report*, 1–48, retrieved from internet: https://matthey.com/media/2024/johnson-matthey-publishes-latest-pgm-market-report-2024

Cossu R. (2013). The urban mining concept. *Waste Management*, **33**(3), 497–498, https://doi.org/10.1016/j.wasman.2013.01.010

Hughes A. E., Haque N., Northey S. A. and Giddey S. (2021). Platinum group metals: a review of resources, production and usage with a focus on catalysts. *Resources*, **10**(9), 1–40, https://doi.org/10.3390/resources10090093

Iavicoli I. and Leso V. (2021). Iridium. In: G. F. Nordberg and M. Costa, M. (Eds.) Handbook on the Toxicology of Metals: volume II: Specific metals (5th ed., pp. 360–390). Academic Press, London, United Kingdom.

Kalavrouziotis I. K. and Koukoulakis P. H. (2009). The environmental impact of the platinum group elements (Pt, Pd, Rh) emitted by the automobile catalyst converters. *Water, Air, and Soil Pollution*, **196**, 393–402, https://doi.org/10.1007/s11270-008-9786-9

MacDonald L., Zhang D. and Karamalidis A. (2024). Platinum group metals: key solid phase adsorption technologies for separation from primary and secondary sources. *Resources, Conservation and Recycling*, ***205***, 107590, https://doi.org/10.1016/j.resconrec.2024.107590

Pereira K., Wabnitz C. C. C., Schildt L., Kuiper J. J., Schmitt R. J. P., Barbour F. and Jouffray J.-B. (2025). Rethinking sand circularity through sufficiency. *One Earth*, ***8***, 101207, https://doi.org/10.1016/j.oneear.2025.101207

Smol M., Adam C. and Presiner M. (2020). Circular economy model framework in the European water and wastewater sector. *Journal of Material Cycles and Waste Management*, **22**, 682–697, https://doi.org/10.1007/s10163-019-00960-z

Part II

Biogeochemistry of platinum group elements

doi: 10.2166/9781789065084_0015

Chapter 2

Platinum group metals: sources, occurrence and applications

Selva Kumar T[1], Rajiv Periakaruppan[2], Aminu Abdullahi[3] and Rajkuberan Chandrasekaran[4]*

[1]Vel Tech Rangarajan Dr. Sagunthala R&D Institute of Science and Technology, Chennai 600062, Tamil Nadu, India
[2]Department of Biotechnology, PSG College of Arts & Science, Coimbatore 641014, Tamil Nadu, India
[3]Department of Biotechnology, Modibbo Adama University, Yola-PMB 2078, Adamawa State, Nigeria
[4]Department of Biotechnology, Karpagam Academy of Higher Education, Coimbatore 641021, Tamil Nadu, India
**Corresponding author: selvathirumalaisamy@gmail.com*

ABSTRACT

This chapter overviews the sources, occurrence, and wide range of applications of platinum group metals (PGMs). It explains how PGMs travel from their natural sources to a variety of technological applications by looking at geological processes and human interventions. The importance of these metals in electronics, clean energy, and catalysis has been highlighted, showcasing their critical role in modern industries. The chapter emphasizes the significance of an in-depth awareness of sustainable resource management and technological advancement as we explore the challenges of sourcing and using PGMs.

Keywords: PGMs, sources, occurrence, catalyst, automotive, electronics, jewellery

2.1 INTRODUCTION

The platinum group metals (PGMs), prized for their exceptional qualities and wide range of uses, have a prominent place in the vast field of contemporary industry and technology. They are the six d-block metals from periods 5 and 6 as well as groups 8, 9, and 10. The elements include platinum (Pt), ruthenium (Ru), rhodium (Rh), palladium (Pd), osmium (Os), and iridium (Ir) as shown in Figure 2.1. These elements share similar chemistry, which is based on either 5d (Os, Ir, and Pt) or 4d electron shells (Ru, Rh, and Pd). PGMs are occasionally

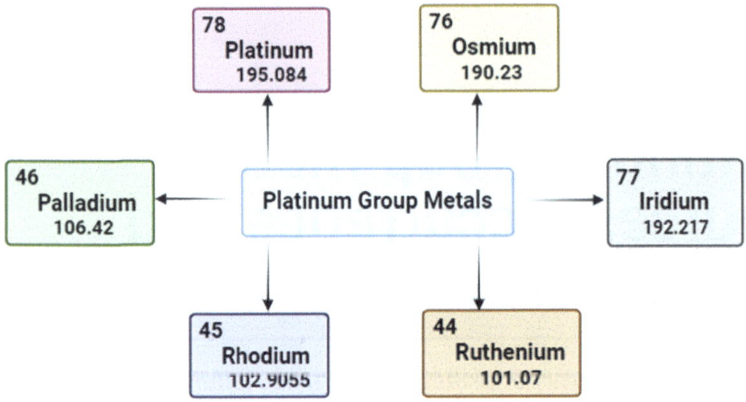

Figure 2.1 Overview of PGMs.

divided into two categories: light PGMs (Ru, Rh, Pd) and strong PGMs (Os, Ir, and Pt), which correspond to their respective atomic numbers (Hughes *et al.*, 2021). PGMs are used as the catalyst's active species in industrial applications because of their catalytic properties and thermal stability. There are several PGM-based catalysts on the market that may be used for different applications. Consequently, PGMs are sometimes referred to as 'high-technology metals' (Dong *et al.*, 2015).

Concerns over the PGM's sustainability have been raised by their high demand and limited supply. For example, PGMs are on the European Commission's list of 'critical raw materials'. All six metals have an extensive number of commercial and scientific applications. The explosive industrial growth is predicted on the quick use of renewable resources, which also produce trash to allow for metal recycling (Zheng *et al.*, 2021). Consumer demand for platinum and palladium has been significant in recent years, but rhodium has been of low interest. The American Mint and other mints throughout the world produce platinum bullion coins in a range of sizes, ranging from 1/10 troy ounce to one ounce, whereas the Royal Canadian Mint has produced three palladium coins. Both metals have traditionally been marketed as ingots for sale (McCown & Shaw, 2017).

The objective of this chapter is to present a thorough analysis of PGM distribution and origin, exploring their geological sources and worldwide abundance. It also aims to clarify the complex web of their uses, explaining their essential function in important sectors, including jewellery, electronics, and automotive development, as well as their growing relevance as catalysts in the field of sustainable chemistry. This chapter also aims to provide an integrated view of the role played by PGMs in influencing modern technological innovations, and their possible consequence for sustainable development and creative industrial practices by examining the interaction between their geological occurrence and the current industrial landscape.

2.2 SOURCES OF PGMs

2.2.1 Natural PGM sources

The origin of PGMs has long captivated the attention of both geologists and scientists. It is essential to comprehend the distribution of these uncommon components for both research and practical uses. The several regions in which PGMs are found due to the geological formations have various extraction techniques with limit to the geology. These regional origins also affect international trade and supply networks. Due to their distinct geological origin, PGMs are rare and have a high market value (Cabri *et al.*, 2022). Mafic and ultramafic igneous rocks, including those found in the United States, Zimbabwe's Great Dyke, and South Africa's Bushveld Complex are frequently associated with these elements. Fractional crystallization and sulfide segregation are two intricate magmatic processes that are strongly related to the production of PGM deposits (Glaister & Mudd, 2010; Oberthür *et al.*, 2013).

One of the most layered igneous intrusions in the world, the Bushveld Complex, is a significant source of PGMs, particularly palladium and platinum. Significant PGM deposits may be found in this layered mafic to ultramafic complex in South Africa together with sulfide ores. Similar to this, the Great Dyke in Zimbabwe is well known for having abundant PGM deposits that are contained in layered mafic intrusions, especially platinum and palladium. Another important source of PGMs is the Still Water Complex in Montana (USA), which has reserves of platinum group elements (Woodall *et al.*, 2021).

2.2.2 Technological developments in extraction

PGM extraction from their individual geological origin requires advanced and intricate methods. To remove ores containing PGM from the Earth's crust, conventional mining techniques including open-pit and underground mining are frequently used. However, because of the complicated mineralogy of PGM ores and their interactions with other precious and non-value minerals, the extraction procedure is sometimes difficult (Sefako *et al.*, 2017). Technological developments have resulted in the creation of novel extraction techniques, such as froth flotation, smelting, and hydrometallurgical processes, in addition to conventional mining methods. By effectively separating and extracting PGMs from complicated ore bodies, these cutting-edge techniques help to lower environmental impact and boost production (Sefako *et al.*, 2019).

2.3 OCCURRENCE OF PGMs

2.3.1 Diversity of PGMs in nature

The presence of PGM in nature is mostly linked to certain geological processes, which frequently includes sedimentary deposits, hydrothermal systems, and magmatic intrusions (Figure 2.2). One of the main sources of PGM is magmatic intrusions, especially those that occur in layered mafic intrusions. PGM concentrations are high in these intrusions, which are found in a few places across the world, such as the Bushveld Complex in South Africa, the Great Dyke in Zimbabwe, and the Stillwater Complex in the United States.

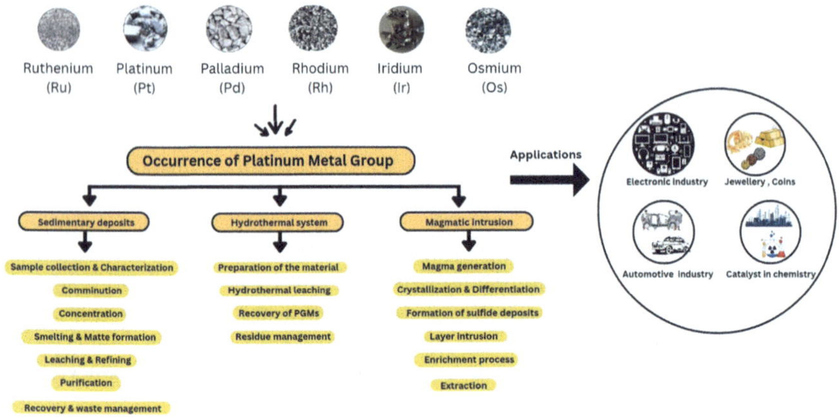

Figure 2.2 Occurrence and applications of PGMs.

These intrusions are the result of the gradual crystallization of magma, which separates PGM-rich alloys and sulfides into separate strata inside igneous rocks (Maier, 2005).

High PGM concentrations are also a result of hydrothermal systems, which are defined by the movement of heated aqueous solutions inside the crust of the Earth. PGMs are frequently found in ore deposits created by hydrothermal processes and are usually connected to tellurides and sulfides of base metals (Shamaila & O'Connor, 2008). Examples of these deposits are the Sudbury Basin in Canada and the Norilsk-Talnakh deposits in Russia. These deposits can be found in a variety of geological settings, including volcanic arcs, rift zones, and subduction zones. Even though they are less frequent, sedimentary deposits are a significant source of PGM. PGMs are frequently found in placer deposits, which are created when heavy minerals are mechanically concentrated in riverbeds and alluvial fans. These deposits are generally linked to gold and are rare; examples can be found in some parts of America and the Russian Ural Mountains (Zaccarini *et al.*, 2018).

The diverse PGM abundances and patterns of dispersion in various geological settings reflect the variety of PGMs in nature. While elements like rhodium, ruthenium, iridium, and osmium are uncommon and frequently found in lesser quantities, platinum and palladium are comparatively more plentiful and broadly dispersed. Furthermore, certain elements exhibit different enrichment or depletion in particular geological environments due to differences in their geochemical activity.

2.3.2 PGMs extraordinary qualities

PGMs are considered to be of very high quality because of their strong physical and chemical properties. PGMs have exceptional catalytic qualities that allow for a wide range of applications in vehicle catalytic converters. These metals efficiently lower the negative effects of exhaust fumes on the environment by facilitating the transformation of toxic pollutants into less toxic or harmless compounds (Murray *et al.*, 2017). This group of metals also has extraordinarily high melting points, which makes them perfect for uses where exposure to high temperatures is expected. Because of this feature, they are essential in the manufacturing of components and machinery that can withstand high temperatures used in scientific research and industrial operations (Couderc, 2010). PGMs, particularly palladium and platinum, exhibit exceptional resistance to oxidation and corrosion, even in adverse environmental circumstances. Because of their extraordinary resilience, they are essential in the production of components that resist corrosion, such lab apparatus, electrical connections, and some industrial gear (Li *et al.*, 2022). Due to their excellent electrical conductivity, platinum and palladium are prized commodities in the electronics sector. Because of their effective electrical conductivity, PGMs are used to make electrodes, electrical contacts, and other electronic devices (Zhao *et al.*, 2023). PGMs may be easily formed into complex structures because of their exceptional malleability and ductility, especially platinum, palladium, and rhodium. Because of this characteristic, they are in great demand in the jewellery business, where they are utilized to build beautiful and long-lasting jewellery. PGMs, especially rhodium, palladium, and platinum, have a uniqueness that adds to their attraction when used to make luxury goods and jewellery. Their desirable beauty and rarity make them sought-after materials for the production of priceless and distinctive decorations. Even in harsh environments like high pressure and severe radiation, PGMs retain their structural integrity and stability. Because of this quality, they are ideal for a wide range of industrial and scientific applications where materials that can survive harsh conditions are required (Rakhtsaum, 2013).

2.3.3 Multidimensional structures

PGMs have an amazing variety of multidimensional structures, which frequently contribute to their special qualities and uses. These structures are essential for characterizing the properties and applications of PGMs in many domains. At the core of PGM's extraordinary features and applicability across several disciplines are their multidimensional architectures and adaptable chemistry. Their capacity to adapt to various structural configurations and build compounds with a broad variety of components makes them important and valuable in the field of science and technology (Cullen *et al.*, 2016).

In their solid state, PGMs frequently form crystalline lattice patterns. Atoms or ions are arranged in regular, repeating patterns within these formations. For instance, rhodium crystallizes in a hexagonal close-packed (HCP) structure, whereas platinum and palladium form a face-centred cubic (FCC) lattice. The metals' high melting points and resistance to deformation – two important characteristics that make them useful in high-temperature and high-stress environments – are a result of the arrangement of atoms in these structures (Junge *et al.*, 2015).

PGMs have a noteworthy ability to combine with other elements to produce alloys. When compared to particular metals, the alloys produced via PGMs frequently have special qualities. For instance, Pt–Rh alloys' consistent temperature-dependent electrical resistance makes them a popular choice for thermocouples. Pd-based alloys are used in hydrogen storage applications as well as dental products. PGM's flexibility is demonstrated by their ability to mix with various metals in a number of forms (Rakhtsaum, 2013).

PGMs are renowned for their capacity to combine with ligands, molecules, or ions to generate coordination complexes. Because of these substances' catalytic qualities, they have been used in a variety of chemical processes. PGMs with stable coordination compounds improve their catalytic efficacy and selectivity in a variety of chemical reactions (Rao & Reddi, 2000). Certain PGMs can display multidimensional crystal forms that change according to the pressure and temperature levels. For example, under high pressure, palladium can change from having a FCC structure to a body-centred cubic (BCC) structure. These phase transitions have consequences for material research and high-pressure applications (Yusenko *et al.*, 2017).

PGMs can be found in layered structures in particular kinds of mineral deposits in geological environments. These layered intrusions, like those found in South Africa's Bushveld Complex, are composed of alternating layers of silicate minerals and sulfides rich in PGMs. The mining and extraction of PGMs are significantly aided by the stratified nature of these deposits (Cawthorn *et al.*, 2002). PGMs and other elements can combine to generate a variety of intermetallic complexes. These substances have distinct structures and qualities, which frequently help explain why they are used in niche markets. PGM–tin intermetallics, for instance, have demonstrated potential in superconductors and electronic materials (Su *et al.*, 2023).

2.3.4 PGMs in alloys and compounds

PGMs are distinctive in their chemical characteristics and occur naturally in a wide range of alloys and compounds. Numerous industries, including the automotive, electronics, ornaments, and chemical production sectors, use these compounds and alloys. In some geological environments, PGMs form alloys combining other elements, such as iron, nickel, and copper, when they occur naturally. For instance, the nickel–copper–PGM ore resources in Canada's Sudbury Basin are well-known. These alloys are essential to the mining and recovery of PGM because they modify the chemical and physical characteristics of the ore, which impacts the methods used by the mining sector to extract PGM (Xiao & Laplante, 2004). PGMs combine with other transition metals including cerium, lanthanum, and zirconium to make alloys that are widely utilized in catalytic converters in the automobile sector. These alloys function as catalysts, helping to convert hazardous contaminants in car emissions into less dangerous ones. These PGM alloys' special catalytic qualities are essential to environmental preservation since they greatly lower hazardous exhaust emissions (Omodara *et al.*, 2019).

In the electronics sector, PGMs, especially palladium and platinum, are used to make electrodes, connections, and electrical contacts. These alloys are useful for a variety of electrical applications because they may be mixed with other metals like copper, silver, and gold to increase their durability and conductivity. These PGM-based alloys' remarkable electrical and thermal conductivity raises the dependability and efficiency of electronic machinery and gadgets (Hughes *et al.*, 2021).

PGM compounds are frequently used as catalysts in a variety of chemical interactions and industrial processes. Examples of these compounds are palladium acetate and platinum chloride. These substances have remarkable selectivity and catalytic activity, which makes it possible to produce specialized chemicals, polymers, and medications with high efficiency. Their importance in the development of contemporary industrial chemistry is highlighted by their application in chemical synthesis and production procedures (Rao & Reddi, 2000).

Precious PGM like platinum, palladium, and rhodium are often utilized to make jewellery of superior quality. These metals are frequently used with other precious metals, such copper, silver, and gold, to make jewellery that is both glossy and long-lasting. The resultant alloys and compounds are widely sought after in the market for luxury products because of their exceptional strength, resistance to corrosion, and stunning visual appeal (Rakhtsaum, 2013).

2.4 APPLICATIONS

Because of the diversified properties of PGMs, they have been used for various applications (Table 2.1).

2.4.1 PGMs for automotive development

Catalytic converters are one of the main uses of PGMs in the automobile industry. In these appliances, palladium, rhodium, and platinum function as catalysts to help transform toxic emissions into less toxic forms. PGMs considerably lower pollution emissions of hydrocarbons, atmospheric nitrogen oxides, and carbon monoxide (CO) by encouraging oxidation and reduction processes. Playing a crucial role in reducing emissions has helped the world's environmental requirements to be met (Moschovi *et al.*, 2021). Proton interaction membrane fuel cells (PEMFCs), a clean and effective replacement for conventional internal combustion engines, rely heavily on platinum as a component. Platinum acts as a catalyst in the fuel cell, facilitating the electrochemical processes that turn oxygen and hydrogen into water and energy that powers the car. Concerns over the condition of the environment and climate change are being addressed by the rise of zero-emission automobiles, which is made possible by the use of platinum in fuel cells (Du *et al.*, 2021).

PGMs are essential to the advancement of hybrid and electric vehicles. Platinum-based catalysts improve combustion in fuel-efficient hybrid engines, maximizing fuel efficiency (Crisostomi *et al.*, 2017). Furthermore, lithium-ion batteries, a vital component of electric car technology, use platinum as one of its electrodes. PGMs are in high demand for battery manufacture, which highlights their contribution to the development of sustainable transportation solutions (Kuzuhara *et al.*, 2021). Platinum-group metals are utilized in a variety of cutting-edge materials and technologies seen in contemporary

Table 2.1 Various applications of PGMs with their properties.

Metal name	Property	Role	Application	Reference
Platinum	Thermal stability	Oxidation of CO	Catalytic converters	Dey and Dhal (2020)
	Electrocatalytic activity	Kinetic oxidation	Fuel cells	Ramli and Kamarudin (2018)
	Hardness	Raw material	Jewellery	Wisniak (2005)
	piezoresistive	fabrication	electronics	Lukas *et al.* (2021)
Palladium	Thermophysical nature	Conduction	Catalytic converters	Cawthorn (2006)
	Thermoelectric property	Photodetectors, sensors	Optolectronics	Wang *et al.* (2021)
	Corrosion resistance	alloy	Dentistry	Wataha *et al.* (2008)
Rhodium	Mutagenic property	Catalyst	Catalytic converters	Bünger *et al.* (1996)
	Electrochromic property	Alloy	Electrochromic devices, glass industry	Wang *et al.* (2001)
	Corrosion resistance	Alloy	Jewellery	Rudolf *et al.* (2013)
Ruthenium	Catalytic property	Catalyst	Chemical catalysts	Over (2012)
	Structural property	Sensitizers	Solar cells	Yin *et al.* (2012)
	Electrochemical properties	Alloy	Electronics	Wenger (2019)
Iridium	Mechanical properties	Alloy	Spark plugs	Kuzuhara *et al.* (2021)
	Photophysical property	Complexes	Electronic devices	Ma *et al.* (2017)
	Chemical property	Precursors	Medical application	Karakovskaya *et al.* (2021)
Osmium	Thermophysical property	Alloy	Electrical contacts	Yamaura (2016)
	Mechanical property	Complexes	Fountain pen tips	Yamaura (2016)
	Corrosion resistance	Alloy	Jewellery	Pandey *et al.* (2009)

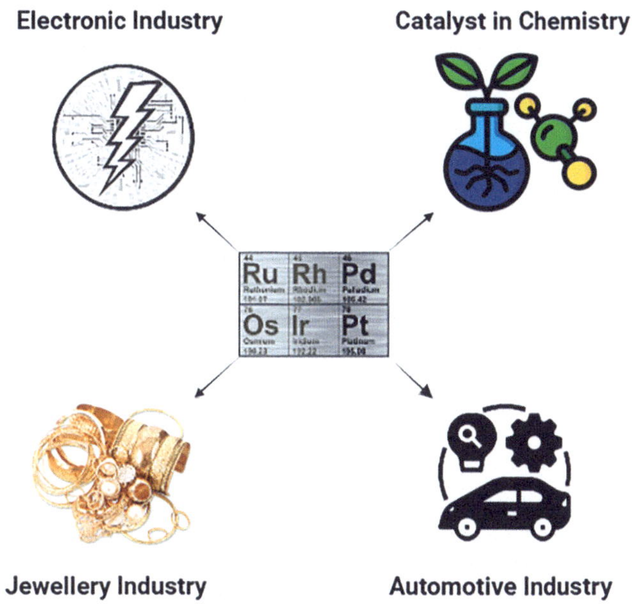

Figure 2.3 General applications of PGMs.

automobiles, in addition to electrification and emission reduction. For example, spark plugs made of platinum help to increase engine performance and ignition (Alonso *et al.*, 2012). Palladium is also used in the production of sensors, which guarantees precise exhaust gas and engine status monitoring (Darmadi *et al.*, 2020).

The corrosion resistance of PGMs, especially rhodium, improves the robustness and capacity of several car parts to withstand corrosion. Exhaust systems use rhodium-coated materials to shield them from corrosive substances and prolong their life. This addresses both financial and environmental issues while extending the general lifespan and dependability of automobiles. In the automobile sector, platinum-group metals are now considered essential, as they promote innovation and tackle issues related to emissions, fuel economy, and ecological sustainability. PGMs are expected to play a bigger role as the automotive industry continues to shift towards more effective and environmental-friendly technologies, confirming their position as essential facilitators of advancement in the quest for a more sustainable and greener future for transportation (Iavicoli & Leso, 2022). Apart from catalytic converters PGMs are also used for other applications as shown in Figure 2.3.

2.4.2 PGMs in sustainable chemistry as catalysts

Platinum and palladium are essential to hydrogenation and dehydrogenation processes. The production of fine chemicals, agrochemicals, and medicines depends on these procedures. PGM catalyst minimize the energy consumption and environmental effects of chemical manufacture by facilitating these reactions at milder conditions (Bellabarba *et al.*, 2023). Particularly good in initiating oxidation and reduction processes is rhodium. The synthesis of many different compounds, including specialized and pharmaceutical intermediates, depends on these interactions. The general sustainability of chemical manufacturing processes is improved by rhodium-based catalysts by several mechanisms like enhanced efficiency, improved selectivity, mild reaction conditions, catalyst generation, atom economy and less hazardous by-products, which provide more efficient and selective paths (Gatte & Phillips, 1987). Significant catalytic activity is shown by ruthenium and iridium in alkylation along with isomerization reactions, which are essential processes in the synthesis of petrochemicals and fuels. PGM catalysts improve yield, specificity, and energy efficiency in these processes, which help to provide more economical and environmental-friendly pathways for the synthesis of important chemical products (Gnanamgari *et al.*, 2009).

2.4.3 Electronic industry

PGMs as catalysts in electronics manufacturing, particularly palladium and platinum, act as catalysts in the semiconductor manufacturing processes. These metals are essential for the accurate development of thin coatings on semiconductor wafers during chemical vapour deposition (CVD) techniques. The development of integrated circuits, the fundamental components of electronic gadgets, depends on this (Ivanov *et al.*, 2021). PGMs are used as catalysts in techniques like electroless plating to produce printed circuit boards (PCBs). Intricate circuit designs on PCBs are made possible by the catalysis of the deposition of conductive materials onto non-conductive surfaces by platinum and palladium (Reuter & van Schaik, 2016). PGMs, like platinum, are important in keeping rooms clean and other similar settings perfect while making electronics components. They accelerate processes that aid in the management and removal of impurities, protecting the integrity of delicate electronic components (Izatt *et al.*, 2015).

Electrodes in electronic devices rely on the unique properties of PGMs for efficient and reliable performance. PGMs, including platinum, palladium, rhodium, ruthenium, iridium, and osmium, are chosen for their exceptional characteristics that enhance the functionality of various electronic components. In electronic applications, platinum's exceptional electrical conductivity is essential. PGMs enable the effective transfer of electrical signals in electronic circuits by acting as electrodes in sensors, capacitors, and resistors. Essentially, the use of platinum-group metals as electrodes in electronic devices capitalizes on their exceptional electrical conductivity, high melting temperatures, corrosion resistance, and catalytic ability. These characteristics support the usefulness and dependability of electronic equipment in electronic applications (Wang *et al.*, 2023). Catalytic converters in electronics recycling using PGMs have emerged as a critical component of ethical and ecological waste management techniques. PGM catalysts in catalytic converters are essential to recovering valuable elements from electronic trash. PGMs are among the many valuable metals found in electronic garbage, or 'e-waste', which is made up of many types of electronic components. To extract and separate PGMs from electronic waste and support sustainable resource management, catalyst converters play a crucial role in the recycling process (Islam *et al.*, 2020). This has led to improvements in semiconductor technology, miniaturization, deposition methods, flexible electronics, durability, stability, efficiency, and overall performance of electronic devices. PGMs have special qualities such as stability, conductivity, biocompatibility, corrosion resistance, optical, mechanical and catalytic properties that make them excellent choices for thin-film technologies (Dutta *et al.*, 2017).

Because of their high magnetic permeability, platinum-group metals make excellent choices for magnetic storage media. This characteristic makes it possible to create magnetic domains that are stable and well-defined, which is essential for encoding and preserving digital data in storage devices. PGMs' strong magnetic coercivity makes it possible to create magnetic storage media with higher data densities. This is necessary for the electronic industry's ongoing trend of downsizing since it enables the storage of more data in fewer physical locations (Mohapatra *et al.*, 2020). PGMs are essential to hard disk drives (HDDs), which use cobalt–platinum alloy thin films as conventional magnetic storage layers. The accuracy and stability of the magnetic recording made possible by these films enhances the dependability and speed of data retrieval in HDDs (Ryan, 2010). Additionally, magnetic tape technologies which are frequently used for data backup and archival storage – use platinum-group metals. PGMs and cobalt are frequently found in the magnetic layers of tapes, which offer long-term stability and large data storage capacities. PGMs help magnetic storage media maintain their temperature stability. These metals are dependable options for applications where temperature swings may occur because of their capacity to retain their magnetic qualities at high temperatures, which guarantees the integrity of data stored in them (Hassan *et al.*, 2022). PGMs are making progress in cutting-edge technology such as magnetic random-access memory (MRAM), which goes beyond conventional storage media. The development of non-volatile memory devices with quicker read and write rates is made possible by their special magnetic properties, which holds promise for improvements in electronic storage. PGMs' stability and corrosion resistance add to the long-term reliability of magnetic storage media. This lessens the need for regular data migration or storage device replacements and guarantees the durability of stored data (Palomino *et al.*, 2021). Heat dissipation is a crucial component that influences the performance and reliability of high-performance electronic components such as transistors and microprocessors. PGMs offer a dependable way to make sure these components work well and last a long time at high temperatures because of their extreme high melting point and oxidation resistance (Hur *et al.*, 2023).

2.4.4 Jewellery industry

PGM's distinct qualities make them extremely sought-after for creating elegant and long-lasting jewellery. PGMs are preferred in jewellery making due to their

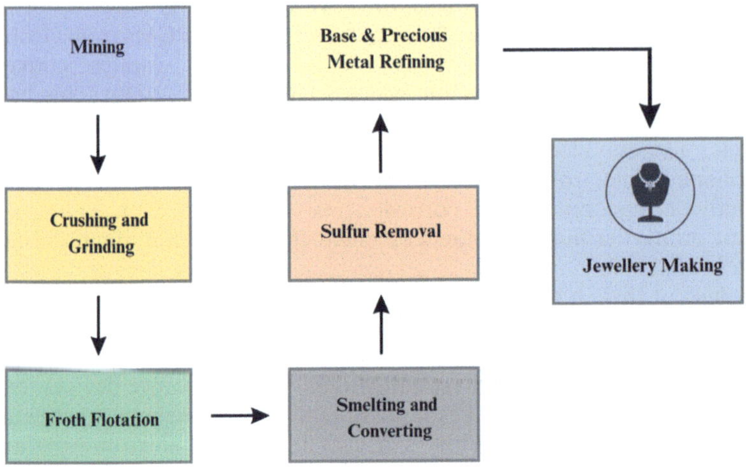

Figure 2.4 Manufacturing of PGM-based jewellery.

scarcity and intrinsic value. Because platinum, for instance, is far rarer than gold, jewellery made of this metal is even more exclusive. The process involved in the usage of PGMs in jewellery making is depicted in Figure 2.4. PGMs are rare resources, which add to their value and increase the price on jewellery crafted from these metals (Artyukh *et al.*, 2021). In addition to being extremely rare, PGMs have outstanding endurance and tarnish as well as corrosion resistance. Because of their resilience, jewellery crafted from metals in the platinum group is guaranteed to hold their shine and beauty throughout time. PGMs are perfect for creating durable and low-maintenance jewellery because they are comparatively untouched by environmental factors, unlike many other metals that might tarnish or react with the environmental factors (Gunn, 2014).

PGMs are frequently used as metals and settings for jewellery that has gemstones even though they are not gemstones themselves. The most popular application is in settings for precious stones, such as diamonds. PGMs are excellent options for protecting and enhancing the beauty of gemstones because of their strength and endurance. Because of its strength, platinum in particular is a popular metal for prongs that firmly hold gemstones in place (Moyer & Dieleman, 2003). By offering a contrasting backdrop to the brilliance of gemstones, PGMs enhance the jewellery's overall aesthetic appeal. For instance, the shine of diamonds is accentuated by the white luster of platinum and palladium, producing an eye-catching contrast. Because of this, PGMs are a great option for engagement rings and other high-end jewellery items where the gemstones' brilliance serves as the main attraction. PGMs can also occasionally be utilized to make distinctive, detailed jewellery designs that resemble gemstones. Without the need for real gemstones, artists can create intricate patterns and textures with these metals' malleability and ductility, giving the item depth and personality (Cabri *et al.*, 2008).

PGMs are inert and have good biocompatibility, in contrast to other base metals that might irritate people's sensitive skin. Because of this, they are the perfect material to use while making jewellery, especially for people who have sensitive skin (Zemelka-Wiacek, 2022). PGMs are hypoallergenic due to their ability to withstand corrosion. PGMs stay stable and do not experience chemical changes that could cause irritation or discoloration, in contrast to certain other metals that might react with the skin's natural oils and acids. Because of this feature, jewellery composed of PGMs is guaranteed to hold up over time, making it a comfortable and long-lasting choice for people who are allergic to certain metals (Nouri & Wen, 2021).

In the jewellery business, rhodium plating is a popular and useful process that is especially used with PGMs like palladium and platinum. To improve the look and functionality of jewellery pieces, a tiny layer of rhodium is applied to their surface. To create a bright and shiny surface, rhodium plating is a common practice in the jewellery industry (Rudolf *et al.*, 2013). Rhodium has a brilliant white colour and is incredibly shiny, which can improve the jewellery's overall appearance. This is especially important for platinum because rhodium plating prevents dulling or discoloration that could come from exposure and wear over time, keeping the platinum's dazzling, silvery luster. In addition, rhodium plating gives PGM jewellery a layer of protection. It serves as a shield, keeping the metal underneath from corroding or tarnishing. This reduces the need for routine cleaning or maintenance and guarantees the jewellery's lifetime. Consumers will find rhodium-plated PGM jewellery to be both appealing and functional due to its increased durability and wear resistance (Corti, 2022).

2.5 CONCLUSION

This chapter gives a thorough overview of PGMs, including their origin, occurrence, and applications. PGMs have undergone an interesting journey from their geological origin to the complex technological landscapes where they find numerous applications. PGMs' importance in a number of fields, such as electronics, sustainable energy, and catalysis, highlights the crucial part they play in forming the modern technological landscape. It is clear from overcoming the difficulties associated with locating and using these metals that advanced technology and sustainable resource management depends on knowledge of their geological occurrence and usage.

REFERENCES

Alonso E., Field F. R. and Kirchain R. E. (2012). Platinum availability for future automotive technologies. *Environmental Science & Technology*, **46**(23), 12986–12993, https://doi.org/10.1021/es301110e

Artyukh T., Hryhorenko I., Ternova A., Yaheliuk S., Verenikin O. and Cernavca M. (2021). Identification of white jewelry alloy based on silver and platinum for testing purposes. *Eastern-European Journal of Enterprise Technologies*, **12**(5), 47–59, https://doi.org/10.15587/1729-4061.2021.243179

Bellabarba R., Johnston P., Moss S., Sievers C., Subramaniam B., Tway C., Wang Z. and Zhu H. (2023). Net zero transition: possible implications for catalysis. *ACS Catalysis*, **13**(12), 7917–7928, https://doi.org/10.1021/acscatal.3c01255

Bünger J., Stork J. and Stalder K. (1996). Cyto- and genotoxic effects of coordination complexes of platinum, palladium and rhodium in vitro. *International Archives of Occupational and Environmental Health*, **69**, 33–38, https://doi.org/10.1007/BF02630736

Cabri L. J., Rudashevsky N. S., Rudashevsky V. N. and Oberthür T. (2008, January). Electric-Pulse Disaggregation (EPD), Hydroseparation (HS) and their use in combination for mineral processing and advanced characterization of ores. Canadian Mineral Processors 40th Annual Meeting, Proceedings, Paper, Vol. **14**, pp. 211–235.

Cabri L. J., Oberthür T. and Keays R. R. (2022). Origin and depositional history of platinum-group minerals in placers – A critical review of facts and fiction. *Ore Geology Reviews*, **144**, 104733, https://doi.org/10.1016/j.oregeorev.2022.104733

Cawthorn R. G. (2006). Centenary of the discovery of platinum in the Bushveld complex. *Platinum Metals Review*, **50**(3), 130–133, https://doi.org/10.1595/147106706X119746

Cawthorn R. G., Lee C. A., Schouwstra R. P. and Mellowship P. (2002). Relationship between PGE and PGM in the Bushveld complex. *The Canadian Mineralogist*, **40**(2), 311–328. https://citeseerx.ist.psu.edu/document?repid=rep1&type=pdf&doi=40a7e6b11288b37ca9c6484246d9a5c4ff7a6c16 https://doi.org/10.2113/gscanmin.40.2.311

Corti C. W. (2022). The evolution of platinum jewellery alloys: from the 1920s to the 2020s: developments in new alloys and techniques. *Johnson Matthey Technology Review*, **66**(4), 418–434, https://doi.org/10.1595/205651322X16390711562364

Couderc C. (2010). Platinum group metals in glass making. *Platinum Metals Review*, **54**(3), 186–191, https://doi.org/10.1595/147106710X514012

Crisostomi E., Shorten R., Stüdli S. and Wirth F. (2017). Electric and Plug-in Hybrid Vehicle Networks: Optimization and Control. CRC Press, Boca Raton. https://doi.org/10.1201/9781315151861

Cullen D. A., Sneed B., More K. L., Brewster J. H., Odgaard M., Chung H. T. and Zelenay P. (2016, September). Resolving Active Sites and Porosity in PGM-Free Catalysts By Electron Microscopy. In Electrochemical Society Meeting Abstracts 230 (No. 38, pp. 2809–2809). The Electrochemical Society, Inc, https://doi.org/10.1149/MA2016-02/38/2809

Darmadi I., Nugroho F. A. A. and Langhammer C. (2020). High-performance nanostructured palladium-based hydrogen sensors – current limitations and strategies for their mitigation. *ACS Sensors*, **5**(11), 3306–3327, https://doi.org/10.1021/acssensors.0c02019

Dey S. and Dhal G. C. (2020). Property and structure of various platinum catalysts for low-temperature carbon monoxide oxidations. *Materials Today Chemistry*, **16**, 100228, https://doi.org/10.1016/j.mtchem.2019.100228

Dong H., Zhao J., Chen J., Wu Y. and Li B. (2015). Recovery of platinum group metals from spent catalysts: a review. *International Journal of Mineral Processing*, **145**, 108–113, https://doi.org/10.1016/j.minpro.2015.06.009

Du L., Zhang G. and Sun S. (2021). Proton exchange membrane (PEM) fuel cells with platinum group metal (PGM)-free cathode. *Automotive Innovation*, **4**(2), 131–143, https://doi.org/10.1007/s42154-021-00146-0

Dutta S., Sankaran K., Moors K., Pourtois G., Van Elshocht S., Bömmels J., Vandervorst W. and Adelmann C. (2017). Thickness dependence of the resistivity of platinum-group metal thin films. *Journal of Applied Physics*, **122**(2), 025107, https://doi.org/10.1063/1.4992089

Gatte R. R. and Phillips J. (1987). Dynamic phase behavior of graphite-supported iron-rhodium catalysts during oxidation-reduction. *The Journal of Physical Chemistry*, **91**(23), 5961–5968, https://doi.org/10.1021/j100307a030

Glaister B. J. and Mudd G. M. (2010). The environmental costs of platinum – PGM mining and sustainability: Is the glass half-full or half-empty? *Minerals Engineering*, **23**(5), 438–450, https://doi.org/10.1016/j.mineng.2009.12.007

Gnanamgari D., Sauer E. L., Schley N. D., Butler C., Incarvito C. D. and Crabtree R. H. (2009). Iridium and ruthenium complexes with chelating N-heterocyclic carbenes: efficient catalysts for transfer hydrogenation, β-alkylation of alcohols, and N-alkylation of amines. *Organometallics*, **28**(1), 321–325, https://doi.org/10.1021/om800821q

Gunn G. (2014). Platinum-group metals. In: Critical Metals Handbook, John Wiley & Sons, Ltd, UK, pp. 284–311, https://www.google.co.in/books/edition/Critical_Metals_Handbook/oLLQAgAAQBAJ?hl=en&gbpv=0

Hassan M., Laureti S., Rinaldi C., Fagiani F., Barucca G., Casoli F., Mezzi A., Bolli E., Kaciulis S., Fix M., Ullrich A., Albrecht M. and Varvaro G. (2022). Thin-film heterostructures based on Co/Ni synthetic antiferromagnets on polymer tapes: toward sustainable flexible spintronics. *ACS Applied Materials & Interfaces*, **14**(45), 51496–51509, https://doi.org/10.1021/acsami.2c14000

Hughes A. E., Haque N., Northey S. A. and Giddey S. (2021). Platinum group metals: A review of resources, production and usage with a focus on catalysts. *Resources*, **10**(9), 93, https://doi.org/10.3390/resources10090093

Hur J., Kang D., Moon D. I., Yu J. M., Choi Y. K. and Yu S. (2023). Cryogenic storage memory with high-speed, low-power, and long-retention performance. *Advanced Electronic Materials*, **9**(6), 2201299, https://doi.org/10.1002/aelm.202201299

Iavicoli I. and Leso V. (2022). Rhodium. In: Handbook on the Toxicology of Metals. Academic Press, Netherlands, pp. 691–728, https://doi.org/10.1016/B978-0-12-822946-0.00025-8

Islam A., Ahmed T., Awual M. R., Rahman A., Sultana M., Abd Aziz A., Monir M. U., Teo S. H. and Hasan M. (2020). Advances in sustainable approaches to recover metals from e-waste – A review. *Journal of Cleaner Production*, **244**, 118815, https://doi.org/10.1016/j.jclepro.2019.118815

Ivanov I. I., Baranov A. M., Talipov V. A., Mironov S. M., Akbari S., Kolesnik I. V., Orlova D E. and Napolskii K. S. (2021). Investigation of catalytic hydrogen sensors with platinum group catalysts. *Sensors and Actuators B: Chemical*, **346**, 130515, https://doi.org/10.1016/j.snb.2021.130515

Izatt R. M., Izatt S. R., Izatt N. E., Krakowiak K. E., Bruening R. L. and Navarro L. (2015). Industrial applications of molecular recognition technology to separations of platinum group metals and selective removal of metal impurities from process streams. *Green Chemistry*, **17**(4), 2236–2245. https://pubs.rsc.org/en/content/articlelanding/2015/gc/c4gc02188f/unauth https://doi.org/10.1039/C4GC02188F

Junge M., Wirth R., Oberthür T., Melcher F. and Schreiber A. (2015). Mineralogical siting of platinum-group elements in pentlandite from the Bushveld Complex, South Africa. *Mineralium Deposita*, **50**, 41–54, https://doi.org/10.1007/s00126-014-0561-0

Karakovskaya K. I., Dorovskikh S. I., Vikulova E. S., Ilyin I. Y., Zherikova K. V., Basova T. V. and Morozova N. B. (2021). Volatile iridium and platinum MOCVD precursors: chemistry, thermal properties, materials and prospects for their application in medicine. *Coatings*, **11**(1), 78, https://doi.org/10.3390/coatings11010078

Kuzuhara S., Ota M. and Kasuya R. (2021). Recovery of platinum group metals from spent automotive catalysts using lithium salts and hydrochloric acid. *Materials*, **14**(22), 6843, https://doi.org/10.3390/ma14226843

Li F. M., Huang L., Zaman S., Guo W., Liu H., Guo X. and Xia B. Y. (2022). Corrosion chemistry of electrocatalysts. *Advanced Materials*, **34**(52), 2200840, https://doi.org/10.1002/adma.202200840

Lukas S., Hartwig O., Prechtl M., Capraro G., Bolten J., Meledin A., Mayer J., Neumaier D., Kataria S., Duesberg G. S. and Lemme M. C. (2021). Correlating nanocrystalline structure with electronic properties in 2D platinum diselenide. *Advanced Functional Materials*, **31**(35), 2102929, https://doi.org/10.1002/adfm.202102929

Ma D., Tsuboi T., Qiu Y. and Duan L. (2017). Recent progress in ionic iridium (III) complexes for organic electronic devices. *Advanced Materials*, **29**(3), 1603253, https://doi.org/10.1002/adma.201603253

Maier W. D. (2005). Platinum-group element (PGE) deposits and occurrences: mineralization styles, genetic concepts, and exploration criteria. *Journal of African Earth Sciences*, **41**(3), 165–191, https://doi.org/10.1016/j.jafrearsci.2005.03.004

McCown J. R. and Shaw R. (2017). Investment potential and risk hedging characteristics of platinum group metals. *The Quarterly Review of Economics and Finance*, **63**, 328–337, https://doi.org/10.1016/j.qref.2016.06.001

Mohapatra J., Xing M., Elkins J. and Liu J. P. (2020). Hard and semi-hard magnetic materials based on cobalt and cobalt alloys. *Journal of Alloys and Compounds*, **824**, 153874, https://doi.org/10.1016/j.jallcom.2020.153874

Moschovi A. M., Giuliano M., Kourtelesis M., Nicol G., Polyzou E., Parussa F., Iakovos Y. and Sgroi M. F. (2021). First of its kind automotive catalyst prepared by recycled PGMs-catalytic performance. *Catalysts*, **11**(8), 942, https://doi.org/10.3390/catal11080942

Moyer I. and Dieleman J. (2003). Miniaturization and the opening of the mouth in a Greek magical text (PGM XII. 270–350). *Journal of Ancient Near Eastern Religions*, **3**(1), 47–72. https://brill.com/view/journals/jane/3/1/article-p47_3.xml https://doi.org/10.1163/1569212031960320

Murray A. J., Zhu J., Wood J. and Macaskie L. E. (2017). A novel biorefinery: biorecovery of precious metals from spent automotive catalyst leachates into new catalysts effective in metal reduction and in the hydrogenation of 2-pentyne. *Minerals Engineering*, **113**, 102–108, https://doi.org/10.1016/j.mineng.2017.08.011

Nouri A. and Wen C. (2021). Noble metal alloys for load-bearing implant applications. In: Structural Biomaterials. Woodhead Publishing, Duxford, UK, pp. 127–156, https://doi.org/10.1016/B978-0-12-818831-6.00003-3

Oberthür T., Weiser T. W., Melcher F., Gast L. and Wöhrl C. (2013). Detrital platinum-group minerals in rivers draining the Great Dyke, Zimbabwe. *The Canadian Mineralogist*, **51**(2), 197–222, https://doi.org/10.3749/canmin.51.2.197

Omodara L., Pitkäaho S., Turpeinen E. M., Saavalainen P., Oravisjärvi K. and Keiski R. L. (2019). Recycling and substitution of light rare earth elements, cerium, lanthanum, neodymium, and praseodymium from end-of-life applications – A review. *Journal of Cleaner Production*, **236**, 117573, https://doi.org/10.1016/j.jclepro.2019.07.048

Over H. (2012). Surface chemistry of ruthenium dioxide in heterogeneous catalysis and electrocatalysis: from fundamental to applied research. *Chemical Reviews*, **112**(6), 3356–3426, https://doi.org/10.1021/cr200247n

Palomino A., Marty J., Auffret S., Joumard I., Sousa R. C., Prejbeanu I. L., Ageron B. and Dieny B. (2021). Evaluating critical metals contained in spintronic memory with a particular focus on Pt substitution for improved sustainability. *Sustainable Materials and Technologies*, **28**, e00270, https://doi.org/10.1016/j.susmat.2021.e00270

Pandey B. D., Singh D. and Yadawa P. K. (2009). Ultrasonic study of osmium and ruthenium. *Platinum Metals Review*, **53**(2), 91–97, https://doi.org/10.1595/147106709X430927

Rakhtsaum G. (2013). Platinum alloys: a selective review of the available literature. *Platinum Metals Review*, **57**(3), 202–213, https://doi.org/10.1595/147106713X668596

Ramli Z. A. C. and Kamarudin S. K. (2018). Platinum-based catalysts on various carbon supports and conducting polymers for direct methanol fuel cell applications: a review. *Nanoscale Research Letters*, **13**(1), 410, https://doi.org/10.1186/s11671-018-2799-4

Rao C. R. M. and Reddi G. S. (2000). Platinum group metals (PGM); occurrence, use and recent trends in their determination. *Trends in Analytical Chemistry*, **19**(9), 565–586, https://doi.org/10.1016/S0165-9936(00)00031-5

Reuter M. and Van Schaik A. (2016). Gold – A key enabler of a circular economy: recycling of waste electric and electronic equipment. In: Gold Ore Processing. Elsevier, Amsterdam, Netherlands, pp. 937–958, https://doi.org/10.1016/B978-0-444-63658-4.00053-0

Rudolf R., Budić B., Stamenković D., Čolić M., Ivanič A. and Kosec B. (2013). Rhodium platings – experimental study. *Metalurgija*, **52**(3), 337–340. https://hrcak.srce.hr/95771

Ryan M. (2010). PGM highlights: platinum in next-generation materials for data storage. *Platinum Metals Review*, **54**(4), 244–249, https://doi.org/10.1595/147106710X525263

Sefako R., Sekgarametso K. and Sibanda V. (2017). Potential processing routes for recovery of platinum group metals from southern African oxidized PGM ores: A review. *Journal of Sustainable Metallurgy*, **3**(4), 797–807, https://doi.org/10.1007/s40831-017-0146-0

Sefako R., Sibanda V. and Sekgarametso K. (2019). PGM extraction from oxidized ores using flotation and leaching. *Journal of the Southern African Institute of Mining and Metallurgy*, **119**(11), 929–, https://doi.org/10.17159/2411-9717/287/2019

Shamaila S. and O'Connor C. T. (2008). The role of synthetic minerals in determining the relative flotation behaviour of Platreef PGE tellurides and arsenides. *Minerals Engineering*, **21**(12–14), 899–904, https://doi.org/10.1016/j.mineng.2008.07.006

Su L., Fan X., Jin Y., Cong H. and Luo W. (2023). Hydroxyl-binding energy-induced kinetic gap narrowing between acidic and alkaline hydrogen oxidation reaction on intermetallic Ru3Sn7 catalyst. *Small*, **19**(11), 2207603, https://doi.org/10.1002/smll.202207603

Wang H., Yan M. and Jiang Z. (2001). Electrochromic properties of rhodium oxide films prepared by a sol–gel method. *Thin Solid Films*, **401**(1–2), 211–215, https://doi.org/10.1016/S0040-6090(01)01604-2

Wang Y., Pang J., Cheng Q., Han L., Li Y., Meng X., Ibarlucea B., Zhao H., Yang F., Liu H., Zhou W., Wang X., Rummeli M H., Zhang Y. and Cuniberti G. (2021). Applications of 2D-layered palladium diselenide and its van der Waals heterostructures in electronics and optoelectronics. *Nano-Micro Letters*, **13**(1), 143, https://doi.org/10.1007/s40820-021-00660-0

Wang X., Zhang W., Zhou Q. and Ran F. (2023). Integrating supercapacitor with sodium hyaluronate based hydrogel as a novel All-In-One wound dressing: self-powered electronic stimulation. *Chemical Engineering Journal*, **452**, 139491, https://doi.org/10.1016/j.cej.2022.139491

Wataha J. C., Lockwood P. E., Messer R. L., Lewis J. B. and Mettenburg D. J. (2008). Brushing-induced surface roughness of nickel-, palladium-, and gold-based dental casting alloys. *The Journal of Prosthetic Dentistry*, **99**(6), 455–460, https://doi.org/10.1016/S0022-3913(08)60107-9

Wenger O. S. (2019). Is iron the new ruthenium? *Chemistry–A European Journal*, **25**(24), 6043–6052, https://doi.org/10.1002/chem.201806148

Wisniak J. (2005). Platinum – from exotic to commodity NIScPR online periodicals repository. *IJCT*, **12**, 601-614, http://nopr.niscpr.res.in/handle/123456789/8567

Woodall C. M., Lu X., Dipple G. and Wilcox J. (2021). Carbon mineralization with North American PGM mine tailings – characterization and reactivity analysis. *Minerals*, **11**(8), 844, https://doi.org/10.3390/min11080844

Xiao Z. and Laplante A. R. (2004). Characterizing and recovering the platinum group minerals – a review. *Minerals Engineering*, **17**(9–10), 961–979, https://doi.org/10.1016/j.mineng.2004.04.001

Yamaura K. (2016). Short review of high-pressure crystal growth and magnetic and electrical properties of solid-state osmium oxides. *Journal of Solid State Chemistry*, **236**, 45–54, https://doi.org/10.1016/j.jssc.2015.06.037

Yin J. F., Velayudham M., Bhattacharya D., Lin H. C. and Lu K. L. (2012). Structure optimization of ruthenium photosensitizers for efficient dye-sensitized solar cells – a goal toward a "bright" future. *Coordination Chemistry Reviews*, **256**(23–24), 3008–3035, https://doi.org/10.1016/j.ccr.2012.06.022

Yusenko K. V., Riva S., Carvalho P. A., Yusenko M. V., Arnaboldi S., Sukhikh A. S., Hanfland M. and Gromilov S. A. (2017). First hexagonal close packed high-entropy alloy with outstanding stability under extreme conditions and electrocatalytic activity for methanol oxidation. *Scripta Materialia*, **138**, 22–27, https://doi.org/10.1016/j.scriptamat.2017.05.022

Zaccarini F., Garuti G., Pushkarev E. and Thalhammer O. (2018). Origin of platinum group minerals (PGM) inclusions in chromite deposits of the Urals. *Minerals*, **8**(9), 379, https://doi.org/10.3390/min8090379

Zemelka-Wiacek M. (2022). Metal allergy: state-of-the-art mechanisms, biomarkers, hypersensitivity to implants. *Journal of Clinical Medicine*, **11**(23), 6971, https://doi.org/10.3390/jcm11236971

Zhao X., Wang Y., Lin P., Li L., Jiao L., Wu Z., Guo Z., Luo G. and Lin M. (2023). Dissolution behavior of palladium and rhodium in glass melts: implications for the viscosity and electrical conductivity of high-temperature melts. *Progress in Nuclear Energy*, **164**, 104886, https://doi.org/10.1016/j.pnucene.2023.104886

Zheng H., Ding Y., Wen Q., Liu B. and Zhang S. (2021). Separation and purification of platinum group metals from aqueous solution: recent developments and industrial applications. *Resources, Conservation and Recycling*, **167**, 105417, https://doi.org/10.1016/j.resconrec.2021.105417

doi: 10.2166/9781789065084_0033

Chapter 3

Recent advances in analytical techniques for the determination of platinum group elements

Nang-Htay Yin, Javier Jiménez-Lamana and Hugues Preud'Homme

Universite de Pau et des Pays de L'Adour, E2S UPPA, CNRS, IPREM, Institut des Sciences Analytiques et de Physico-chimie pour L'Environnement et Les Matériaux, Technopole Hélioparc, 2 avenue du Président Pierre Angot, 64053 Pau, France

Corresponding author: nangsao87@gmail.com

ABSTRACT

This chapter focuses on state-of-the-art advancements in analytical methodologies aimed at precisely determining platinum group elements (PGEs). It explores key aspects of sample pretreatment, chromatographic separation, purification, and recent innovations in mass spectrometry. The chapter elucidates the roles of various sample pretreatment methods in improving the accuracy and efficiency of PGE analysis. Subsequently, it outlines the significance of chromatographic separation, extraction, and purification techniques in isolating PGEs from complex matrices for subsequent analytical procedures. Furthermore, the chapter evaluates the single vs. double-spiking techniques for concentration and isotopic analysis of PGEs, shedding light on the advantages and limitations of both techniques. Finally, the chapter comprehensively addresses the various instrumentation employed in PGE analysis, providing detailed insights into instrumental setup and operational conditions to optimize sensitivity, precision, and overall efficiency. The chapter offers a comprehensive overview of state-of-the-art analytical techniques applied in the determination of PGEs, providing valuable insights for researchers, analysts, and professionals involved in the exploration of these economically and environmentally significant elements.

Keywords: Concentration, isotopes, analysis, precision, mass spectrometry

3.1 INTRODUCTION

3.1.1 Importance of PGEs in modern society

Platinum group elements (PGEs) consist of six chemically similar elements: platinum (Pt), palladium (Pd), rhodium (Rh), ruthenium (Ru), iridium (Ir), and osmium (Os). These elements are rare and economically valuable, found in various geological sources such as mafic to ultramafic intrusions, alluvial deposits, chromite deposits, impact craters, and meteorites (Cabri *et al.*, 2022). PGEs play a crucial role in diverse industrial applications due to their unique properties. The automotive industry is a major consumer, particularly for catalyzing exhaust gases (Omran *et al.*, 2020). Their catalytic roles extend to the chemical and petroleum sectors, contributing to processes like hydrogenation, isomerization, oxidation, and refining (Antos & Aitani, 2004). Additionally, PGEs are essential in manufacturing high-quality glass, electrical devices, and heat-resistant equipment. Platinum-based chemotherapeutic agents, namely cisplatin and carboplatin, are instrumental in the treatment of various cancers, notably gynecological cancers, germ cell tumors, head and neck cancer, thoracic cancers, and bladder cancer (Ho *et al.*, 2016). They are also widely employed in biomedical implants (pacemakers and stents) and dental applications, showcasing their diverse utility in modern medical practices (Woodward, 2014).

The extensive use of PGEs in various sectors has led to their release into the environment through exhaust fumes, fertilizers, and pharmaceutical products, making them emerging anthropogenic contaminants (Birke *et al.*, 2018; Savignan *et al.*, 2021). PGEs are released into the air through exhaust fumes, into the soil through PGE-based fertilizers (Savignan *et al.*, 2021), and into hospital wastewater from PGE-based pharmaceuticals (Vyas *et al.*, 2014). Studies show a significant rise in PGE concentrations across various environmental matrices, in several animal tissues (Pawlak *et al.*, 2014), in human hair (Lo Medico *et al.*, 2023) indicating under-studied biogeochemical transformations and posing potential health risks (Ravindra *et al.*, 2004). With increasing PGE production and use, understanding their sources, distribution, and biogeochemical behavior in different environmental media and associated environmental and health risks becomes essential.

In biogeochemical transformation studies of PGEs, rigorous sample pretreatment and advanced analytical techniques are essential for achieving precise concentration and isotopic measurements. PGEs have potential to undergo intricate transformations in environmental matrices due to their diverse oxidation states and having multiple stable/radiogenic isotopes (Coursey & Dragoset, 2000), necessitating accurate data to comprehend their behavior and potential impacts. Sample pretreatment involves extracting and processing environmental samples to enhance sensitivity by eliminating interferences. The development of sophisticated analytical techniques ensures reliable measurements, providing crucial insights into PGE biogeochemical processes. The pivotal role of stringent sample preparation and cutting-edge methods lies in unraveling the complexities of PGE biogeochemistry, contributing to a more comprehensive understanding of their environmental fate, transport and associated risks.

3.1.2 Environmental and geological relevance of PGEs

The analysis of PGEs holds significant environmental and geological relevance. The analysis of PGEs contributes to a comprehensive understanding of geological processes, aids in the exploration of valuable mineral deposits, provides insights into the Earth's history and evolution, and enables the monitoring of environmental changes and anthropogenic inputs to this.

3.1.2.1 Indicator of geological processes

PGEs, are often associated with specific geological processes. Examining their concentrations and isotopic composition in rocks and minerals can provide valuable insights into the geological history and the processes that have shaped the Earth's crust. Studying the distribution and behavior of PGEs in different types of magmatic rocks helps in understanding the dynamics of magmatic systems, such as magma evolution, differentiation, and crystallization (Ely & Neal, 2003). PGEs are also considered valuable tracers of mantle processes. The study of PGE patterns in mantle-derived rocks contributes to our understanding of mantle composition, melting, and mantle–crust interactions. PGEs can act as geochemical fingerprints, helping to identify specific geological environments or the nature of weathering processes associated with brief climatic excursions (Barnes & Lightfoot, 2005). Their unique signatures in rocks can be used to distinguish between different geological settings, providing information on the tectonic history and geological evolution of an area.

3.1.2.2 Significance in cosmochemistry

PGEs are abundant in meteorites, which are remnants from the early solar system. Analyzing the PGE content of meteorites provides insights into the processes that occurred during the formation of the solar system. While other aspects, such as the study of PGEs in planetary differentiation, exploration of celestial bodies, and their role in stellar nucleosynthesis are also highly significant, the understanding of solar system formation has profound implications for our broader comprehension of the cosmos (Palme, 2000, 2001).

3.1.2.3 Environmental indicators

Environmental monitoring of PGEs is critical for understanding their presence and impact on ecosystems (Zereini & Alt, 2006). Identification of sources, including emissions from vehicle catalytic converters and industrial processes, is essential. Air quality monitoring assesses airborne PGE concentrations, especially in areas with high vehicular traffic or industrial activities (Rauch *et al.*, 2001). Water quality monitoring focuses on surface water and sediment samples to evaluate aquatic ecosystem impact. Soil contamination studies near emission sources or PGE-rich areas help assess soil quality and contamination extent (Rauch & Morrison, 2008). Biological monitoring, examining PGE accumulation in plants and aquatic organisms, provides insights into bio-availability and bio-accumulation in the food chain. Overall, environmental monitoring contributes to ecosystem health assessments, regulatory compliance, and the identification of long-term trends in PGE concentrations

(Ravindra *et al.*, 2004). It also aids in assessing human exposure, especially in occupational settings, and informs the development and assessment of remediation strategies to mitigate environmental impact and protect ecosystems and human health.

3.1.2.4 Economic implications

PGEs have substantial economic value due to their use in various industrial applications, particularly in catalytic converters for automobiles and in jewelry (Gunn, 2014; Chapter 2). Understanding the distribution and availability of PGEs has economic implications for mining and resource exploration (Mudd & Jowitt, 2018; Chapters 7 and 8). PGE prices are influenced by factors such as industrial demand, geopolitical events, and supply constraints. The economic viability of mining and processing PGE-rich ores is closely tied to market dynamics (Roskill Information Services Ltd., 2017). As global industries evolve, the demand for PGEs continues to rise, creating opportunities for investments, technological innovations, and economic growth (Bloxham *et al.*, 2022), while also posing challenges related to resource sustainability and environmental impacts.

3.2 SAMPLE PRETREATMENT FOR PGEs ANALYSIS

Accurate determination of the elemental and isotopic composition of geological samples is fundamental in basic geochemical and geochronological studies. The low concentrations of naturally occurring PGEs pose a significant challenge for most existing analytical methods, often pushing them to their limits of detection. Moreover, ensuring the accurate and precise quantification of elements like Ru, Rh, Pd, Re, Os, Ir, and Pt within natural matrices continues to present a persistent challenge in analytical practice. The choice of an appropriate sample preparation method depends on several critical factors, including the sample's composition, pre-concentration needs, the blanks, the target analytes, the chosen analytical technique, and the specific research goals. Additionally, each analytical approach necessitates a rigorous focus on sensitivity and careful considerations regarding quality assurance and quality control. This section outlines the latest advancements in sample preparation methods, as illustrated in Figure 3.1, to address these challenges effectively.

3.2.1 Closed digestion (for liquid mode analysis)

In general, geological samples such as rocks, ores, nodules, and soils are ground into fine powder in an agate grinding machine to minimize the risk of contamination. It is imperative to achieve complete dissolution of samples to ensure the precise determination of both elemental and isotopic composition.

3.2.1.1 Microwave-assisted digestion

The microwave pressure-dissolution technique involves placing a sample vessel (Teflon or quartz) with acids into a high-pressure chamber filled with nitrogen gas (at 40 bar) and subjecting it to microwave irradiation at frequencies between 300 MHz and 300 GHz, reaching temperatures of 220–260°C. This method facilitates rapid and efficient decomposition of samples for geochemical analysis

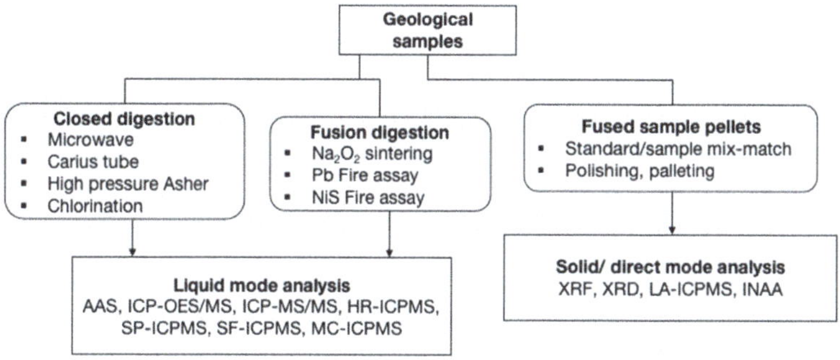

Figure 3.1 Different sample pre-treatment techniques and mode of analysis for PGEs. AAS: atomic absorption spectroscopy, MS: mass spectrometry, ICP-OES: inductively coupled plasma-optical emission spectrometry, HR: high resolution, SP: single particle, SF: sector field, MC: multi-collector.

in the laboratory. It is crucial to ensure that the sample quantity and selection of acid mixtures are in accordance with the anticipated concentration levels and the size of the vessel for optimal results. Typically, up to 0.5 g of sample in a mixture of 2 mL of HNO_3, 1 mL of HCl, 2 mL of H_3PO_4, 2 mL of HBF_4 in 15 mL Teflon vials is used for the complete digestion of BCR-2 (Basalt) as given in application guidelines for microwave digestion methods by Milestone Srl (Milestone Helping Chemists, available online).

3.2.1.2 Carius tube digestion

The Carius tube (CT) method stands as an alternative high-pressure digestion technique widely employed for the precise analysis of PGEs and accurate isotopic ratio measurement. This method involves dissolving samples in aqua regia ($2HNO_3$:1HCl) at elevated temperatures (220–260°C) with intermittent agitation for efficient sample digestion over 48 h (Hopp *et al.*, 2016). Following digestion, solutions undergo centrifugation to eliminate silicate particles, then evaporation at 100°C. The residues are subsequently redissolved in the desired acids for relevant analyses. Recognized for its minimal blank and prevention of volatile OsO_4 loss, quantitative recovery of highly siderophile elements like PGEs, the CT method holds particular significance in stable isotope analyses of PGEs.

3.2.1.3 High-pressure asher (HPA-S)

This technique is similar to CTs digestion with capability of achieving higher temperatures. The method involves placing samples and acids in closed vessels (highly pure quartz glass) within a pressure vessel, which is then pressurized with nitrogen up to 130 bar and heated up to a maximum of 320°C with a predetermined temperature program. The technique has limitations on sample size, typically 2 g. For example, chromite samples ranging from 0.5 to 4 g were dissolved using an HNO_3/HCl (5:2) acid combination through the application of HPA-S at 300°C and 125 bar pressure within an HPA-S apparatus (Paliulionyte *et al.*, 2006).

3.2.1.4 Chlorination

The oldest technique involves converting precious metals into water-soluble chloro-complexes by heating the samples with chlorine (Perry *et al.*, 1993). It was developed to pre-concentrate PGEs from geological samples and it accommodates large sample sizes (up to 250 g). Powdered samples, mixed with sodium chloride, undergo heating at 580°C in a chlorine gas current. Precious metal salts are separated by dissolving them in water or weak HCl, and filtering out insoluble matrix components. Combined with inductively coupled plasma-mass spectrometry (ICP-MS), it achieves detection limits below 1 ng/g and is adaptable to other techniques based on desired sensitivity. While effective for native metals and sulfide minerals, some minerals like silicates resist chlorination, requiring alternative digestion methods.

3.2.2 Fusion digestion (for liquid mode analysis)

Fusion digestion techniques are employed in geological material analysis when close digestion methods (section 3.2.1) struggle to fully dissolve refractory minerals like chromite or rare earth element (REE) ores. Fusion techniques effectively address incomplete digestion issues by utilizing both chemical and thermal energy to disintegrate mineral phases, forming water-soluble PGE salts. Fusion decompositions involve subjecting samples to elevated temperatures (500–1000°C) with alkali fluxing agents ($Na_2B_4O_7$, Na_2CO_3, CaF_2, SiO_2, Na_2O_2, NaOH, and $LiBO_2$), converting them into more soluble forms for subsequent dissolution by acids or water. However, the high sample-to-flux ratio can introduce impurities, leading to elevated total dissolved solids (TDS) in sample solutions, potentially hindering instrument functionality, especially for ICP-MS. Therefore, careful consideration of flux properties is essential for accurate and reliable geochemical analysis.

3.2.2.1 Lead-fire assay

The conventional lead-fire assay method involves the fusion of the sample at high temperatures (around 1100°C) using a flux containing sodium carbonate (Na_2CO_3), borax ($Na_2B_4O_7$), silica (SiO_2), and lead oxide (PbO) acting as a collector and reducing agent (flour/starch) (Juvonen *et al.*, 2007; Ni *et al.*, 2021). Within this mixture, lead oxide is transformed into metallic lead, which effective extracts of gold, silver, and PGEs, ultimately settling at the crucible's base. Simultaneously, the remaining matrix elements react with the flux components, forming slag. After cooling and slag removal, the lead button is heated in a bone ash cupel in an open furnace, known as cupellation. The lead is oxidized to PbO, which is absorbed by the cupel, leaving behind just metal bead. This residue metal bead is dissolved using aqua regia and subsequently analyzed using instrumental techniques such as atomic absorption spectroscopy (AAS), inductively coupled plasma-optical emission spectrometry (ICP-OES) and ICP-MS.

3.2.2.2 Nickel-fire assay

The nickel sulfide fire-assay involves fusing the sample with flux reagents such as $Na_2B_4O_7$, Na_2CO_3, CaF_2, and SiO_2, along with collectors like nickel and sulfur in a clay crucible at a temperature of approximately 1100°C (Juvonen *et al.*, 2007; Ni *et al.*, 2019). As PGEs are siderophile and charcophile elements, they are known to be more attractive to the molten nickel sulfide phase, serving as a more effective collector compared to the lead-fire assay. Similar to the lead-fire assay, the resulting nickel sulfide button after separation from the slag is ground into a fine powder and treated with hot concentrated HCl. The insoluble sulfides of precious metals, together with the dissolved precious metals, are collected on tellurium. This Te co-precipitation step efficiently captures all PGE sulfides, improving the overall PGEs recoveries. The final black precipitate is then filtered and subjected to analysis using methods such as ICP-MS, HR-ICP-MS, ICP-OES.

3.2.2.3 Sodium peroxide (Na_2O_2) sintering

Sodium peroxide (Na_2O_2) sintering presents an alternative approach to conventional alkali flux methods (Na_2CO_3, NaOH, and $LiBO_2$) for dissolving PGE-containing samples at lower temperatures (480°C) as given in Meisel *et al.* (2002) and Bokhari and Meisel (2017). This technique offers minimal blank values and reduces TDS levels in comparison to conventional alkali fluxes. Sintering with a sample-to-Na_2O_2 ratio of 1:4 is widely used in this method, enabling rapid mineral attack and the production of an easily soluble residue. Lower temperatures help in minimizing corrosion of the vessels (glass, carbon, zirconium, or nickel). Corundum crucibles have been successfully employed in the determination of PGE and Au via this technique, offering the advantage of minimal PGE blanks, along with the cost-effectiveness of the crucibles. The use of zirconium crucibles should be avoided as it can create $^{90}Zr^{16}O^+$ and $^{179}Hf^{16}O^+$ ions, causing interference with $^{106}Pd^+$ and $^{195}Pt^+$ ions in the ICP-MS analysis. Once the sample is fused into a bead, it is cooled down and dissolved by a drop-by-drop addition of 6 M HCl, and subsequently prepared for analysis using ICP-MS.

3.2.3 Fusion techniques (for solid mode analysis)

Fused sample pellets are prepared using the borate fusing technique, where the sample combines with a mixture of $LiBO_2$ (35.3%) and $Li_2B_4O_7$ (64.7%) at high temperatures (1000–1200°C). The sample dissolves in the flux, and the resulting molten mixture is shaped into disks using a mold. These disks offer a nearly complete homogenization, effectively mitigating the impact of particle size, surface roughness, and matrix effects. Glass pellets, disks, or beads created with fluxes like $LiBO_2$, $Li_2B_4O_7$, and $Na_2S_2O_7$ fusions are commonly employed for X-ray fluorescence (XRF) and in laser ablation ICP-MS (LA-ICP-MS).

While qualitative and semi-quantitative analyses are generally routine on LA-ICP-MS, the calibration process poses a challenge. Although certified reference materials exist, they do not cover all sample types, leading to many laboratories creating their own matrix-matched standards as external calibration. Techniques include adding a reference element to a powdered matrix, mixing, pressing, or co-precipitation of PGEs. Additionally, the dual sample introduction method by Thompson *et al.* (1989) provides an alternative by utilizing two channels for ablated material and nebulized aqueous solution standards in LA-ICP-MS calibration. Hence, special care and attention are crucial during the preparation of solid calibration standards tailored for the targeted PGEs elements to ensure optimal accuracy in quantitative analysis.

3.3 CHROMATOGRAPHIC SEPARATION, EXTRACTION, AND PURIFICATION OF PGEs

Following comprehensive sample pre-treatment, the subsequent critical step involves chromatographic column separation to enhance the purification of target PGEs and refine the analytical process. PGEs exhibit distinctive behavior in mineral acids, forming robust anionic chloro complexes; a behavior distinct from the majority of transition metals, alkali metals, and alkaline earth metals that typically form weaker anionic or stable cationic complexes. This unique feature presents an opportunity for effective separation of PGEs from matrix elements. Various elution methods have been explored to optimize analyte recovery, facilitate matrix separation, minimize background interference, and isolate each PGE from one another.

3.3.1 Cation-exchange chromatography (pre-concentrate PGEs)

PGEs have a tendency to form anionic complexes in mineral acids (<1 M HCl). These complexes exhibit limited adherence to cation resins, allowing them to pass directly through, while other metals quantitatively adsorb to the resin. Consequently, the cation-exchange resin selectively avoids absorbing Ru, Re, Ir, Pt, Rh, Pd, and Mo due to the formation of negatively charged chloro-complexes (Lommelen *et al.*, 2019). In contrast, the resin efficiently retains major elements like Fe, Mg, Zn, and Cu. Notably, PGE yields through a cation column can achieve up to 95%, especially when utilizing a mixed organic-solvent-acid medium. Table 3.1 outlines the steps to pre-concentrate PGEs from geological materials using cation-exchange resins (Hopp *et al.*, 2016, 2018).

The PGE fractions obtained from the cation-exchange column require a process of evaporation to dryness, followed by re-dissolving in a few mL of concentrated HCl and HNO_3, and undergo another drying step at 80°C. This critical procedure is essential to eliminate organic compounds from the cation resin, enhancing the purity of the PGEs in preparation for concentration analysis or subsequent isolation steps.

Distillation to purify Ru: Following the pre-concentration of all PGEs by a cation resin, ruthenium (Ru) can be further isolated from the rest of the PGEs through a distillation process described in Fischer-Gödde *et al.* (2015). The dried PGEs samples are redissolved in a solution of H_2SO_4-CrO_3 in a 2:1 ratio and heated in three steps of 2 h each at 70°C, 80°C, and 90°C. The oxidation will result in the formation of volatile RuO_4 which is then collected in vials containing 10 mL of 3% H_2O_2 where the volatile RuO_4 is reduced and recovered as RuO_2. Os is the only PGE oxidized alongside Ru as OsO_4; however, the volatile OsO_4 remains unreduced by H_2O_2, ensuring the Ru fraction trapped in the H_2O_2 is exceptionally pure. The purified Ru is subsequently subjected to concentration and isotopic composition analysis.

Table 3.1 Elution steps to pre-concentrate PGEs in cation-exchange resin (Hopp *et al.*, 2016, 2018).

Steps	Reagents
Resin	AG50W-X8 (100–200 mesh)
Resin form	Cations
Resin volume	10 mL
Resin cleaning	40 mL of 6 M HCl
	40 mL of 6 M HCl + 1 M HF
	40 mL of Milli-Q
Resin conditioning	20 mL 0.2 M HCl
Sample loading_Step 1	2.5 mL 0.2 M HCl
Sample loading_Step 2+ To collect for Ru, Re, Ir, Pt, Rh, Pd and Mo	7.5 mL 0.2 M HCl
Resin rinsing+ To collect for Ru, Re, Ir, Pt, Rh, Pd and Mo	6.5 mL 0.2 M HCl

(*Remarks:* Further separation or isolation of one PGE from another is still required for precise isotopic ratio measurement).

3.3.2 Anion-exchange chromatography

Anion-exchange resins are widely used for their ability to effectively isolate individual PGEs from one another. In contrast to cation-exchange resins, anion resins exhibit robust retention of stable PGE anionic chloro-complexes in mineral acids. The key factors influencing PGE adsorption are their valence states, and the potential formation of stable ion pairs between chloro-complexes and quaternary ammonium groups on the resins.

3.3.2.1 Pt and Pd separation

Various combinations of eluants are possible in order to elute specific PGEs from resins, for instance, the purification of Pt from the pre-concentrated PGEs is given in Table 3.2 (Creech *et al.*, 2014, 2017).

3.3.2.2 Anion-exchange chromatography for Pt, Pd, and Ir separation

An alternative approach, outlined in Table 3.3, introduces consecutive separation in two steps to effectively isolate Pt and Ir from matrix elements and the remaining PGEs (Hunt *et al.*, 2017). The initial step of column separation involves Pt or Pd separation, using Br water as an oxidant to keep Ir in its oxidized (Ir^{4+}) state. The obtained Pt or Pd fractions undergo treatment with a mixture of concentrated HNO_3 and HCl (1.5 mL in a 2:1 ratio), which is then evaporated to dryness. The residue is subsequently redissolved either in ascorbic acid or peroxide acid, undergoing further purification detailed in Table 3.3. The role of ascorbic acid and peroxide acid is to serve as reducing agents for the reduction of Ir. Procedural blanks for the entire chemical separation procedure fall within the range of 0.1–0.3 ng of Pt per 0.3 g of the initial solid sample.

Table 3.2 Elution steps to purify Pt or Pd from other PGEs using anion exchange chromatography.

Steps	Reagents
Element	Pd and Pt target
Resin	AG1-X8 (100–200 mesh, Cl)
Resin form	Anions
Resin volume	1 mL
Resin cleaning	10 mL of 11 M HCl
	20 mL of Milli-Q
	10 mL 13.5 M HNO_3
	40 mL of 6 M HCl
Resin conditioning	40 mL 0.5 M HCl
Sample loading	10 mL 0.5 M HCl
Matrix elution	40 mL 0.5 M HCl
Elution–1 (Rinsing)	20 mL 1 M HCl
Elution–2 (transitional metals)	20 mL 0.8 M HNO_3
Elution–3 (Pd)	60 mL 11 M HCl
Elution–4 (Pt)	40 mL 13.5 M HNO_3

Specific choice between anion- and cation-exchange resins for PGE purification will depend on the nature of the samples, the chemical forms of the PGEs, and the overall analytical objectives. Resin selection, optimization of the elution and purification efficiency is crucial to achieving accurate and reliable results in PGE analysis.

3.3.3 Liquid–liquid extraction of Os, Ru, and Re

Osmium (Os) displays excellent solubility in liquid bromine (Br_2) and efficiently undergoes reduction to $Os(Br_6)^{2-}$ when exposed to concentrated HBr acid (>5 M to 12 M). More than 99.5% of OsO_4 stays in solution, avoiding penetration into the walls of PFA (perfluoroalkoxy alkane or Teflon) and plastic vials, therefore preventing losses in preparation of an Os spike solution. The extraction of osmium as OsO_4 from the aqua regia digest solution involves chloroform solvent extraction, as outlined in Cohen and Waters (1996). Subsequently, OsO_4 is back-extracted using concentrated HBr, followed by purification through micro-distillation, as described by Birck *et al.* (1997). In brief, the samples in Carius digestion tubes are cooled down to –10°C in a freezing MeOH/water mixture to prevent potential loss of volatile OsO_4 upon opening. 12 mL of H_2O and 30 mL of extractant (chloroform, $CHCl_3$), both pre-chilled to –4°C, are added, and the tubes are resealed with a propane-oxygen flame. A brief agitation at room temperature is ensured to mix the acid and extractant well, where the tubes are stored overnight upright to promote the separation of organic extractant, mineral acids, and solid phases.

Table 3.3 Consecutive elution steps to purify Pt, Pd or Ir from the remaining PGEs.

Steps (First Column)	Reagents
Element	Pt or Pd target
Resin	AG1-X8 (200–400 mesh, Cl)
Resin form	Anions
Resin volume	1.25 mL
Resin cleaning	20 mL 0.8 M HNO_3
	10 mL Concentrated HCl
	25 mL Concentrated HNO_3
	40 mL of 6 M HCl
Resin conditioning	8 mL 0.5 M HCl + 10% Br water
Sample loading	10 mL 0.5 M HCl + 10% Br water
Matrix elution	12 mL 1 M HCl +10% Br water
Matrix elution	5.5 mL 0.5 M HCl + 10% Br water
Matrix elution (Ag, Ru)	12 mL Concentrated HCl
Elution–1 (Pd, Ru)	10 mL 8 M HNO_3 (80–90°C)
Elution–2 (Pt)	14 mL 13.5 M HNO_3
Steps (Second Column)	**Reagents**
Resin conditioning	8 mL 0.5 M HCl
Sample loading	5 mL 0.5 M HCl + 0.1 M ascorbic acid
Resin rinsing	10 mL 0.5 M HCl
Elution–1 (Ir)	20 mL 2 M HCl
Elution–2 (Pt)	30 mL 13.5 M HNO_3
OR	
Resin conditioning	8 mL 0.5 M HCl
Sample loading (Fe, Ni elution)	5 mL 0.5 M HCl
Resin rinsing (Fe, Ni elution)	10 mL 0.5 M HCl
Ir reduction	20 mL 0.5 M HCl + 1% H_2O_2
Elution–1 (Ir)	20 mL 2 M HCl
Elution–2 (Pt)	30 mL 13.5 M HNO_3

The upper acidic layer contains Re and the denser organic layer at the bottom contains Os where the two layers are extracted using a pipette. Approximately 5 mL of concentrated HBr is added to the dense organic layer with Os and are agitated for a few minutes and left to settle overnight. The upper HBr layer containing the reduced Os is transferred and subjected to heating at 70°C for overnight to facilitate the complete reduction of OsO_4 to $Os(Br_6)^{2-}$ after which HBr is evaporated to dryness. The sample is purified further by micro-distillation where the dry residue is redissolved in 10 μl of CrO_3 in 12 N H_2SO_4 at 8% m/v (10 μl of a solution containing chromium trioxide (CrO_3) dissolved in 12 N sulfuric acid (H_2SO_4) with a mass/volume concentration of 8%), heated at 80°C for approximately 3 h where the volatile OsO_4 is collected in 25 μl of concentrated HBr. Extraction with chloroform ($CHCl_3$) or tetrachloromethane (CCl_4) can also be applied to RuO_4 separation from an aqua regia solution as both of these organic solvents have high sensitivity toward Os and Ru with a high partition coefficient in an organic-acid mixture.

It is possible to recover Re using the same technique. Unlike Os, the direct addition of extractant into the aqua regia solution is avoided due to the poor recovery of Re. Therefore, the upper acidic layer containing Re from the previous step is evaporated to dryness. The residue is redissolved in 2 M H_2SO_4 and Re is extracted into a 1% solution of tribenzylamine (TBA) in $CHCl_3$, then back-extracted into $NH_{3(aq.)}$ and further purified on an anion-exchange column.

Ion-exchange techniques require substantial quantities of resin to adsorb non-PGE metals, necessitating an extensive cleaning process with large volumes of clean acid. On the other hand, liquid–liquid extraction methods offer a simpler and cleaner alternative to traditional distillation and ion-exchange techniques, encompassing overall blank values and total recovery factors.

3.4 SINGLE VS. DOUBLE-SPIKING TECHNIQUES FOR CONCENTRATION AND ISOTOPIC ANALYSIS

Two distinct techniques are employed in studies requiring high precision and robust data for understanding geochemical processes and environmental behavior of PGEs. Single spiking or isotopic dilution serves as a powerful method for the concentration measurement of PGEs, meanwhile double spiking is employed for the precise measurement of stable isotopic ratios of PGEs.

3.4.1 Isotopic dilution technique

The isotopic dilution technique is a method used in analytical chemistry to determine the concentration of a specific element in a sample. It involves introducing a known amount of a stable isotope of the element of interest (the spike) into the sample. The ratio of the spiked isotope to the natural isotope in the sample is then measured. By comparing this ratio to a standard calibration curve, the concentration of the element in the original sample can be accurately determined. The isotopic dilution technique is particularly valuable for its ability to correct for various analytical interferences and improve the accuracy of concentration measurements in complex sample matrices. Challenges in the isotope dilution technique include the complexity of identifying two isotopes without isobaric interferences, as detailed in Table 3.4.

Isotopic dilution techniques can be applied to all PGE, except for the mono-isotopic elements, Rh. Single isotope spikes of $^{102,104}Ru$, $^{104,110}Pd$, ^{185}Re, $^{188,189,190}Os$, ^{191}Ir and ^{198}Pt can be the relevant candidates to prevent significant isobaric interferences and in determination of PGEs concentrations. Optimizing the spike solution is a crucial process for minimizing potential errors and obtaining reliable results in the analysis of PGEs. Minimizing excessive spiking in samples is crucial for mitigating the influence of the spike on measured isotopic ratios.

Table 3.4 Potential interferences of the isotopic dilution technique on each isotope of PGE.

PGE Isotopes	Interferences (Chlorides, Oxides, Argides)
^{99}Ru	$^{62}Ni^{37}Cl$, $^{64}Ni^{35}Cl$, $^{64}Zn^{35}Cl$, $^{61}Ni^{38}Ar$, $^{63}Cu^{36}Ar$
^{100}Ru	^{100}Mo, $^{65}Cu^{35}Cl$, $^{66}Ni^{40}Ar$, $^{62}Ni^{38}Ar$
^{101}Ru	$^{64}Ni^{37}Cl$, $^{64}Zn^{37}Cl$, $^{66}Zn^{35}Cl$, $^{61}Ni^{40}Ar$, $^{65}Cu^{36}Ar$
^{105}Pd	$^{89}Y^{16}O$, $^{65}Cu^{40}Ar$
^{106}Pd	^{106}Cd, $^{90}Zr^{16}O$, $^{90}Mo^{16}O$, $^{66}Zn^{40}Ar$
^{108}Pd	^{108}Cd, $^{92}Zr^{16}O$, $^{68}Zn^{40}Ar$
^{192}Os	$^{176}Yb^{16}O$, $^{176}Lu^{16}O$, $^{176}Hf^{16}O$
^{191}Ir	$^{175}Lu^{16}O$
^{193}Ir	$^{177}Hf^{16}O$
^{194}Pt	$^{178}Hf^{16}O$
^{195}Pt	$^{179}Hf^{16}O$
^{196}Pt	^{196}Hg, $^{180}Hf^{16}O$

The challenges in spiking arise from the diverse PGE distributions found in natural rocks across broad concentration ranges. Preparation of various mixed PGE spikes is necessary to optimize spiking for samples displaying diverse PGE distribution patterns.

3.4.2 Double-spiking technique

3.4.2.1 Working principle

The double-spike technique is a highly effective approach for determining the stable isotopic composition of PGEs possessing a minimum of four stable isotopes and predominantly employed especially in conjunction with multi-collector ICP-MS (MC-ICP-MS). By measuring the isotopic composition in both the double spike and the sample-spike mixture, the actual isotopic composition of the sample can be calculated using algebraic inversion, as demonstrated by Klaver and Coath (2019).

The advantage of the double-spike technique over the traditional standard-sample method lies in its ability to distinguish natural mass fractionation from any mass-dependent isotope fractionation induced during chemical purification and mass spectrometry (Zhang *et al.*, 2019). Consequently, this technique is less susceptible to potential analytical artifacts stemming from variations in sample matrices and incomplete chemical yields during the purification of the element of interest. In double-spike techniques, it is commonly assumed that any isotopic difference between the sample and the standard is purely mass-dependent, which is the case for most terrestrial samples. However, it is essential to conduct a distinct isotope measurement on an unspiked aliquot in order to determine mass-independent isotope anomalies due to nucleosynthetic anomalies and nuclear field shift effects in the case of extraterrestrial samples (meteorites).

The double-spike technique involves the use of four different isotopes of an element of interest. Two of which are enriched in the double spike, whereas the other two are used in the inversion of measured isotopic data to derive the natural mass fractionation factor, α. Rudge *et al.* (2009) compiled a comprehensive list of double-spike compositions along with their corresponding theoretical uncertainties on determined α for elements with more than four isotopes. This compilation is grounded in the isotopic compositions of individual spikes sourced from the Oak Ridge National Laboratory (USA). In this section, the emphasis is on the double-spike cocktail list used in many studies to determine the isotopic composition for PGEs, following the guidelines provided by Rudge *et al.* (2009), minimizing the theoretical error on α.

3.4.2.2 Double-spiking technique for PGEs

Ruthenium (Ru): Ruthenium has seven stable isotopes, therefore providing numerous possibilities for designing a double spike. The optimal double spike for Ru would involve a combination of ^{98}Ru-^{102}Ru spike pair, with the isotopes ^{96}Ru, ^{98}Ru, ^{102}Ru, and ^{104}Ru used in the inversion process (Hopp *et al.*, 2016). However, using the isotopes ^{96}Ru, ^{98}Ru, ^{102}Ru, and ^{104}Ru is not ideal due to isobaric interferences from Mo, Zr, and Pd affecting all four isotopes. Additionally, the cup configuration of the Neptune Plus MC-ICP-MS complicates matters as it does not allow the simultaneous measurement of Zr, Mo, and Pd as to monitor these interferences. As an alternative, ^{98}Ru-^{101}Ru is the double-spike pair with ^{98}Ru, ^{99}Ru, ^{101}Ru, and ^{102}Ru to be used in the inversion process. Potential isobaric interferences are only relevant for ^{98}Ru and ^{102}Ru by ^{102}Pd where the impact of Pd interferences is to be minimal due to its low abundance (1.02%).

Platinum (Pt): Having five stable isotopes and one with an exceptionally lengthy half-life (^{190}Pt, with a half-life of approximately 10^{11} years), Pt is exceptionally well-suited for the double-spike technique. There are two potential pairs of Pt double spikes such as ^{192}Pt–^{198}Pt and ^{196}Pt–^{198}Pt (Creech *et al.*, 2014). The ^{192}Pt–^{198}Pt pair comprises a mixture of 38.9% ^{192}Pt and 61.1% ^{198}Pt spikes with the smallest predicted error propagation, where the inversion is performed using ^{192}Pt, ^{194}Pt, ^{196}Pt, and ^{198}Pt. However, a potentially large direct isobaric interference arises from the high abundance of ^{192}Os (40.8%) on the low abundance of ^{192}Pt (0.78%). Even trace amounts of Os in a sample can significantly impact the Pt signal on these masses, leading to inaccurate ^{192}Pt–^{198}Pt double-spike corrected ratios. On the other hand, the ^{196}Pt–^{198}Pt pair consists of a mixture of 27.7% ^{196}Pt and 72.3% ^{198}Pt spikes, and the inversion is performed using ^{194}Pt, ^{195}Pt, ^{196}Pt, and ^{198}Pt. The hypothetical error propagation on the ^{196}Pt–^{198}Pt pair is slightly larger than that of the ^{192}Pt–^{198}Pt pair. However, the ^{196}Pt–^{198}Pt pair presents potential advantages over the ^{192}Pt–^{198}Pt pair as it is unaffected by the presence of Os, making it an ideal double-spike pair for measuring Pt isotopes.

Palladium (Pd): Palladium has six stable isotopes: ^{102}Pd, ^{104}Pd, ^{105}Pd, ^{106}Pd, ^{108}Pd, and ^{110}Pd, offering a variety of possibilities for Pd double-spike techniques (Creech *et al.*, 2017). Following the minimum predicted errors using the double-spike toolbox by Rudge *et al.* (2009), the optimal pair is ^{106}Pd-^{110}Pd, with ^{105}Pd, ^{106}Pd, ^{108}Pd, and ^{110}Pd used in the inversion process. The ^{106}Pd-^{110}Pd double spikes are prepared by blending the two spikes in the optimal proportions of 47.17% ^{106}Pd and 52.83% ^{110}Pd. The inversion process excludes ^{102}Pd or ^{104}Pd to prevent potential interference from ^{102}Ru and ^{104}Ru.

Osmium (Os): Osmium has five stable isotopes (^{184}Os, ^{188}Os, ^{189}Os, ^{190}Os, and ^{192}Os), therefore, there is possibility of employing ^{188}Os, ^{189}Os, ^{190}Os, and ^{192}Os in a double-spike approach. A promising double-spike pair is the ^{188}Os-^{190}Os, pair, consisting of 27.61% ^{188}Os and 39% ^{190}Os with the optimal spike-to-sample ratio to be 55:45. The double-spike deconvolution approach for Os is based on the geometric iterative resolution method as outlined by Nanne *et al.* (2017) and cross-checked using an algebraic resolution method by Millet *et al.* (2016) and it involves resolving the following non-linear equation:

$$R_m = [(1-f)R_{\text{standard}}(i_x/i_n)^{\alpha} + fR_{\text{spike}}] * (i_x/i_n)^{\beta}$$

where R_{m}, R_{standard}, and R_{spike} stands for the measured, standard, and spike isotope ratios, respectively. i_n denotes the atomic weight of the normalizing isotope (^{188}Os), while i_x signifies the atomic weight of one of the other isotopes used in resolving the equation (^{189}Os, ^{190}Os, and ^{192}Os). f represents the relative proportion of ^{188}Os originating from the spike in the sample–spike mixture, while α and β correspond to the natural and processing/instrumental exponential fractionation factors, respectively.

Unlike Ru, Pt, and Pd, the error model used for Os deviates from the one presented by Rudge *et al.* (2009) in two key aspects: (i) errors on the natural fractionation factor (α) are computed with a constant intensity for the most abundant isotope in the natural Os-double-spike (OS-DS) mixture, as opposed to being calculated based on a constant total Os ion beam, and (ii) the determination of errors is achieved through Monte Carlo modeling.

3.4.2.3 Double-spike solution preparation

The preparation and calibration of double spike for PGEs analysis can be prepared in the following way (Ru is taken as an example). The powder single spikes of ^{98}Ru and ^{101}Ru are digested in aqua regia ($3HCl{:}1HNO_3$) inside CTs at 200°C for 14 days (Hopp *et al.*, 2016, 2018). Following complete digestion, the solutions are evaporated to dryness and re-dissolved in 6 M HCl. The Ru isotopic composition and concentration are determined through isotopic dilution using multi-collector ICP-MS (MC-ICP-MS) for calibration. Portions of these single spikes are then combined to create a ^{98}Ru-^{101}Ru double spike with optimal proportions (47.87% ^{98}Ru spike and 52.13% ^{101}Ru spike) to minimize error propagation in the double-spike inversion. This double spike undergoes equilibration for 48 h at 120°C on a hot plate using 6 M HCl, then goes through multiple drying cycles in concentrated HCl, and is ultimately re-dissolved in 6 M HCl. In the subsequent step, a small aliquot of the double spike is used to determine its concentration (via isotopic dilution) and isotopic composition.

3.5 ANALYTICAL TECHNIQUES FOR PGEs ANALYSIS

3.5.1 Inductively coupled plasma

Recent advanced techniques, such as ICP-MS in MS/MS mode and neutron activation analysis, excel in detecting PGEs at ultra-low concentrations: μg/L, ng/L, up to pg/L.

3.5.1.1 Working principle of ICP-MS

In ICP-MS, an argon plasma is generated using a torch with three concentric gas flows: the main gas flow supplies argon, an auxiliary flow controls the plasma's position, and the central channel introduces the sample (Figure 3.2). An electric discharge initiates the plasma, sustained by an electromagnetic field from a copper load coil. This field accelerates electrons, causing collisions and plasma formation. The sample undergoes vaporization, atomization, and ionization in the plasma, with resulting ions extracted into a mass spectrometer. Two (three in some case) interface cones guide ions through ion lenses into the mass filter, where the commonly used quadrupole mass filter isolates ions based on their mass-to-charge ratio (m/z). Other mass filters, like sector field and time-of-flight mass spectrometers, can also be used with ICP-MS. Samples, introduced as gas, droplets, or solid particles, use pneumatic, ultrasonic, and micro-concentric nebulizers for liquid samples, while laser ablation or electrothermal vaporization transforms solid samples.

3.5.1.2 Sector field SF-ICP-MS

Interference poses a significant challenge in the determination of trace elements using ICP-MS, manifesting in two distinct forms. The first is spectral interference, where multiple ions share a similar mass-to-charge ratio and remain indistinguishable by the mass filter. The second is non-spectral interference or matrix effects. Spectral interferences can be further categorized into three types: isobaric overlap, polyatomic ions, and double-charged ions. Sector field ICP-MS, an early and conventional method, stands out as a powerful analytical technique used to precisely measure the concentration of PGEs. It addresses spectral interferences associated with analytes in solutions, offering enhanced resolutions that are effective in mitigating interference challenges. SF-ICP-MS instruments offer a range of resolution settings from low ($m/\Delta m = 300$) to high ($m/\Delta m = 10\,000$). Table 3.5 provides a compilation of minimum required resolution to analyze PGEs using SF-ICP-MS. The effectiveness of signal separation between interferences and analytes depends on the mass differences between interfering ions and the analyte, as well as the ratios of signal intensity and the instrument's resolution capabilities (Moldovan *et al.*, 2004). A resolution of 10 000 is often adequate for eliminating certain interferences, including those from polyatomic and double-charged ions affecting analyte ions. The resolution ($m/\Delta m$) required to overcome ZrO^+ on ^{106}Pd is 26 000, and $ArCu^+$ on ^{103}Rh is 8000.

3.5.1.3 Multi-quadrupole 4Q-ICP-MS

A recent breakthrough in the realm of ICP-MS is the introduction of the NexION® 5000 Multi-quadrupole ICP-MS, operating in MS mode, MS/MS mode or Mass shift mode, enabling highly accurate measurements at ultra-low concentrations (Figure 3.2). The third cone allow to work in 'Extraction' mode (200 V) or 'Focused' softer mode (4 V) in combination with the narrow mass band width of Q0 in order to maximize the signal to noise. With four sets of quadrupoles, the ion beam is shaped and directed within Q0 (quadrupole ion deflector) and filtered in Q1 (first transmission analyzer quadrupole). The reaction is controlled in Q2

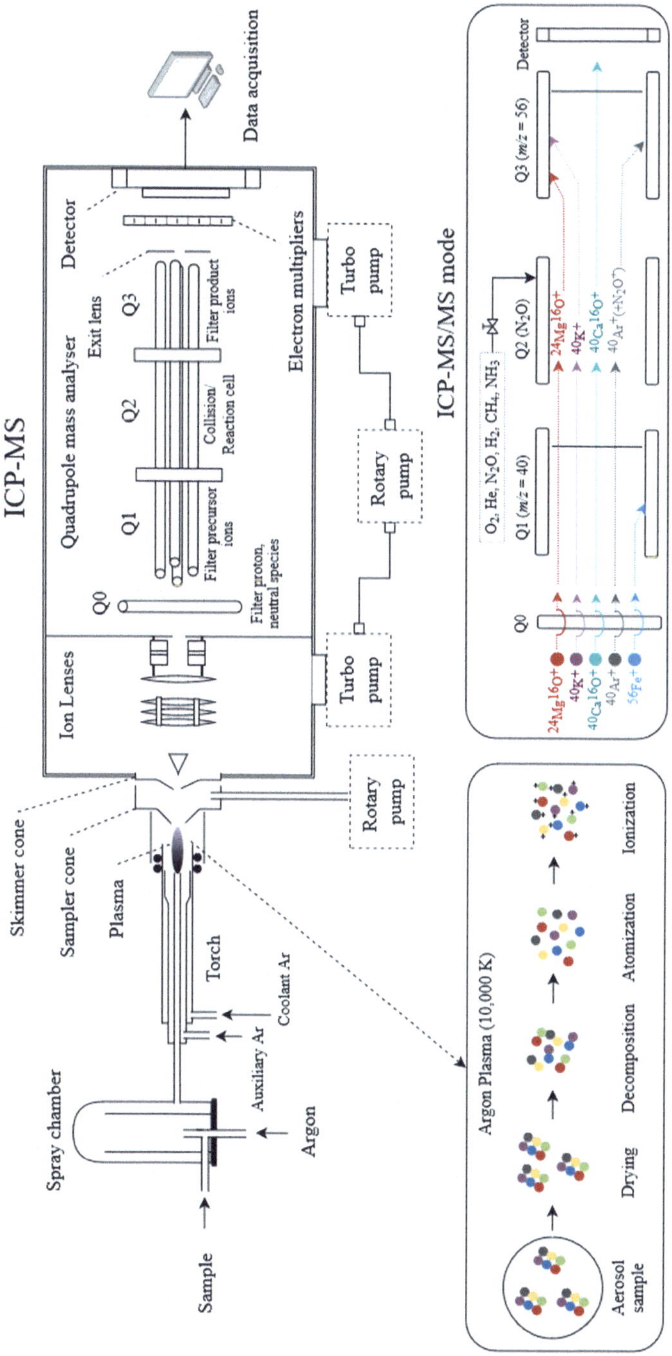

Figure 3.2 Schematic diagram of ICP-MS.

Table 3.5 Interferences and minimum required resolution to analyse PGEs in mass spectrometry.

Element	Interfering Species	Minimum Resolution Required (m/Δm)
^{103}Rh	$ArCu^+$	8040
	$ArZn^+$	10 100
	Pb^{2+}	1248
	SrO^+	102 900
	RbO^+	147 000
^{105}Pd	$ArCu^+$	7300
	YO^+	27 600
	SrO^+	1 000 000
	RbO^+	28 400
^{106}Pd	ZrO^+	26 000
^{195}Pt	HfO^+	8100

(quadrupole universal cell), and the resulting ions are separated in Q3 (second transmission analyzer quadrupole). The universal multichannel cell allows analysis to be linked in standard mode (MS Q_1 or Q_3), MS/MS (kinetic energy discrimination-KED or dynamic reaction cell-DRC), mass shift and ion scan. The system offers flexibility for simultaneous analysis of high and low concentration analytes in a single run, delivering background equivalent concentrations (BECs) for Ca and K in hot plasma of less than 1 ppt. Additionally, it provides ultra-low detection limits for ultra-trace elements like PGEs.

An evaluation on a set of PGE standards confirmed the blank response, the background equivalent concentrations (BECs), and the analytical detection limit (D.L.) in three acquisition modes. These modes encompassed the standard MS/MS mode, the single quad mode with Q3 exclusively, and the MS/MS mode integrated with a kinetic energy discrimination (KED) system utilizing helium gas to mitigate potential interferences. The results, presented in Table 3.6, indicate that the blank concentration in a 1% HCl matrix is less than one count per second. The background equivalent concentration (BEC) ranges between 0.008 and 0.025 parts per trillion (ppt), while the analytical detection limit (D.L.) spans from 0.105 to 0.424 ppt for the corresponding PGEs. Notably, these results showcase the most exceptional blank response achieved to date, coupled with the highest sensitivity of the analyte in the history of ICP-MS. It is evident that the utilization of 1/Q3 mode is the most suitable approach for the analysis of PGEs based on the BECs and D.L. (Table 3.6).

3.5.1.4 Single particle inductively coupled plasma mass spectrometry

Single particle SP-ICP-MS emerges as a powerful tool, facilitating the examination of individual nanoparticles (defined as objects with at least one characteristic dimension in the range 1–100 nm) in diverse and complex

Table 3.6 Blank response, background equivalent concentrations (BECs), and the analytical detection limit (D.L.) of PGEs measurement on NexION 5000 ICP-MS/MS in three acquisition modes (Cavalheiro *et al.*, 2014; Preud'Homme *et al.*, 2019).

Mode	MS/MS Mode			1/Q3 Mode			MS/MS Mode with Helium KED Cooled		
PGEs	Blank (cps)	BEC (ppt)	DL (ppt)	Blank (cps)	BEC (ppt)	DL (ppt)	Blank (cps)	BEC (ppt)	DL (ppt)
Ni	713.18	0.8050	2.8210	1350.286	0.6550	1.3340	200.279	1.1980	5.9090
Ru	2.8580	0.0220	0.1580	4.9480	0.0040	0.0640	0.5560	0.0250	0.1970
Rh	3.0770	0.0030	0.0390	17.3330	0.0080	0.0340	0.8330	0.0080	0.1050
Pd	3.0770	0.0310	0.2750	9.6480	0.0440	0.1110	0.2780	0.0190	0.1890
Os	4.7390	0.0410	0.1710	18.9360	0.0400	0.0690	0.8140	0.0230	0.1460
Ir	0.6840	0.0090	0.0730	3.1110	0.0090	0.0440	0.5560	0.0110	0.2500
Pt	2.0510	0.0230	0.3250	7.5560	0.0060	0.2100	0.8330	0.0090	0.4240

Table 3.7 Instrumental set-up and operational parameters to measure nanoparticles of PGEs (Jiménez-Lamana *et al.*, 2016; Kińska *et al.*, 2018; Gupta *et al.*, 2023).

Instrumental Parameters	NexION 300X		NexION 5000	Agilent 7900
RF Power (W)	1450		1600	1550
Analog stage voltage (V)	−1787		–	–
Interface cones – Sampler	Ni		Pt	
Interface cones – Skimmer	Ni		Pt	
Nebulizer	Meinhard® glass microcentric		Inert concentric PFA	Meinhard® glass microcentric
Spray chamber	Glass cyclonic		Ultra inert cyclonic chamber	Scott chamber
Plasma gas flow rate (L/min)	15	15	–	–
Auxiliary gas flow rate (L/min)	1	0.9	–	–
Nebulizer gas flow rate (L/min)	0.98	1.1	1.03	1.05
Operational parameters				
Analyte	Pt	Pd	Au, Ag, Pt	Au, Ag, Pt
Mass (amu)	195	105	197, 107, 195	
Density (g/cm^3)	21.45	12.023	19.3, 10.5, 21.45	
Dwell time (μs)	100	100	10/25/50/100	100
Acquisition time (s)	100	60	100	100
Transport efficiency (TE%) Au NP equivalent	8–10%	3.5–4%	11–14	7–8

samples (Meermann & Nischwitz, 2018). SP-ICP-MS is able to provide the following information after very short (few minutes 60–100 sec) measurements and at very low concentrations: the presence of nanoparticles and/or dissolved species of the same element in a nanoparticle suspension, the nanoparticle size, the nanoparticle number and mass concentration as well as the mass of the element per particle (Laborda *et al.*, 2023).

PGEs nanoparticles (NPs) find extensive use and consequently end up in the environment (Jiménez-Lamana *et al.*, 2016; Chapter 10). Notably, platinum nanoparticles (PtNPs), palladium nanoparticles (PdNPs), and rhodium nanoparticles (RhNPs) are integral components of automotive catalytic converters (Omran *et al.*, 2020) and serve various applications in fields such as fuel cells (Muthuswamy *et al.*, 2013), immunosensors (Tang & Tang, 2015), and biomedical technologies (Gao *et al.*, 2007). SP-ICP-MS has been used to detect, characterize, and quantify PGE-NPs in diverse environmental compartments such as in road dust (Folens *et al.*, 2018), where the presence of Pt is shown to be in nano-particulated form, with sizes ranging from 9 to 21 nm. In addition, SP-ICP-MS offers comprehensive insights into the behavior of PGE-NPs in various plant tissues during the interaction of PGE-NPs with plants, involving uptake, bioaccumulation, and translocation.

The instrumental setup and operational parameters for detecting PGE-NPs in plant tissues using the NexION 300X as well as Agilent 7900 are detailed in Table 3.7 (Jiménez-Lamana *et al.*, 2016; Kińska *et al.*, 2018). The study on single particle analysis utilizes the NexION 5000 (ICPMultiQuadMS/MS) and Agilent 7900 (ICP-CRC-MS), assessing the performance of analyzing a range of nanoparticle types, including spherical, core/shell, rod-shaped, cubic, and those with unknown shapes (Gupta *et al.*, 2023). This evaluation involves a comparison across various mass spectrometry modes, instrumental setups, and conditions.

3.5.1.5 Single particle ICP time-of-flight mass spectrometry

Single particle inductively coupled plasma time-of-flight mass spectrometry (SP-ICP-TOF-MS) is a cutting-edge analytical technique used for the characterization of nanoparticles in various sample matrices. This method combines the capabilities of single particle ICP-MS and TOF-MS, offering improved sensitivity, resolution, and elemental specificity for individual nanoparticles. In SP-ICP-TOF-MS, nanoparticles are introduced into the plasma, where the ions are subjected to a time-of-flight mass analyzer, which measures the time it takes for each ion to travel a known distance.

SP-ICP-TOF-MS has applications in fields such as nanotoxicology, environmental monitoring, and the characterization of engineered nanoparticles. Its advanced capabilities contribute to a deeper understanding of the behavior and fate of nanoparticles in various systems, providing crucial insights for both research and regulatory purposes. Key features of SP-ICP-TOF-MS include its ability to provide rapid acquisition of mass spectra and simultaneous monitoring of all nuclides at high speed, simultaneous measurement of multiple elements, high sensitivity for low-concentration samples, and the capability to analyze a wide range of particle sizes (Hendriks *et al.*, 2023), therefore addressing the technical issues faced in SP-ICP-MS. A study by Taskula *et al.* (2022) demonstrated that employing a time-of-flight ICP-MS instrument in single particle mode (SP-ICP-TOF-MS) enables the differentiation of anthropogenic PtNPs from the natural background in sediments.

In comparison, SP-ICP-MS is the sequential scanning-based ICP-MS where the dwell time or the duration of transient events can be as long as 100 ms (Table 3.7), resulting in an unavoidable loss of signal. Despite advancements like SP-ICP quadrupole mass spectrometry (SP-ICP-QMS) with dwell times <1 ms designed for fast transient signal monitoring, sequential mass analyzers are limited to observing signals from one or a maximum of two nuclides with different mass-to-charge (m/z) ratios (Zhuravlev *et al.*, 2016). Consequently, the observation of only a fraction of a particle event per nuclide hinders accurate quantification.

3.5.1.6 Multi-collector inductively coupled plasma mass spectrometry

Multi-collector inductively coupled plasma mass spectrometry (MC-ICP-MS) is an advanced technique tailored for precise isotopic analysis (Figure 3.3).

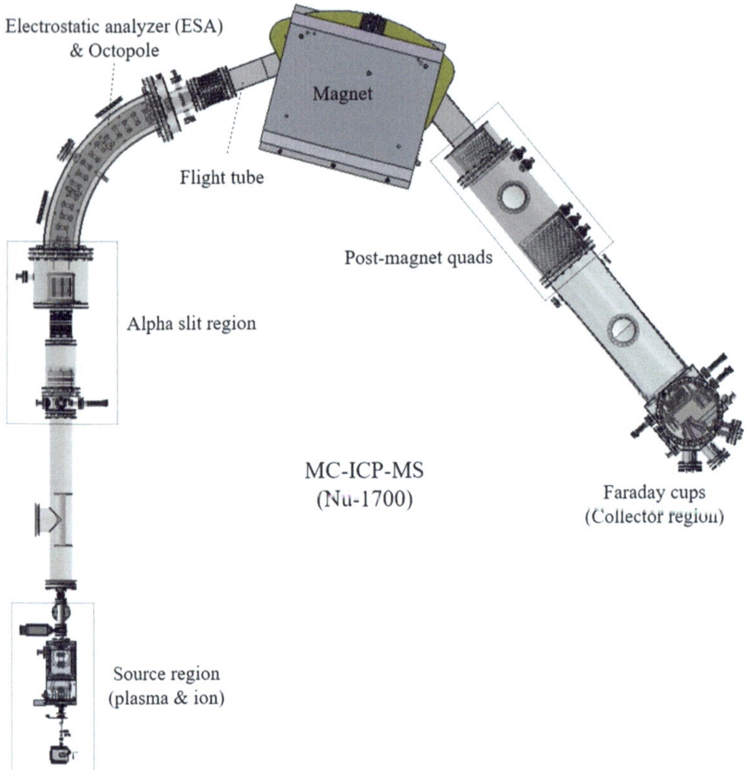

Figure 3.3 Schematic diagram of multi-collector ICP-MS (MC-ICP-MS).

Distinguishing itself from single-collector systems, MC-ICP-MS employs multiple detectors, allowing simultaneous measurement of isotopes to enhance both efficiency and accuracy. In this method, the sample undergoes ionization within the plasma, and the resulting ions are accelerated through a magnetic field toward multiple detectors strategically positioned to collect ions with specific mass-to-charge ratios. The concurrent collection of isotopic signals facilitates precise determination of isotopic ratios. The versatility of MC-ICP-MS finds applications in various fields such as geology, environmental science, archaeology, and nuclear forensics. Continuous advancements in instrumentation and methodologies continually bolster the capabilities of MC-ICP-MS, making it a widely utilized tool across diverse scientific disciplines.

3.5.1.6.1 Measurement of stable ruthenium (Ru) isotopic ratios

Ru isotopes are measured using a ThermoScientific Neptune+ where Ru in 0.28 M HNO_3 is introduced via ESI Apex-IR desolvating system coupled with a MicroMist glass nebulizer (100 μL/min uptake rate) as described in Hopp

Table 3.8 Instrumental set-up and measurement parameters for PGEs on MC-ICP-MS.

	Ru	Pt		Pd
Operating Conditions	**Neptune+**	**Nu Plasma**	**Neptune**	**Neptune+**
RF power (W)	1300	1300	1200	1300
Plasma gas flow rate (L/min)	16	14	15	16
Interface cones – Sampler	Ni	Ni	Jet cone	Jet cone
Interface cones – Skimmer	Ni	Ni, H-type	X-cone	Ni, H-type
Acceleration voltage (kV)	10	6	10	10
Instrumental resolution	Low resolution	~1840	~1600	~1700
Mass analyser pressure (mbar)	$5–9 \times 10^{-9}$	$5–9 \times 10^{-9}$	8×10^{-9}	8×10^{-9}
Sample introduction system	ESI ApexIR	Nu DSN–100	ESI ApexIR w/ACM	ESI ApexHF
Sample uptake rate (μL/min)	~100	~80	~30	~50–100
Solution concentration (ppb)	200	60_75	100	100
Sensitivity (V/ppm)	100	280	100	900
Sample measurement time (s)	50×4.2	100×8	100×8.4	100×8.4
Washout time (min)	30	30	30	20
Background measurement time (s)	40×4.2	100×8	100×8.4	8.4

et al. (2016). Ion beams are simultaneously collected in static mode for all seven stable Ru isotopes (^{96}Ru, ^{98}Ru, ^{99}Ru, ^{100}Ru, ^{101}Ru, ^{102}Ru, ^{104}Ru) as well as ^{97}Mo and ^{105}Pd, serving as interference monitors. Oxide formation is monitored as Ce/CeO and kept below 1% by introducing N_2 into the sample gas. Baseline measurements are done on peaks at 40×4.2 s integration on a solution blank, followed by each sample/standard measurement at 50×4.2 s integration of the ion beams, consuming around 90 ng of Ru. Concentrations of 200 ppb Ru yielded a total ion beam intensity of about 2×10^{-10} A, corresponding to a sensitivity of ~100 V/ppm. A spiked Alfa Aesar standard solution ($RuCl_3$; Lot # 61300952) is used as a bracketing standard (Hopp *et al.*, 2016). The detailed instrumental set-up is given in Table 3.8.

The raw isotope data are processed using the double-spike toolbox developed by Rudge *et al.* (2009), utilizing MATLAB's non-linear equation solving routine. Raw ion beams measured at ^{98}Ru*, ^{99}Ru, ^{101}Ru, and ^{102}Ru* are employed for data deconvolution, correcting for isobaric interferences from ^{98}Mo and ^{102}Pd on ^{98}Ru* and ^{102}Ru*. The initial estimation of β is performed to calculate mass bias-corrected ^{97}Mo and ^{105}Pd intensities, subsequently used for isobaric interference correction. These interference-corrected raw beams are then input into the double-spike toolbox to derive a new estimate of β. This process iterates until the β values converge, typically achieved after two iterations (Hopp *et al.*, 2016). The Ru isotopic composition of the samples is then calculated using the natural fractionation factor α obtained from the double-spike inversion.

$$\delta^{102/99}Ru_{meas} = -1000 \times (\alpha_{sample} - \alpha_{standard}) \times \ln(m_{102} / m_{99})$$

where m_{102} and m_{99} represent the atomic weights of ^{102}Ru and ^{99}Ru, and 'standard' refers to the mean composition of the spiked Alfa Aesar standard solution acquired in each analytical session.

3.5.1.6.2 Measurement of stable osmium (Os) isotopic ratio

Both stable and radiogenic Os isotopic ratios can be measured using both a plasma source ThermoFisher Scientific Neptune (MC-ICP-MS) or a negative thermal ionization mass spectrometry Triton + (N-TIMS) (Nanne *et al.*, 2017). The advantage of MC-ICP-MS lies in its ability to measure Os as Os^+, simplifying the mass spectrum and enabling easy corrections for isobaric interferences. With a relatively low ionization efficiency (0.08%), it is particularly well-suited for the analysis of larger samples and for calibration with standards. In contrast, N-TIMS boasts a much higher ionization efficiency for Os (1–5%), making it ideal for samples with a low Os abundance or limited sample size. Os is measured as the tri-oxide ion OsO_3^-, introducing complexity to the mass spectrum and necessitating multiple oxide corrections that contribute to the overall uncertainty.

The development of methods on both MC-ICP-MS and N-TIMS enables the analysis of samples across a wide range of Os concentrations (>1 ppb) with precisions of 0.01–0.02‰ on δ^{190}Os (2 s; 80 cycles for MC-ICP-MS and 280 cycles for N-TIMS). Additionally, the method allows for the simultaneous collection of stable isotopes and radiogenic isotope ratios, specifically ^{187}Os/^{188}Os and ^{186}Os/^{188}Os, with a precision of <40 ppm (2 s; 80 cycles for MC-ICP-MS and 280 cycles for N-TIMS).

Measurement on MC-ICP-MS: The sample is introduced in 3 M HCl using an ESI PFA-50 micro-flow nebulizer and GE micro-cyclonic 'Cinnabar' spray-chamber with uptake of ~80 μL/min, corresponding to a sensitivity of ~30 V/ppm (Nanne *et al.*, 2017). All Os isotopes except ^{184}Os are measured simultaneously and ^{183}W, ^{185}Re, and ^{194}Pt are monitored to correct isobaric interferences. The baseline was measured on-peak zero (OPZ) at 20 × 8.4 s integrations in a solution blank. Measurements are conducted in static mode in a single block of 80 cycles, each with an integration time of 8.4 s. Washout between samples is done in clean 3 M HCl for 5 min, leading to ^{188}Os signal below 2 mV. The total analysis time including OPZ and wash-out lasts approximately 15 min for each sample with the consumption of 0.8 mL of sample solution (400 ng total Os).

Measurement on TIMS: The sample is loaded onto Pt single filaments in concentrated HBr or 3 M HCl (the same sample matrix measured in MC-ICP-MS by Nanne *et al.* (2017). Following the deposition of the sample, approximately 0.5 μL of NaOH-Ba$(OH)_2$ activator is added to enhance the ionization of Os. The quantity of Os loaded can range from 2.3 to 45 ng of natural Os, where the intensity obtained does not correlate with the variation in the quantity of Os, but rather with the activator and the method of loading the sample on N-TIMS. In general, Os masses from 232 to 242 are analyzed in a two-sequence static multi-collection routine. Masses 241 and 242 are collected to determine the precise oxygen isotopic compositions (specifically, ^{17}O/^{16}O and ^{18}O/^{16}O), which will later be used for oxygen interference corrections. A single analysis comprises 28 blocks, each consisting of 10 cycles with 8.4 s integration per cycle, followed by 3 s of idle time. The presence of interferences, such as PtO_2^-, ReO_3^-, and WO_3^-, can be quantified by measuring masses 228, 230, and 233 using an ion counter before and after each Faraday measurement for 4–6 cycles, resulting in a total analysis time of 110–160 s.

3.5.1.6.3 Measurement of stable Pt isotopic ratios

Stable isotope measurements for Pt are conducted using a Nu Plasma MC-ICP-MS, in low-resolution mode with a mass resolution of ~1840 (M/ΔM; peak edge width from 5 to 95% full peak height). The analysis employs a DSN-100 desolvation nebulizer system with an ESI PFA micro-flow nebulizer, measuring samples containing 60 μg/L of natural Pt in a 0.5 M HCl matrix (Creech *et al.*, 2023). The sample uptake rate is around 80 mL/min, and the sensitivity is approximately 280 V/ppm of Pt. Each analysis comprises 100 × 8 s integrations, totaling 13.3 min of data acquisition and using roughly 80 ng of natural Pt. Before each analysis, baselines are established by measuring on-peak for 100 × 8 s integrations. A 30-min washout time is applied between analyses and baseline measurements, resulting in background levels typically <1 mV on all Pt isotopes. Data reduction can be done offline using the free software Iolite, operating within Igor Pro (Paton *et al.*, 2011). The double-spike (DS) calculations can be algebraically performed based on the equations outlined in Rudge *et al.* (2009) using the procedure integrated into Iolite (Paton *et al.*, 2011). Pt stable isotope results are presented as δ^{198}Pt, indicating the permil deviation of the $^{198}Pt/^{194}Pt$ ratio in the sample from the corresponding ratio for the IRMM-010 Pt isotope standard. This technique has demonstrated external reproducibility in δ^{198}Pt measurements of Pt solution standards a ±0.04 (i.e., ±0.01/amu; Creech *et al.*, 2014).

Alternatively, platinum isotopes can be analyzed using a Thermo Scientific Neptune Plus instrument featuring a Cetac Aridus II desolvating nebulizer (uptake rate of 100 μl/min) in low resolution mode. This approach allows for the simultaneous collection of all Pt isotopes, along with $^{188}Os^+$ and $^{200}Hg^+$, facilitating corrections for isobaric interferences, and $^{191}Ir^+$ to evaluate Ir tailing effects on Pt isotopes. Each measurement involves 60 × 8.4 s integrations, with a 30 s baseline and peak centering step preceding each measurement (Berglund & Wieser, 2011). The NIST SRM 3140 Pt reference solution is used as bracketing standard, and the ion beam intensity of both the sample and bracketing solution is adjusted to achieve a match within 10%. Approximately 1.2 mL of solution with 300 ng of Pt per analysis is required to achieve an instrumental sensitivity of 200 V/µg Pt. Mercury interferences are first corrected on $^{198}Pt^+$ using $^{200}Hg/^{198}Hg = 2.31695$ (Berglund & Wieser, 2011). The corrected $^{198}Pt^+$ is then used for internal normalization and interference corrections. The intensity of $^{196}Hg^+$ was also subtracted from $^{196}Pt^+$. Correction for Os interference on $^{192}Pt^+$ is carried out using $^{188}Os/^{192}Os = 0.32467$ (Berglund & Wieser, 2011). Subsequently, a 2 s rejection criterion was applied to the 60 corrected cycles. All data are presented in epsilon notation, denoted as $\varepsilon^{191}Pt/^{195}Pt$).

3.5.1.6.4 Measurement of stable palladium (Pd) isotopic ratios

Palladium stable isotope measurements are conducted using a Thermo-Scientific Neptune+ (MC-ICP-MS). The operational and measurement conditions of this instrument are outlined in Table 3.8. Samples are introduced into the mass spectrometer using an ESI Apex-HF desolvator with a 100 μL/min PFA nebulizer. Sample and standard solutions are typically analyzed at concentrations of 100 ppb Pd in 0.5 M HCl matrix at low resolution mode with a mass resolution of approximately 1700 (Creech *et al.*, 2017). During Pd isotope measurements, ion beams at ^{102}Pd, ^{103}Rh, ^{104}Pd, ^{105}Pd, ^{106}Pd, ^{107}Ag, ^{108}Pd, ^{109}Ag, and ^{110}Pd were simultaneously collected in a single cycle using Faraday cups connected to amplifiers with $10^{11}\Omega$ resistors for all Pd channels. This configuration resulted in a combined ion beam intensity for all Pd isotopes of around 90 V at an uptake rate of approximately 100 μL/min, corresponding to a sensitivity of about 900 V/ppm (Hunt *et al.*, 2017). Pd stable isotope results are presented as δ^{106}Pd notation, the corresponding ratio for the in-house solution standard due to the absence of a certified Pd isotopic standard. This technique provides an internal precision of approximately $\pm$0.003‰ (2 SE, $n = 100$), and the long-term apparent reproducibility is $\pm$0.017‰ (2 SD, $n = 84$) over a period of six months.

Table 3.9 Instrumental set-up of laser ablation system coupled to ICP-MS (From: Ni *et al.*, 2021).

Laser set-up	
Laser ablation system	ArF excimer laser
Type specification	GeoLas HD
Wavelength	193 nm
Energy density	7 J/cm^2
Repetition rate	6 Hz
Spot size	60 μm
Pulse number	200
Carrier gas (He) flow rate	0.6 L/min
ICP set-up	
ICP-MS	Agilent 7700x
RF power	1550 W
Auxiliary gas (Ar) flow rate	1.05 L/min
Plasma gas (Ar) flow rate	15 L/min
Sampling depth	7.5 mm
Measurement mode	STD
Survey runs	Jump
Dwell time per isotope	20 ms
Detector mode	Dual
Measured isotopes	^{197}Au, ^{195}Pt, ^{105}Pd and ^{109}Ag

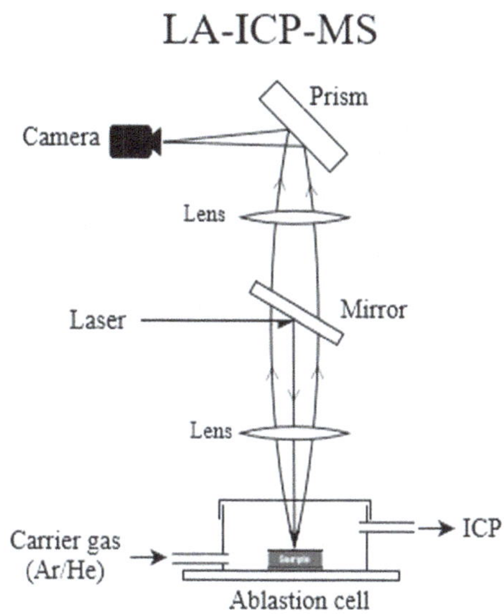

Figure 3.4 Schematic diagram of laser ablation inductively coupled plasma-mass spectrometry (LA-ICP-MS).

3.5.1.7 Laser ablation ICP-MS (LA-ICP-MS)

Laser ablation stands out as a versatile solid-sampling technique employed for both qualitative and quantitative analyses (Barefoot, 1998). This method involves delivering a brief laser pulse to the sample's surface, dislodging a small volume of particles which are then transported by an inert gas to plasma. Compared to sample introduction via a nebulizer solution, laser ablation sampling presents numerous advantages. Notably, it allows for the direct analysis of solid samples, eliminating the need for chemical dissolution that can introduce risks of contamination and analyte loss (Barefoot, 2004). It also enables the analysis of minute areas within solid matrices and facilitates the determination of spatial element distributions in samples. Table 3.9, extracted from Ni *et al.* (2021), complies the instrumental conditions used in the PGEs analysis via laser ablation coupled with ICP-MS (Figure 3.4).

For homogeneous samples, no specific sample preparation is required for quantitative bulk analysis. In cases of non-homogeneous samples, the preparation of disks involves pulverization and pressing, with or without a binder (Barefoot, 2004). The selection of the laser wavelength significantly influences the ablation process. Research suggests that shorter wavelengths result in higher ablation rates and reduced size fractionation of the ablated material. The applications discussed in this review predominantly utilize wavelengths of 1064, 266, and 213 nm (Campbell & Humayun, 1999; Reddi & Rao 1999). Shorter wavelengths, such as 266, 213, and 197 nm, provide higher photon energies for effective bond breaking and solid ionization (Rauch *et al.*, 2001, 2002).

However, calibration presents a notable challenge in LA-ICP-MS, relying on the development of solid, matrix-matched standards. While certified reference materials are available to facilitate matrix matching, they may not precisely correspond to every sample type, necessitating the preparation of customized matrix-matched standards. Beyond external standards, incorporating one or more internal standards is crucial to account for variations in instrumental operating conditions during analysis and discrepancies in masses ablated from samples and standards. A comprehensive review regarding technical details on laser ablation is given by Di Russo *et al.* (2002), and offers valuable insights.

This technique allows for the determination of PGEs with estimated detection limits in the lower ng/g range. Interferences are managed by assessing the isotope or concentration ratios in model samples. Unlike solution analysis, dry sample ablation results in lower oxide formation due to reduced oxygen concentration. A novel approach has been devised for the LA-ICP-MS analysis of PGE in road dust, involving the use of uncontaminated road dust spiked with PGE from both solution and solid certified reference material (NIST 2557).

3.5.2 Neutron activation analysis

Neutron activation analysis (NAA) stands as a highly sensitive and non-destructive analytical technique utilized for the quantitative determination of PGEs across diverse sample matrices. In the NAA process, samples are subjected to irradiation with neutrons, inducing the formation of radioactive isotopes within the material. The subsequent decay of these isotopes produces characteristic gamma rays that are unique to each element. By precisely measuring these gamma rays emitted during the decay process, the PGE concentrations can be accurately determined. The typical irradiation protocol consists of a brief exposure of 5 min to a thermal neutron flux of 10^{16} m/^{2} /s

Table 3.10 Nuclear data for thermal neutron activation of platinum group elements.

Synthetic Radiogenic Isotopes of PGEs					
Element	**% Abundance**	**Reaction and Nuclide**	**Half-life**	**Cross-section**	**γ-energy (keV)**
Ru	31.6	^{102}Ru(n,γ)^{103}Ru	39.26 d	1.3 ± 0.15 b	497.1
Rh	100	^{103}Rh(n,g)^{104m}Rh*	4.36 m	11 ± 1 b	51.4, 555.8
Pd	26.7	^{108}Pd(n,γ)^{109m}Pd*	13.43 h	8–12 ± 2 b	88.1
Os	26.4	^{190}Os(n,γ)^{191}Os	15.3 d	13 ± 0.3 b	64, 129.4
Ir	37.4	^{191}Ir(n,γ)^{192}Ir	74.3 d	630–924 ± 53 b	316.5, 468, 308.4, 296
	62.6	^{194}Ir	19.4 h	112.5 ± 7.5 b	328.5,293.6
Pt	7.2	^{198}Pt(n,γ)^{199}Pt (β^-)^{199}Au	30.8 m	3.7 ± 0.2 b	542.7,317
	7.2	^{199}Au	3.13 d		158.4, 208.2

*Excited nuclear isomer and data are adapted from Li and Ebihara (2003) and international atomic energy agency (IAE, 2005).

after which the sample is allowed for 5 min decay and ^{104m}Rh and ^{109m}Pd are counted at 51 and 188 keV, respectively. Subsequently, a lengthier irradiation of approximately 10 h is administered, followed by measurements after an overnight decay of 16 h and once again after 5 days.

This process enables the optimized detection of various PGEs and nuclear data for the radioisotopes used are listed in Table 3.10 (IAE, 2005; Li & Ebihara, 2003).

NAA offers a significant advantage by providing highly sensitive and accurate results without requiring sample destruction. This makes NAA particularly valuable in fields such as forensic analysis, where sample preservation is crucial. The versatility of NAA extends to its capability to analyze complex sample matrices such as catalysts, pharmaceuticals, iron meteorites, geological and environmental samples. However, the main nuclear interference arises from generation of ^{199}Au during Pt determination and relevant correction should be made. Elevated background elements such as Al, Si, Na, Ca, Ti, and Sc could also lead to additional nuclear interferences during reaction with Ca resulting in ^{47}Sc close to the Pt peak of ^{199}Au at 158 keV. Moreover, NAA does not typically require extensive sample preparation, making it a preferred method for the analysis of geological, environmental, and biological samples. However, the requirement for specialized facilities and the potential for interference from other elements are some factors that need to be carefully considered in the application of neutron activation analysis for the determination of PGEs.

3.6 CONCLUSION

The growing demands for PGEs in various industries necessitates precise trace-level determinations of PGEs for exploration, biological evaluation, and environmental impact monitoring. Speciation and isotopic studies play a crucial role in understanding the biogeochemical transformation of PGEs in the environment, in food chains, and health risk assessment. ICP-MS with possibility to different sample introduction systems stands out as a leading technique globally for PGEs determination, owing to its unmatched sensitivity and selectivity compared to other methods. The decreasing costs and increased availability of instruments ensure that ICP-MS remains indispensable for PGEs analysis in the foreseeable future. Moreover, the prudent use of certified reference materials (CRMs) when available, and the adoption of rigorously validated methods, including inter-laboratory comparisons, will further enhance the reliability and consistency of generated data.

REFERENCES

Antos G. J. and Aitani A. M. (2004). Catalytic Naphtha Reforming, Revised and Expanded, 2nd edn. CRC Press, Boca Raton, USA, https://doi.org/10.1201/9780203913505.

Barefoot R. R. (1998). Determination of the precious metals in geological materials by inductively coupled plasma mass spectrometry. *Journal of Analytical Atomic Spectrometry*, **13**(10), 1077–1084, https://doi.org/10.1039/A803216E

Barefoot R. R. (2004). Determination of platinum group elements and gold in geological materials: a review of recent magnetic sector and laser ablation applications. *Analytica Chimica Acta*, **509**(2), 119–125, https://doi.org/10.1016/j.aca.2003.12.031

Barnes S.-J. and Lightfoot P. C. (2005). Formation of magmatic nickel-sulfide ore deposits and associated platinum-group elements. In: Exploration for Platinum-Group Element Deposits, J. A. Mungall (ed.), Mineralogical Association of Canada Short Course 35, Oulu, Finland, pp. 179–213.

Berglund M. and Wieser M. E. (2011). Isotopic compositions of the elements 2009 (IUPAC technical report). *Pure and Applied Chemistry*, **83**(2), 397–410, https://doi.org/10.1351/PAC-REP-10-06-02

Birck J. L., Barman M. R. and Capmas F. (1997). Re-Os isotopic measurements at the femtomole level in natural samples. *Geostandards Newsletter*, **21**, 19–27, https://doi.org/10.1111/j.1751-908X.1997.tb00528.x

Birke M., Rauch U., Stummeyer J., Lorenz H. and Keilert B. (2018). A review of platinum group element (PGE) geochemistry and a study of the changes of PGE contents in the topsoil of Berlin, Germany, between 1992 and 2013. *Journal of Geochemical Exploration*, **187**, 72–96, https://doi.org/10.1016/j.gexplo.2017.09.005

Bloxham L., Brown S., Cole L., Cowley A., Fujita M., Girardot N., Jiang J., Raithatha R., Ryan M., Tang B. and Wang A. (2022) The Johnson Matthey PGM market report, May 2022. https://matthey.com/documents/161599/509428/PGM-market-report-May-2022.pdf/542bcada-f4ac-a673-5f95-ad1bbfca5106?t=1655877358676

Bokhari S. N. H. and Meisel T. C. (2017). Method development and optimization of sodium peroxide sintering for geological samples. *Geostandards and Geoanalytical Research*, **41**(2), 181–195, https://doi.org/10.1111/ggr.12149

Cabri L. J., Oberthür T. and Keays R. R. (2022). Origin and depositional history of platinum-group minerals in placers–a critical review of facts and fiction. *Ore Geology Reviews*, **144**, 104733, https://doi.org/10.1016/j.oregeorev.2022.104733

Campbell A. J. and Humayun M. (1999). Trace element microanalysis in iron meteorites by laser ablation ICPMS. *Analytical Chemistry*, **71**(5), 939–946, https://doi.org/10.1021/ac9808425

Cavalheiro J., Preud'Homme H., Amouroux D., Tessier E., Monperrus M. (2014). Comparison between GC-MS and GC-ICPMS using isotope dilution for the simultaneous monitoring of inorganic and methyl mercury, butyl and phenyl tin compounds in biological tissues. *Analytical and Bioanalytical Chemistry*, **406**(4), 1253–1258. https://doi.org/10.1007/s00216-013-7373-4

Cohen A. S. and Waters F. G. (1996). Separation of osmium from geological materials by solvent extraction for analysis by thermal ionisation mass spectrometry. *Analytica Chimica Acta*, **332**(21), 269–275, https://doi.org/10.1016/0003-2670(96)00226-7

Coursey J. and Dragoset R. (2000). Atomic Weights and Isotopic Compositions (Version 2.0), Electronic Publication, https://physics.nist.gov/cgi-bin/Compositions/stand_alone.pl

Creech J. B., Baker J. A., Handler M. R. and Bizzarro M. (2014). Platinum stable isotope analysis of geological standard reference materials by double-spike MC-ICPMS. *Chemical Geology*, **363**, 293–300, https://doi.org/10.1016/j.chemgeo.2013.11.009

Creech J. B., Moynier F. and Bizzarro M. (2017). Tracing metal–silicate segregation and late veneer in the Earth and the ureilite parent body with palladium stable isotopes. *Geochimica et Cosmochimica Acta*, **216**, 28–41, https://doi.org/10.1016/j.gca.2017.04.040

Creech J., Baker J., Handler M., Schiller M. and Bizzarro M. (2023). Platinum stable isotope ratio measurements by double-spike multiple collector ICPMS. *Journal of Analytical Atomic Spectrometry*, **28**, 853–865, https://doi.org/10.1039/C3JA50022E

Di Russo F., Martínez A., Sereno M. I., Pitzalis S. and Hillyard S. A. (2002). Cortical sources of the early components of the visual evoked potential. *Human Brain Mapping*, **15**(2), 95–111, https://doi.org/10.1002/hbm.10010

Ely J. C. and Neal C. R. (2003). Using platinum-group elements to investigate magmatic processes in the Earth's mantle: a review of recent results. *Chemical Geology*, **196**(1–4), 233–250, https://doi.org/10.1016/S0009-2541(02)00415-1

Fischer-Gödde M., Burkhardt C., Kruijer T. S. and Kleine T. (2015). Ru isotope heterogeneity in the solar protoplanetary disk. *Geochimica et Cosmochimica Acta*, **168**, 151–171, https://doi.org/10.1016/j.gca.2015.07.032

Folens K., Van Acker T., Bolea-Fernandez E., Cornelis G., Vanhaecke F., Du Laing G. and Rauch S. (2018). Identification of platinum nanoparticles in road dust leachate by single particle inductively coupled plasma-mass spectrometry. *Science of the Total Environment*, **615**, 849–856, https://doi.org/10.1016/j.scitotenv.2017.09.285

Gao J., Liang G., Zhang B., Kuang Y., Zhang X. and Xu B. (2007). FePt@CoS_2 yolk–shell nanocrystals as a potent agent to kill HeLa cells. *Journal of the American Chemical Society*, **129**(5), 1428–1433, https://doi.org/10.1021/ja067785e

Gunn G. (ed.) (2014). Critical Metals Handbook. Wiley-Blackwell, Chichester, West Sussex, United Kingdom, ISBN: 978-0-470-67171-9

Gupta S., Allal A. and Preud'Homme H. (2023). Multi-screening analysis of single nanoparticles by the first multi-quadrupole ICPMS/MS. *Green Technology, Resilience, and Sustainability*, **3**(1), 1, https://doi.org/10.1007/s44173-023-00011-9

Hendriks L., Brünjes R., Taskula S., Kocic J., Hattendorf B., Bland G., Lowry G., Bolea-Fernandez E., Vanhaecke F., Wang J., Baalousha M., von der Au M., Meermann B., Holbrook T. R., Wagner S., Harycki S., Gundlach-Graham A. and von der Kammer F. (2023). Results of an interlaboratory comparison for characterization of Pt nanoparticles using single-particle ICP-TOFMS. *Nanoscale*, **15**, 11268–11279, https://doi.org/10.1039/D3NR00435J

Ho G. Y., Woodward N. and Coward J. I. G. (2016). Cisplatin versus carboplatin: comparative review of therapeutic management in solid malignancies. *Critical Reviews in Oncology/Hematology*, **102**, 37–46, https://doi.org/10.1016/j.critrevonc.2016.03.014

Hopp T., Fischer-Gödde M. and Kleine T. (2016). Ruthenium stable isotope measurements by double spike MC-ICPMS. *Journal of Analytical Atomic Spectrometry*, **31**(8), 1515–1526, https://doi.org/10.1039/C6JA00041J

Hopp T., Fischer-Gödde M. and Kleine T. (2018). Ruthenium isotope fractionation in protoplanetary cores. *Geochimica et Cosmochimica Acta*, **223**, 75–89, https://doi.org/10.1016/j.gca.2017.11.033

Hunt A. C., Ek M. and Schönbächler M. (2017). Separation of platinum from palladium and iridium in iron meteorites and accurate high-precision determination of platinum isotopes by multi-collector ICP-MS. *Geostandards and Geoanalytical Research*, **41**, 633–647, https://doi.org/10.1111/ggr.12176

International Atomic Energy Agency. (2005). Nuclear Analytical Methods for Platinum Group Elements (IAEA Report No. IAEA-TECDOC-1443). IAEA, Vienna, Austria.

Jiménez-Lamana J., Wojcieszek J., Jakubiak M., Asztemborska M. and Szpunar J. (2016). Single particle ICP-MS characterization of platinum nanoparticles uptake and bioaccumulation by *Lepidium sativum* and *Sinapis alba* plants. *Journal of Analytical Atomic Spectrometry*, **31**, 2321–2329, https://doi.org/10.1039/C6JA00201C

Juvonen M. R., Bartha A., Lakomaa T. M., Soikkeli L. A., Bertalan E., Kallio E. I. and Ballók M. (2007). Comparison of recoveries by lead fire assay and nickel sulfide fire assay in the determination of gold, platinum, palladium and rhenium in sulfide ore samples. *Geostandards and Geoanalytical Research*, **28**(1), 123–130, https://doi.org/10.1111/j.1751-908X.2004.tb01048.x

Kińska K., Jiménez-Lamana J., Kowalska J., Krasnodębska-Ostręga B. and Szpunar J. (2018). Study of the uptake and bioaccumulation of palladium nanoparticles by *Sinapis alba* using single particle ICP-MS. *Science of the Total Environment*, **615**, 1078–1085, https://doi.org/10.1016/j.scitotenv.2017.09.203

Klaver M. and Coath C. D. (2019). Obtaining accurate isotopic compositions with the double spike technique: practical considerations. *Geostandards and Geoanalytical Research*, **43**, 5–22, https://doi.org/10.1111/ggr.12248

Laborda F., Abad-Álvaro I., Jiménez M. S. and Bolea E. (2023). Catching particles by atomic spectrometry: benefits and limitations of single particle – inductively coupled plasma mass spectrometry. *Spectrochimica Acta Part B: Atomic Spectroscopy*, **199**, 106570, https://doi.org/10.1016/j.sab.2022.106570

Li X. L. and Ebihara M. (2003). Determination of all platinum-group elements in mantle-derived xenoliths by neutron activation analysis with NiS fire-assay preconcentration. *Journal of Radioanalytical and Nuclear Chemistry*, **255**(1), 131–135, https://doi.org/10.1023/A.1022296201902

Lo Medico F., Varrica D. and Alaimo M. G. (2023). Occurrence of palladium and platinum in human scalp hair of adolescents living in urban and industrial sites. *Science of the Total Environment*, **892**, 164777, https://doi.org/10.1016/j.scitotenv.2023.164777

Lommelen R., Vander Hoogerstraete T., Onghena B., Billard I. and Binnemans K. (2019). Model for metal extraction from chloride media with basic extractants: a coordination chemistry approach. *Inorganic Chemistry*, **58**(18), 12289–12301, https://doi.org/10.1021/acs.inorgchem.9b01782

Meermann B. and Nischwitz V. (2018). ICP-MS for the analysis at the nanoscale – a tutorial review. *Journal of Analytical Atomic Spectrometry*, **33**(9), 1432–1468, https://doi.org/10.1039/C8JA00037A

Meisel T., Schöner N., Paliulionyte V. and Kahr E. (2002). Determination of rare earth elements, Y, Th, Zr, Hf, Nb, and Ta in geological reference materials G-2, G-3, SCo-1, and WGB-1 by sodium peroxide sintering and inductively coupled plasma-mass spectrometry. *Geostandards and Geoanalytical Research*, **26**(1), 53–61, https://doi.org/10.1111/j.1751-908X.2002.tb00623.x

Milestone Helping Chemists. UltraWAVE Application Book Manual. https://www.scribd.com/document/488141931/UltraWAVE-Application-Book-pdf

Millet M.-A., Dauphas N., Greber N.D., Burton K.W., Dale C.W., Debret B., Macpherson C.G., Nowell G.M. and Williams H.M. (2016). Titanium stable isotope investigation of magmatic processes on the Earth and Moon. *Earth and Planetary Science Letters*, **449**, 197–205, https://doi.org/10.1016/j.epsl.2016.05.039

Moldovan M., Krupp E. M., Holliday A. E. and Donard O. F. X. (2004). High resolution sector field ICP-MS and multicollector ICP-MS as tools for trace metal speciation in environmental studies: a review. *Journal of Analytical Atomic Spectrometry*, **19**(7), 815–822, https://doi.org/10.1039/B403128H

Mudd G. M. and Jowitt S. M. (2018). A detailed assessment of global nickel resource trends and endowments. *Economic Geology*, **113**(4), 655–689, https://doi.org/10.2113/ECONGEO.109.7.1813

Muthuswamy N., de la Fuente J. L., Ochal P., Giri S., Raaen S. and Sunde S. (2013). Towards a highly-efficient fuel-cell catalyst: optimization of Pt particle size, supports and surface-oxygen group concentration. *Physical Chemistry Chemical Physics*, **15**(12), 3803–3813, https://doi.org/10.1039/C3CP43659D

Nanne J. A. M., Millet M.-A., Burton K. W., Dale C. W., Nowell G. M. and Williams H. M. (2017). High precision osmium stable isotope measurements by double spike MC-ICP-MS and N-TIMS. *Journal of Analytical Atomic Spectrometry*, **32**, 749–765, https://doi.org/10.1039/C6JA00406G

Ni W., Mao X. and Zhang H. (2019). Determination of ultra-trace platinum, palladium, ruthenium, rhodium, and iridium in rocks and minerals by inductively coupled-plasma mass spectrometry following nickel sulfide fire assay preconcentration and open mixed acid digestion. *Analytical Letters*, **52**(11), 1699–1710, https://doi.org/10.1080/00032719.2019.1566348

Ni W., Mao X., Zhang H., Liu L., Guo X. and Zhang L. (2021). Matrix-matched multi-external standards combined internal standard calibration strategy for the simultaneous determination of ultra-trace Au, Pt and Pd in geochemical samples by LA-ICP-MS after lead fire assay preconcentration. *Microchemical Journal*, **170**, 106724, https://doi.org/10.1016/j.microc.2021.106724

Omran M., Goriaux M., Liu Y., Martinet S., Jean-Soro L. and Ruban V. (2020). Platinum group elements study in automobile catalysts and exhaust gas samples. *Environmental Pollution*, **257**, 113477, https://doi.org/10.1016/j.envpol.2019.113477

Paliulionyte V., Meisel T., Ramminger P. and Kettisch P. (2006). High pressure Asher digestion and an isotope dilution-ICP-MS method for the determination of platinum-group element concentrations in chromitite reference materials CHR-Bkg, GAN Pt-1 and HHH. *Geostandards and Geoanalytical Research*, **30**(1), 87–96, https://doi.org/10.1111/j.1751-908X.2006.tb00915.x

Palme H. (2000). Are there chemical gradients in the inner solar system? *Space Science Reviews*, **92**, 237–262, https://doi.org/10.1023/A:1005247329412

Palme H. (2001). Chemical and isotopic heterogeneity in protosolar matter. *Philosophical Transactions of the Royal Society, London*, **359**, 2061–2075, https://doi.org/10.1098/rsta.2001.0897

Paton C., Hellstrom J., Paul B., Woodhead J. and Hergt J. (2011). Iolite: freeware for the visualization and processing of mass spectrometric data. *Journal of Analytical Atomic Spectrometry*, **26**, 2508–2518, https://doi.org/10.1039/C1JA10172B

Pawlak J., Łodyga-Chruścińska E. and Chrustowicz J. (2014). Fate of platinum metals in the environment. *Journal of Trace Elements in Medicine and Biology*, **28**(3), 247–254, https://doi.org/10.1016/j.jtemb.2014.03.005

Perry B. J., Speller D. V., Barefoot R. R. and Van Loon J. C. (1993). A large sample, dry chlorination, ICP-MS analytical method for the determination of platinum group elements and gold in rocks. *Canadian Journal of Applied Spectroscopy*, **38**, 131, https://doi.org/10.1134/S1061934820030065

Preud'Homme H., Mahmoud K., Shomar B., Jiménez Lamana J., Mansour S., Reynaud S. and Grassl B. (2019). Non target monitoring tracers and emerging tracers for enhanced reservoir, well integrity and risk management. In: Proceedings 8th International Conference on Tracers and Tracing Methods (Tracer 8), Da Nang, Vietnam, 26–28 February 2019, pp. 126–127. https://tracers-conferences.com/wp-content/uploads/2022/06/Abstract-book-TRACER-8-final.pdf

Rauch S. and Morrison G. M. (2008). Environmental relevance of the platinum-group elements. *Elements*, **4**(4), 259–263, https://doi.org/10.2113/GSELEMENTS.4.4.259

Rauch S., Lu M. and Morrison G. M. (2001). Heterogeneity of platinum group metals in airborne particles. *Environmental Science & Technology*, **35**(3), 595–599, https://doi.org/10.1021/es000048c

Rauch S., Morrison G. M. and Moldovan M. (2002). Scanning laser ablation-ICP-MS tracking of platinum group elements in urban particles. *Science of the Total Environment*, **286**(1–3), 243–251, https://doi.org/10.1016/S0048-9697(01)00988-3

Ravindra K., Bencs L. and Van Grieken R. (2004). Platinum group elements in the environment and their health risk. *Science of the Total Environment*, **318**(1–3), 1–43, https://doi.org/10.1016/S0048-9697(03)00372-3

Reddi G. S. and Rao C. R. M. (1999). Analytical techniques for the determination of precious metals in geological and related materials. *Analyst*, **124**(9), 1531–1540, https://doi.org/10.1039/A904839A

Roskill Information Services Ltd. (2017). Platinum Group Metals: Market Outlook to 2026, 15th edn. Roskill, London, United Kingdom.

Rudge J. F., Reynolds B. C. and Bourdon B. (2009). The double spike toolbox. *Chemical Geology*, **265**, 420–431, https://doi.org/10.1016/j.chemgeo.2009.05.010

Savignan L., Faucher S., Chéry P. and Lespes G. (2021). Platinum group elements contamination in soils: review of the current state. *Chemosphere*, **271**, 129517, https://doi.org/10.1016/j.chemosphere.2020.129517

Tang J. and Tang D. (2015). Non-enzymatic electrochemical immunoassay using noble metal nanoparticles: a review. *Microchimica Acta*, **182**, 2077–2089, https://doi.org/10.1007/s00604-015-1567-8

Taskula S., Stetten L., von der Kammer F. and Hofmann T. (2022). Platinum nanoparticle extraction, quantification, and characterization in sediments by single-particle inductively coupled plasma time-of-flight mass spectrometry. *Nanomaterials*, **12**(19), 3307, https://doi.org/10.3390/nano12193307

Thompson M., Chenery S. and Brett L. (1989). Calibration studies in laser ablation microprobe-inductively coupled plasma atomic emission spectrometry. *Journal of Analytical Atomic Spectrometry*, **4**(1), 11–16, https://doi.org/10.1039/JA9890400011

Vyas N., Turner A. and Sewell G. (2014). Platinum-based anticancer drugs in wastewaters of a major UK hospital and predicted concentrations in recipient surface waters. *Science of the Total Environment*, **493**, 324–329, https://doi.org/10.1016/j.scitotenv.2014.05.127

Woodward B. K. (2014). Platinum group metals (PGMs) for permanent implantable electronic devices. In: Precious Metals for Biomedical Applications, N. Baltzer and T. Copponnex (eds), Woodhead Publishing, Sawston, United Kingdom, pp. 130–147, https://doi.org/10.1533/9780857099051.2.130

Zereini F. and Alt F. (eds) (2006). Anthropogenic Platinum-Group Element Emissions: Their Impact on Man and Environment. Springer, Heidelberg, Germany, https://doi.org/10.1007/978-3-642-59678-0

Zhang Q., Liu J., Zhang Y., Yu H., Qin L. and Shen J. (2019). Factors affecting chromium isotope measurements using the double-spike method. *Rapid Communications in Mass Spectrometry*, **33**, 1390–1400, https://doi.org/10.1002/rcm.8483

Zhuravlev A., Zacharia A., Arabadzhi M., Turetta C., Cozzi G. and Barbante C. (2016). Comparison of analytical methods: ICP-QMS, ICP-SFMS and FF-ET-AAS for the determination of V, Mn, Ni, Cu, As, Sr, Mo, Cd and Pb in ground natural waters. *International Journal of Environmental Analytical Chemistry*, **96**(4), 332–352, https://doi.org/10.1080/03067319.2016.1160380

Part III

Recovery of platinum group elements from wastewaters

doi: 10.2166/9781789065084_0069

Chapter 4

Adsorptive removal of platinum group elements

Shinjini Sen[1], *Talari Madhumanohar*[1], *Piet N. L. Lens*[2] *and Mohanakrishnan Logan*[1]*

[1]*Environmental Engineering Division, Department of Civil Engineering, Indian Institute of Technology Madras, Chennai 600036, India*
[2]*University of Galway, H91 TK33 Galway, Ireland*
**Corresponding author: mohanlogan@iitm.ac.in*

ABSTRACT

This chapter is a general overview of adsorptive removal of platinum group elements (PGEs) from different sources. It deals with the challenges in drawbacks of traditional PGE removal and recovery techniques. The chapter also points to the necessity for sustainable and efficient recovery methods and delves into the mechanism of adsorptive removal of PGEs. The role of environmental conditions on PGE adsorption, and traditional and novel trends in adsorbent materials for PGEs are comprehensively reviewed. The chapter further provides case studies on the recovery of PGE, as well as kinetics and thermodynamics of PGE adsorption. It explores the effectiveness of various adsorbents and analyses the cost and scalability of adsorption processes for PGE recovery. It further addresses desorption and recovery of adsorbed PGEs and investigates the environmental and economic implications of adsorption-based PGE recovery processes. Finally, a conclusion on overview of the future directions in research and development for industrial application is drawn.

Keywords: Adsorbents, adsorption, desorption, kinetics, platinum group elements, recovery, thermodynamics

4.1 INTRODUCTION

Platinum group elements (PGEs) are six rare, noble and precious metallic elements including platinum (Pt), palladium (Pd), rhodium (Rh), iridium (Ir), osmium (Os), and ruthenium (Ru). As such, there are no distinct ores for PGEs;

they mostly occur in trace quantities within nickel and copper ores. PGEs are usually recovered as by-products of base metal mining operations. PGEs have similar chemical and physical properties, especially high melting points, highly catalytic and corrosion resistance properties. Their rarity, coupled with their versatile applications, has made PGEs very popular for industrial purposes such as the automotive industries, electronics, and catalysis (Macheyeki *et al.*, 2020).

The uniqueness in distinguishing the six PGEs lies in their particular properties, which enable their industrial applicability. The most popular is Pt, mainly used in catalytic converters, electronics, and jewellery. It has a very high catalytic efficiency, especially regarding the processes referring to abatement of automobile emission. Meanwhile, Pd has wider applications in catalytic converters for petrol vehicles. Interestingly, owing to the traction in recent years, Pd has emerged as a cheap alternative to Pt in certain uses (Chidunchi *et al.*, 2024).

Of the PGEs, Rh is the rarest and most expensive element. However, its outstanding catalytic properties for reducing NO_x emissions in catalytic converters make it inevitable in the automotive industry. Besides this, the hardness of Ir is well known, along with its resistance to high temperatures. Hence, it has an application in equipment and spark plugs operated at high temperatures. Most commonly, Os and Ru are used as an alloying addition to increase hardness and improve corrosion resistance in some other PGEs or base metals (Nouri & Wen, 2021).

When focusing on sector-wise applications and significance of PGEs, automotive industry enjoys the prominence. Pt, Pd, and Rh are increasingly used as components in catalytic converters. In these catalytic reactors, through the oxidation, the poisonous emissions of internal combustion engines such as carbon monoxide, hydrocarbons, and nitrogen oxides are transformed into relatively harmless products (Zheng & Greedan, 2003). It is evident that the automotive sector has seen an increase in demand for PGEs due to an upsurge in the call for stricter environmental regulations, particularly in countries with strict vehicle emission standards.

The electronics industry employs PGEs in silicon wafers, connectors, and electroplating processes on account of their excellent conductivity and durability. Platinum and palladium are applied in hard disk drives, multilayer ceramic capacitors, and hybrid integrated circuits. In line with technological progress, as the need for smaller and faster electronic equipment continuously grows, PGE use in manufacturing has increased (Zientek *et al.*, 2017). In catalysis, the PGEs are indispensable in petroleum refining, chemicals and fertilizer production. Pt-based catalysts find wide application in hydrogen fuel cells – a plausible solution to the world's energy requirements. Similarly, Pd is used during the course of hydrogenation and dehydrogenation reactions in chemical industries, turning out to be an indispensable industrial catalyst (United States Department of Energy, 2022).

While PGEs are a part of several industries, their extraction, usage, and disposal pose environmental and economic challenges. The environmental footprint associated with mining and refining is huge due to scarce primary sources of PGEs. Apart from ecological destruction and water pollution, mining of these metals generally results in the generation of harmful emissions. Improper disposal of PGE is detrimental to the environment, and their bioaccumulation potential exhibits deleterious effects on aquatic and human life (Karim & Ting, 2021). Recovery and recycling from secondary sources, such as end-of-life vehicles and waste electronic devices, has hence gained critical importance (Buchspies *et al.*, 2017).

Recovery of PGEs from wastes is not only economically viable but has also an environmental role. Products such as waste catalytic converters, electronic devices, and other commodities containing PGEs at their end of life could form valid resources for their secondary recovery. However, because of the generally complex procedures to extract PGEs from wastes, their processing is bound to be costly and power consuming. Moreover, advances in the recovery technologies through hydrometallurgical and pyrometallurgical processes help to reduce costs and improve efficiency (Yakoumis *et al.*, 2021).

The economic value of the PGEs is also an incentive in conducting recovery efforts from secondary sources. In a time when demand around the world is surging, especially in the backdrop of tightening environmental regulations, the supply chain of the PGEs is under considerable pressure. For this reason, recycling offers a more circular and sustainable path that secures these critical metals in continuous supply with less environmental harm from mining and refining processes. Therefore, the recycling of PGEs, which mostly originates from automotive and electronic waste, is turning into an area of growing interest, with a number of countries continuously investing in the optimization of recovery technologies (Schulte, 2019).

4.2 CHALLENGES IN PGE RECOVERY

Recovery of PGEs faces greater challenges because of their complicated chemical properties and diverse sources from where they have to be extracted. The first challenge is the low concentration of PGEs in both ores and waste streams, which makes their extraction economically and technologically difficult. Recovery from industrial wastes and spent catalysts normally involves handling huge volumes to recover small quantities of precious metals.

This problem is exacerbated by environmental hazards from the conventional methods of extraction, including high-energy usage and hazardous by-product generation. Recovery is further made difficult by the chemical similarities between PGEs, thus demanding cumbersome separation techniques for individual PGEs. This further increases cost and environmental impact of the recovery operations. Despite this, there is an ever-growing need for more effective and less environmentally damaging recovery techniques (Taylor, 2012).

4.2.1 Sources of PGE contamination

The main sources of environmental contamination from PGEs include industrial effluents, vehicle catalysts, and waste electronics. In most instances, industrial wastes are generated from refining and manufacturing of PGE-rich materials due to their potential use in many sectors. These effluents can therefore contaminate the soil and aquatic environments, thus posing risks to ecosystems and human health. Automotive catalysts using PGEs reduce emissions, but there is also environmental contamination contributed by road runoff and atmospheric deposition. With the abrasion of such catalysts over time, the PGEs are released into the environment and tend to accumulate in the soil and aquatic systems (Ek *et al.*, 2004).

Another significant source of PGE contamination includes electronic waste, which commonly features PGEs in components such as connectors and circuit boards. If electronic waste is disposed of or recycled incorrectly, it may release metals into the environment, therefore leading to contamination problems. This wide use of PGEs in a number of electrical devices, together with the complex pathways of their disposal, underlines the urgent need for effective recovery and recycling strategies (Weiss *et al.*, 2016).

4.2.2 Limitations of conventional PGE removal methods

Conventional PGE removal methods include a number of limitations related to hydrometallurgical and pyrometallurgical processes. Hydrometallurgical methods, such as solvent extraction and ion exchange, have normally been characterized by low efficiencies with high operating cost, especially when dealing with low-grade ores or complex matrices of waste. Nevertheless, these processes do have vast potential for huge volumes of chemicals. Similar disadvantages are also applicable in the case of the pyrometallurgical techniques, which are involved with high-temperature processing, with the scope of separating the PGEs from other materials, implying high energy use and burdens to the environment. The high temperatures of these processes lead to high energy use and greenhouse gas emissions. Besides, in certain types of waste, such as those with fine particulate matter or complex matrices, pyrometallurgy seems less effective (Compagnone *et al.*, 2023). These associated inefficiencies with conventional methods highlight the importance of innovative approaches towards developing more effective, inexpensive, and environment-friendly methods for the recovery of PGEs from various sources. Nevertheless, it should be noted that of the different techniques employed for recovering PGEs (such as ion exchange, separation, chemical precipitation and solvent extraction) from the liquid phase are inevitable preceded by leaching processes to solubilize PGEs from the solid to the liquid phase.

4.2.3 Need for sustainable and efficient recovery techniques

Given that most of the conventional modes of recovery have some degree of deficiency to be addressed, it is envisaged that enhanced feasibility by sustainable and efficient techniques should be identified, developed, and deployed with regard to PGE recoveries. Recent development in green technologies presents promising alternatives to conventional methods. Biotechnological approaches, like bioleaching and biosorption, for example, have been receiving recent attention due to their ability to open up possibilities for recovering PGEs in a more environmental-friendly way. Different biological procedures are opted for the extraction of metals from ores and wastes, which probably have lower environmental impacts coupled with lower energy use than conventional techniques (Saitoh *et al.*, 2017).

Further research is conducted towards the use of more advanced materials, including functionalized carbon composites and new sorbents that may be utilized in order to further enhance the efficiency of the processes utilized for PGE recovery. These materials can enhance selectivity and increase the capacity for PGE adsorption, hence increasing the rates of recovery while reducing costs (Mosai *et al.*, 2020). Indeed, in light of the ever-pressing demand for these metals, innovative eco-friendly methods must be found to meet the supply requirements, while environmental impacts are minimized concurrently (Compagnone *et al.*, 2023).

4.3 ADSORPTIVE REMOVAL MECHANISM FOR PGEs

The adsorptive removal of PGEs becomes important with the increasing importance of efficient extraction of these metals from wastes for environmental and industrial needs. Adsorption is considered a primary means of recovering PGEs through the interaction of these element ions with the surface of adsorbent materials. It works through a mechanism in which PGEs bind onto the surface of the adsorbent, where the efficiency and effectiveness depend on multiple variables. Recent research has tried to explain how different adsorbents, metal-organic frameworks, and functionalized polymers interact with PGEs, and the basic principles behind those interactions (MacDonald *et al.*, 2024).

4.3.1 Theory of adsorption and its application to PGEs

The theory of adsorption bears utmost importance in understanding how PGEs are removed from solution. There are many models included in the adsorption theory. These include the Langmuir and Freundlich isotherms, which describe the mode with which adsorbates interact on the surfaces of adsorbents. The Langmuir isotherm is based on monolayer adsorption onto a surface made up of a finite number of identical sites. On the other hand, the Freundlich isotherm takes into consideration heterogeneous surfaces having distributed affinities for the adsorbate (Kalam *et al.*, 2021). Theories are essential in developing the capacity and efficiency of different adsorbents in the capture of PGEs from aqueous solutions. Such theories have been used, for instance, to optimize adsorbent materials with the view to enhance the selective removal of PGEs, apparently an indication of the practical relevance of the adsorption theory for the recovery of PGEs.

4.3.2 Types of adsorption

In general, adsorption can be divided into two broad categories: physisorption and chemisorption. Each of these types has different characteristics that will affect the removal of PGEs. Physisorption is physical adsorption characterized by weak Van der Waals forces or electrostatic interactions between adsorbate–adsorbent with low binding energies. Generally, physisorption interactions are reversible. On the other hand, chemisorption is characterized by the formation of strong chemical bonds between the adsorbate and the surface of the adsorbent, mostly presenting a higher binding energy and frequently irreversible adsorption (Aljamali *et al.*, 2021). In the case of PGEs, it is important to highlight the role of chemisorption for metals featuring strong interactions with adsorbent surfaces, for instance thiol groups, on functionalized materials (Woods, 1996). This particular property accordingly improves the efficiency of metal recovery processes. It implies that chemisorption can be a promising mechanism to be used for removing PGE from different complex waste streams.

4.3.3 Influence of environmental factors on PGE adsorption

Similar to heavy metals, the environmental factors that strongly influence the PGE adsorption efficiency are pH, temperature, and ionic strength. pH plays an important role in relation to the ionization state, not only of the adsorbent but also of the adsorbate. In fact, for example, the adsorption of PGEs mostly increases at lower pH values due to more positively charged metal ions that can interact strongly with negatively charged adsorbent surfaces. Temperature also affects adsorption: the higher the temperature, the greater the kinetic energy of the molecules and hence faster the adsorption rate. Ionic strength may affect adsorption by influencing the solubility of metal ions and the competition of metal ions with other dissolved species for active adsorption sites. In general, an understanding of all these factors and their relationship is very important to optimize the PGE adsorption process through efficient recovery of elements with minimal cost and environmental impact. Studies and refinement in the field of adsorption techniques will enable increased capabilities in effectively managing and recovering valuable PGEs from diverse sources.

4.4 CONVENTIONAL AND RECENT DEVELOPMENTS IN ADSORBENT MATERIALS FOR PGEs

The principle of recovery of PGEs based on adsorption involves many types of adsorbent materials with their different advantages and limitations. These can be categorized into three wide divisions: natural adsorbents, synthetic adsorbents, and functionalized or composite adsorbents. Figure 4.1 showcases various adsorbents, emphasizing their features which enhance their efficiency in PGE recovery processes.

Consequently, adsorbent materials – from natural through synthetic to functionalized – open up a range of different options for PGE recovery in an effective way. Each of these material types has its merits, and each has certain worthwhile applications, which is important for choosing an appropriate material according to the requirements of the PGE recovery process.

4.4.1 Natural adsorbents

4.4.1.1 Inorganic materials

Clays, zeolites, and other biosorbents are natural adsorbents that have been utilized ubiquitously in PGE removal from aqueous solutions. Due to their porous structure with high surface area, clays and zeolites exhibit great potential in metal ion adsorption. These materials represent a cheap and eco-friendly

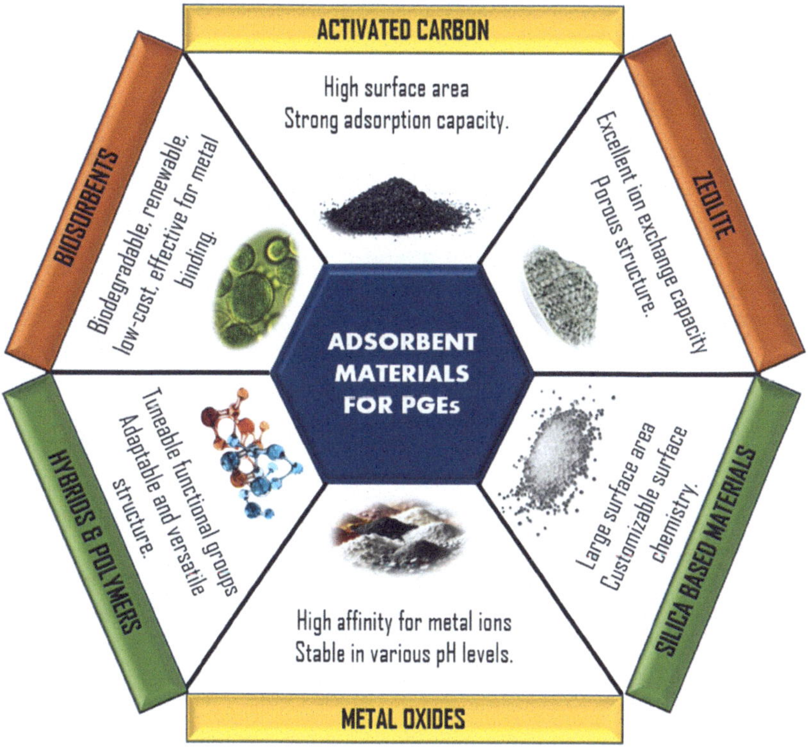

Figure 4.1 Various adsorbents and their features facilitating PGE recovery.

alternative for PGE recovery. As an example, natural clays may have very high ion-exchange capacities and surface reactivity and, hence, can adsorb a range of metal ions, including PGEs (Uddin, 2017). Likewise, zeolites are crystalline in nature with ion-exchange capacity that can efficiently adsorb the PGEs from industrial effluents (Mosai *et al.*, 2019).

Sustainability is also at the forefront of developing adsorbents for PGE recovery. Therefore, green adsorbents, such as natural and biosorbents, provide a more ecological material in comparison with synthetic materials. Focus on the use of natural materials such as agricultural byproducts have lately been observed since such materials not only are economical but are also environmentally benign (Mosai *et al.*, 2019; Uddin, 2017). The sustainability of sourcing these materials is easy to ensure, and because they are natural, they usually need little processing, thereby generally limiting the overall environmental impact.

4.4.1.2 Organic materials

Biosorbents, such as algae and other biomass types, are promising because they are biodegradable and exhibit a high efficiency for the removal of metal ions.

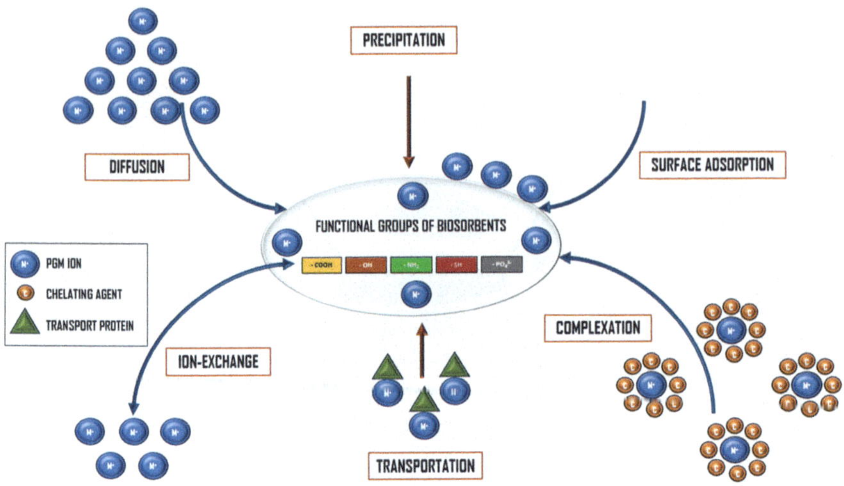

Figure 4.2 Illustration of functional groups in biosorbents and their mechanisms of interaction with PGE.

The materials have functional groups that can bind metal ions, hence suitable for recovery of PGE in a sustainable way (Tofan *et al.*, 2022). Biosorbents are, therefore, useful because these are biodegradable and relatively inexpensive. Figure 4.2 shows how functional groups like COOH, OH, NH_2, or others in biosorbents facilitate processes like ion exchange, complexation, and surface adsorption for effective metal binding.

4.4.2 Synthetic adsorbents

Artificial adsorbents including activated carbon, metal-organic frameworks (MOF) (Furukawa *et al.*, 2013), and few other polymers are widely known due to their tailored properties for high efficiency of PGE recovery. Activated carbon is among the most widely used man-made adsorbents due to its large surface area and extended porosity. It is effective in adsorbing a range of PGEs, because it can support a large number of adsorption sites (Aktas & Morcali, 2011). Modifications of surface properties of the activated carbon can further improve selectivity and capacity (Jha *et al.*, 2021).

MOFs are another advanced class of synthetic adsorbents. Broadly, MOFs are the kind of advanced materials formed by the coordination of metallic ions with organic ligands, resulting in a structure characterized by high porosity that enables them to effectively and selectively adsorb PGEs. Recent studies have demonstrated that MOFs can be engineered for targeting specific PGEs, thus enabling high recovery rates and treatment of complicated waste streams (Maponya *et al.*, 2023). The MOFs' huge degree of flexibility and efficiency render them promising materials towards future developments in the recovery of PGEs.

Polymers, especially those with functional groups designed for the attachment of metal ions, are also employed as adsorbents. Some of the reported synthetic polymers capable of PGEs adsorption include resin-based amberlite and thiol-group-based polymers (MacDonald *et al.*, 2024). These can be prepared in such a way as to maximize their interaction with PGEs, thereby increasing their adsorption capacity and selectivity (Sharma *et al.*, 2017). Since polymers are flexible materials, this allows for the creation of various types of adsorbents with specified properties for use in various applications regarding PGE removal.

4.4.3 Functionalized and composite adsorbents

The area of adsorbents of PGEs has progressed a lot in recent times, particularly with the incorporation of nanotechnology and developing green adsorbents. These developments reflect the trend towards more efficient, effective, and eco-friendly PGE recovery processes more broadly.

Nanotechnology has changed the perception of PGE adsorption by embracing nanomaterials developed with superior features. The dominant area of nano-application is using different types of nanomaterials like nanoparticles and nanocomposites, which offer greater surface area and higher reactivity for the efficient recovery of PGEs. Recent reviews point out that nanomaterials have demonstrated outstanding performance in adsorbing PGEs because of their relatively high surface-to-volume ratios, with tailored properties to meet specific adsorption requirements (McNeice & Mahandra, 2023).

Magnetic nanoparticles, in particular, have gained prominence due to their dual functionality-efficient adsorption of PGEs and easy recovery by magnetic separation. Functionalization of such materials with various ligands maximizes their selectivity towards PGEs, enabling them to exhibit great efficiency in targeted removal processes (Mokhodoeva *et al.*, 2020). Thus, the latter enables fast separation of materials from solution, thereby simplifying recovery and lowering costs in operation. This advanced nanomaterial development marked an important step in the PGE recovery technologies.

Advanced functionalized and composite adsorbents include metal nanoparticle-based and hybrid materials. In this regard, nanoparticle-based adsorbents represent a good example of materials with high surface areas, including metal nanoparticles and nanoscale materials. Such materials generally tend to show the best performance in the adsorption of PGEs owing to their large surface-to-volume ratio and high reactivity. According to McNeice and Mahandra (2023), the use of nanoparticles can substantially improve the recovery efficiency of PGEs.

Other new trends relate to hybrid materials, in which nanoparticles are blended with other adsorbents, such as activated carbon or polymers. Such hybrid adsorbents can avail themselves of the strengths of various materials to improve the overall performance. For example, the nanoparticles may be mixed with activated carbon for the development of materials that have both a high surface area and high reactivity, thereby resulting in an increase in the recovery efficiencies of PGEs (Trieu, 2018). Functionalized adsorbents, on the other hand, primarily look at modifying the surface of the materials to include certain binding sites for PGEs, hence giving a more effective and specific way of removal (Zhang *et al.*, 2024).

4.4.4 Ion exchange processes

Poor selectivity and capacity of the adsorption process can be overcome by using highly selective ion exchange resins. Developments in recovery methods based on ion exchange have focused attention on the high efficiency of silica-based ion exchangers for the extraction of PGEs, especially rhodium, from hydrometallurgical wastes and homogeneous catalysts (Kramer *et al.*, 2001, 2005). Silica-based ion exchangers are promising materials for this application, due to their high kinetic performance, mechanical stability, and thermal resistance, properties that render them suitable for industrial use. Also, ion-exchange resins like Puromet and Lewatit exhibit good selectivity and efficiency in metal recovery and can be used in large-scale operations (Table 4.1). Nevertheless, the dynamic nature of industrial waste streams might necessitate further optimization. On the other hand, sorption methods with anion exchange resins and hydrated zirconium dioxide (HZH) are economically viable alternative, especially in acidic solutions, but issues such as sluggish sorption kinetics and intricate recovery mechanisms remain.

Reusability of ion exchange resins for PGEs is a key consideration in their industrial use. Ion exchange resins can withstand several adsorption–desorption cycles with little loss of performance, especially under mild acidic conditions. Irreversible degradation occurs, however, upon exposure to strong oxidizing conditions at high temperatures. Although certain ion exchangers show a slight decrease in adsorption efficiency (~15%) after the first cycle due to ligand removal, most have consistent recovery rates for repeated cycles.

4.4.5 Case studies on PGE recovery

4.4.5.1 Urban mining

Cases of advanced adsorbents are described with regards to PGE recovery from industrial wastewater, and urban mining of electronic wastes. Case studies of industrial wastewater treatment have shown that conventional and novel adsorbents can be effectively used for the recovery of PGEs from complex effluents (Fu & Wang, 2011). The studies show there is a challenge in high concentration levels of competing ions and a requirement for highly selective and high-capacity adsorbents (Nakhjiri *et al.*, 2022).

Another important area to emphasize is the recovery of PGEs from electronic waste through urban mining. A lot of adsorbents have been shown to extract precious metals from waste electronics, which could be a source of PGEs. Among the promising technologies established in these fora regarding effective recoveries and as a tool for combating electronic waste, leaching processes followed by selective adsorption stand out (Sronsri *et al.*, 2022). The recent research developments in adsorbents for PGEs could imply a development trend towards more efficient, sustainable, and practical solutions. There is integration of nanotechnology, green adsorbents, and lessons from diverse case studies such as improved leaching from catalytic converters (de Oliveira Demarco *et al.*, 2020), enhanced froth flotation in mining (Cilek & Uysal, 2018) and selective extraction from e-waste (Cui & Zhang, 2008), that channel advancement into the PGE recovery field. These innovations (e.g., nanomaterial adsorbents, biosorbents, ionic liquids, and process intensification

Table 4.1 Performance of various ion exchange and adsorption processes for PGE recovery.

Process	Adsorbent /Ion Exchanger	Target Pollutants/ Compounds	Key Performance Metrics	Operating Conditions	Advantages	Disadvantages	Reference
Ion exchange	Puromet MTS9200, Puromet MTS9850, and Lewatit MonoPlus MP600	Pt, Pd, Rh, Au	Sorption efficiency, Elution efficiency, Operational capacity, Total capacity	Contact time: Sorption – 5 min, Elution – 10–30 min; Flow rate: 10 cm^3 min^{-1}; pH: ~7 for Rhodium sorption	Low operating cost, low energy consumption, high selectivity, high efficiency, possibility for large volumes with low-concentration, resins can be regenerated and reused	Tests must be redone in dynamic conditions	Goc *et al.* (2024)
Ion exchange	Puromet MTS9600	Pt (IV), Pd (II), Rh (III)	Impact of hydrochloric acid, temperature, PGM concentration, contact time, PGM adsorption onto the resin and loading isotherms	Acidic chloride media (0.1–9.0 M HCl)	Chelating ion exchange resin used for separation of PGM	N.A.	Hosseinzadeh and Petersen (2024)
Ion exchange	Polyethylenimine (PEI) resin beads	Rh	Metal recovery profiling using ICP-OES	HCl and water considered to recover Rh from PEI	N.A.	PEI would not be a suitable method to recover the Rh from this process	Conroy (2017)

(Continued)

Table 4.1 Performance of various ion exchange and adsorption processes for PGE recovery. (*Continued*)

Process	Adsorbent /Ion Exchanger	Target Pollutants/ Compounds	Key Performance Metrics	Operating Conditions	Advantages	Disadvantages	Reference
Ion exchange	Ion exchange resin	Pt, Rh, Ir, Os, Pd, Ru	Efficiency of PGM recovery, resin selectivity	Acidic and high chloride effluent streams; laboratory-scale experiments	High recovery potential	Resin performance hampered by wastewater composition	Siwela *et al.* (2022)
Sorption	Hydrated zirconium dioxide (HZH)	Trinuclear Ir (IV, IV, IV) and Ir (III, III, IV) oxosulfates	Static exchange capacity, Sorption kinetics, Temperature, Sorbent grain size, pH	Static conditions, 0.01 M H_2SO_4 solutions, Temperature 20–90°C, pH 0.5–7.0, Sorbent grain size 0.125–0.800 mm	HZH selectively and irreversibly sorbs microscopic amounts of PM from acid solutions	Sorption kinetics is slow compared to iridium chloro complexes, Complex mechanism of recovery	Abovskaya *et al.* (2006)
Sorption	Anionites (AH–31, AB–17–8, Purolite S985)	Pt, Ir	Sorption isotherms, Apparent sorption equilibrium constant, Standard Gibbs energy	Temperature: 20°C, Duration: 24 h	Relatively simple, cost-effective and environmental friendly	N.A.	Petrov *et al.* (2021)
Biosorption	Lyophilized cells of unicellular red alga *Galdieria sulphuraria*	Ir	Ir recovery efficiency	Up to 0.2 M acid	Comparable Ir recovery to ion-exchange resin, selective adsorption	Ir can be eluted with HCl, ethylenediaminetetraacetic acid, and potassium hydroxide solutions, but not with a thiourea-HCl solution	Minoda *et al.* (2023)
Bioleaching	Biogenic thiosulfate	Pd	Leaching yield	pH 8, 60°C, NaS_2O_3 1.2 M, $CuSO_4$ 0.03 M, $(NH_4)_2SO_4$ 1.5 M, Na_2SO_3 0.2 M	Sustainable alternative to conventional methods	N.A.	Compagnone *et al.* (2023)

N.A.: Not available.

techniques) improve the efficiency of the PGE recovery process and play into broader strategies for environmental sustainability and resource conservation.

4.4.5.2 Silica-based ion exchangers for rhodium recovery

The most important parameters that define the applicability of an ion exchanger for recovery of PGEs (such as rhodium) are uptake capacity, recovery efficiency, reusability, kinetics, selectivity, and synthesis cost. The adsorption capacity of ion exchangers is determined largely by the nature of ligands attached to them. Phosphorus (P)-containing exchangers exhibit relatively moderate uptake of rhodium (up to 12 mg/g), while nitrogen (N) and nitrogen-sulfur (N, S)-derived ligands exhibit much higher adsorption capacities (>40 mg/g) (Kramer *et al.*, 2001). The hard and soft acids and bases (HSAB) principle is an important factor in the recovery of rhodium, determining the affinity of ligands according to the oxidation state of rhodium. Ion exchangers with both N and S donor atoms have strong affinity for rhodium in various oxidation states, while those with just N ligands prefer certain rhodium species. Multiple structural and chemical parameters affect rhodium adsorption efficiency, such as ligand donor type, N-alkylation, chelate ring size, and tendencies for complex formation (Kramer *et al.*, 2002). For example, five-membered rings provide higher stability than six-membered rings, and didentate/tridentate coordination increases binding affinity. The conditions for extraction, especially pH and solvent selection, also influence rhodium recovery (Kramer *et al.*, 2005). In low pH conditions, rhodium adsorption is dominated by ion-exchange processes, while for high pH, chelation effects prevail. Additionally, alcohol-based solvents enhance adsorption efficiency but impede desorption due to rhodium ion reduction.

4.4.5.3 Biosorption processes for iridium recovery

In the recent years, biosorption and bioleaching are being considered sustainable alternatives for PGE recovery (Table 4.1). *Galdieria sulphuraria* cell lyophilysate efficiently adsorbs iridium, equalling conventional ion-exchange resins in performance. Biogenic thiosulfate is efficient for palladium leaching in controlled alkaline medium. Nevertheless, these biological processes need to be optimized further to achieve better metal recovery and minimize selectivity concerns (Chapter 5). Even with their drawbacks, further advances in these methods could provide cleaner and more economical PGE recovery solutions.

4.5 KINETICS AND THERMODYNAMICS OF PGE ADSORPTION

Adsorption isotherms play a vital role in analysing the interaction of PGEs with various adsorbents. They depict the relationship between the amount of adsorbate that is adsorbed onto an adsorbent and its equilibrium concentration in the solution at a constant temperature.

4.5.1 Adsorption isotherms

4.5.1.1 Langmuir isotherm

The Langmuir isotherm model is based on the premise of monolayer adsorption occurring on a uniform surface with a limited number of identical sites. When utilized for PGEs, this model aids in assessing the maximum adsorption capacity and the strength of interaction between the adsorbent and adsorbate (in this case, PGE ions). It is especially beneficial for examining PGE adsorption in controlled settings, where the adsorption sites are consistent and interactions between adsorbed molecules are negligible (Dada *et al.*, 2019).

The linear equation of the Langmuir isotherm is expressed as: $1/q_e = (K_L/Q_m)(1/C_e) + 1/Q_m$.

Where q_e is the amount of PGE adsorbate adsorbed per unit mass of adsorbent at equilibrium, C_e is the equilibrium concentration of the PGE adsorbate in solution, K_L is the Langmuir constant related to the affinity of the binding sites and Q_m is the maximum adsorption capacity of the adsorbent.

4.5.1.2 Freundlich isotherm

The Freundlich isotherm model is an empirical framework that addresses the presence of heterogeneous surface energies. It is particularly effective for analysing the adsorption of PGEs in complex environmental systems with irregular surfaces. By providing insights into both the intensity and capacity of adsorption across various PGE concentrations, the Freundlich model is valuable for exploring PGE behaviour under different environmental conditions (Dada *et al.*, 2019)

The linear equation of the Freundlich isotherm is expressed as: $\log(q_e) = \log(K_F) + (1/n)\log(C_e)$, where q_e is the amount of PGE adsorbate adsorbed per unit mass of adsorbent at equilibrium, C_e is the equilibrium concentration of the PGE adsorbate in solution, K_F is the Freundlich constant related to the adsorption capacity and n is a heterogeneity factor, indicating the deviation from linearity.

4.5.2 Kinetics of PGE adsorption

The study of the kinetics of PGEs adsorption is essential for understanding how quickly and effectively these elements adhere to adsorbent materials.

4.5.2.1 First-order kinetics

The first-order kinetic model indicates that the rate of adsorption is influenced by the concentration of the PGE adsorbate present in the solution. Specifically, the model can be expressed with the equation: $(dq_t/dt) = k_1(q_e - q_t)$.

Where q_t is the amount of PGE adsorbed at time t, q_e is the equilibrium adsorption capacity, and k_1 is the rate constant for the first-order process. This model is especially applicable in the early phases of adsorption, when the concentration of PGEs is relatively high (Musah *et al.*, 2022).

4.5.2.2 Second-order kinetics

The second-order kinetic model assumes that the rate of PGE adsorption is proportional to the square of the amount of PGE adsorbate remaining in solution.

The rate equation is given by $(dq_t/d_t) = k_2(q_e - q_t)^2$.

Where k_2 is the rate constant for the second-order reaction. This model is frequently more relevant in systems where multiple adsorption sites engage with the PGE adsorbate, making it well-suited for characterizing the kinetics of PGE adsorption over time (Wang & Guo, 2023).

4.5.3 Thermodynamic parameters

The enthalpy change (ΔH) is an important thermodynamic parameter that represents the heat absorbed or released during PGE adsorption. Positive ΔH values signify that the PGE adsorption process is endothermic, necessitating heat input. Conversely, negative values indicate an exothermic reaction, in which heat is released as PGEs are adsorbed onto the surface of the adsorbent. Comprehending the enthalpy change is essential for evaluating the energy demands and stability of the PGE adsorption process (Mphahlele *et al.*, 2024).

The change in entropy (ΔS) represents the level of disorder in a system before and after the adsorption of PGEs. A positive ΔS signifies an increase in disorder, which can happen if the PGE adsorbate molecules gain mobility upon adsorption. Conversely, a negative ΔS indicates that the system becomes more ordered, typically due to strong interactions between the PGE adsorbate and the adsorbent. This parameter is vital for evaluating the spontaneity and feasibility of the PGE adsorption process under different conditions (Prasetyo & Anderson, 2020).

The Gibbs free energy (ΔG) is an important measure of the thermodynamic spontaneity of the PGE adsorption process and is calculated using the equation: $\Delta G = \Delta H – T\Delta S$, where H is the enthalpy, T is the absolute temperature, and S is the entropy. A negative ΔG value indicates that the adsorption process is spontaneous, suggesting that PGEs could naturally adhere to the adsorbent material. By assessing ΔG, the effectiveness of PGE adsorption conditions can be evaluated and adjustments as needed to optimize PGE recovery methods can be made (Chandrasekaran & Karunasagar, 2016).

4.6 COMPARATIVE EFFICIENCY OF ADSORBENTS

4.6.1 PGE removal efficiency

The relative efficiency of PGE adsorbents is essential for enhancing PGE recovery processes from both economic and environmental perspectives. The effectiveness of adsorbents in recovering PGEs can differ considerably depending on their chemical composition and structural properties (Table 4.2).

4.6.2 Economic evaluation and scalability of adsorption methods for PGE recovery

Carrying out an economic assessment of adsorption methods is essential for evaluating their feasibility in PGE recovery. These evaluations analyse costs associated with materials, operational processes, and the anticipated return on investment.

Investing in advanced adsorption technologies can yield significant returns in the long run. By achieving high recovery rates and optimizing material usage, these technologies can enhance profit margins. For instance, innovative approaches that boost the adsorption efficiency may reduce overall material expenses, positively impacting the financial performance of mining operations (Sefako *et al.*, 2017).

Table 4.2 Performance of various adsorbents based on adsorption capacity and selectivity.

Adsorbent	PGE Studied	Adsorption Efficiency (%) and/or Capacity (mg g^{-1})	Experimental Conditions	Reference
Chitosan (biopolymer) (0.01 g)	Pd, Pt (50 mg L^{-1})	80.8 mg g^{-1} (Pt) 89.4 mg g^{-1} (Pd)	15 mL, pH 2.0, 300 min	Asere *et al.* (2019)
Fungi *Aspergillus* sp. (13 mg)	Pd (0.075 mg L^{-1}) Pt (0.1 mg L^{-1})	5.49 mg g^{-1} (Pt) 4.28 mg g^{-1} (Pd)	5 mL, pH 3.5, 45 min	Godlewska-Żyłkiewicz *et al.* (2019)
Yeast-functionalized bentonite (100 mg)	Pt (2 mg L^{-1})	98.5%	10 mL, pH 2.0, 180 min	Mosai *et al.* (2020)
Grape stalk (0.5 g for Pt and 0.8 g for Pd) and activated grape stalk (0.3 g for Pt and 0.2 g for Pd)	Pd, Pt (10 mg L^{-1})	For Grape stalk: 100% (Pt, Pd) For activated grape stalk: 93% (Pt) and 100% (Pd)	200 mL, 200 rpm, 298 K, 24 h	Mavhungu *et al.* (2013)
Granular expanded vermiculite (100 mg)	Pt (0.2 mg L^{-1})	0.07301 mg g^{-1}	2 mL, pH 2, 298 K (RT), 5 min	Treber *et al.* (2024)
Activated carbon (400 mg)	Pt (57.6 mg L^{-1})	99%	200 rpm, 3 h, 80°C	Aktas and Morcali (2011)
MOR–2-QAS (ion-exchange adsorbent) (10 mg)	Pd, Pt (200 mg L^{-1} $Pt(CN)_4^{2-}$) (100 mg L^{-1} $Pd(CN)_4^{2-}$)	>90% 242.6 mg g^{-1} (Pt) 119.3 mg g^{-1} (Pd)	20 mL, pH 9.5, 150 rpm, 10 min	Chen *et al.* (2021)
Fe_3O_4@PEG@ Cyphos® IL 101 (magnetic NPs) (0.02 g)	Pd, Pt (1.5*10^{-4} mM)	99% (Pd) 98% (Pt)	5 mL, 250 rpm, 30 min	Mokhodoeva *et al.* (2020)
Poly(N-(hydroxymethyl) methacrylamide–1-allyl–2-thiourea) hydrogels (0.01 g)	Pt and Pd (10 mg L^{-1})	>90% 477 mg g^{-1} (Pt) 407 mg g^{-1} (Pd)	20 mL, pH 0.5, 400 rpm, 48 h	Döker *et al.* (2005)
Hydrazine-functionalized zeolite (200 mg)	Pt (2 mg L^{-1})	93%, 112 μg g^{-1}	10 mL, pH 2.0, 150 rpm, 360 min	Mosai *et al.* (2019)

Scalability refers to the ability of a process to be effectively expanded from laboratory or pilot-scale to industrial-scale operations. Recent advancements in technology, such as the development of high-performance adsorbent materials, are enhancing the scalability potential of PGE recovery methods. Techniques like functionalization of materials by adding specific chemical groups, for example thiols for platinum (Kang *et al.*, 2004), or amines for palladium (Zhou *et al.*, 2010), that increase adsorption of target PGEs via mechanisms such as coordination and electrostatic interactions, or the use of nanomaterials can lead to significant improvements in adsorption capacity and selectivity. Developing tailored adsorbents using novel materials or modifying existing ones can enhance the adsorption performance while facilitating larger-scale operations (Grilli *et al.*, 2023).

4.7 DESORPTION AND RECOVERY OF ADSORBED PGEs

The desorption and recovery of adsorbed PGEs are crucial for effective recycling and resource recovery. The methods utilized in the desorption process, including the use of acids, alkalis, or organic solvents, significantly influence the overall efficiency and recovery rates.

4.7.1 Techniques for desorbing PGEs from adsorbents

Acid desorption is among the most common methods for recovering PGEs from adsorbents (Suoranta, 2016). Strong acids like hydrochloric acid (HCl) or nitric acid (HNO_3) are used to break the bonds between the adsorbed metals and the adsorbent material. The acidic conditions can effectively protonate the metal ions, aiding their release back into the solution. While concentrated HCl may yield better results than dilute solutions, it requires careful handling to minimize risks associated with strong acids (Mishra, 2014). Alkali desorption utilizes alkaline solutions, such as sodium hydroxide (NaOH) or potassium hydroxide (KOH), to extract PGEs from adsorbents (Wu *et al.*, 2024). The alkaline conditions facilitate the ionization of the adsorbed metal ions, reducing their attachment to the adsorbent surface. By forming soluble complexes or modifying electrostatic interactions, alkali desorption can effectively release PGEs, making them available for recovery processes.

Organic solvents provide an alternative method for desorbing PGEs from different adsorbent materials. Solvents like ethanol, acetone, or specialized extractants can dissolve the adsorbed metal ions, facilitating their recovery. These organic solvent extraction techniques may be preferred for PGEs that are less suitable for acid or alkali desorption, or when a specific selectivity is needed during the recovery process (Iqbal & Ashraf, 2018). Moreover, these methods can be especially valuable when reducing the use of acids or bases is a priority due to environmental or safety concerns.

4.7.2 Adsorbent regeneration and reuse potential

A variety of techniques are utilized for regenerating adsorbents used in PGE recovery. The effectiveness of each method can differ depending on the type of adsorbent material and the specific PGE being recovered, impacting decisions in industrial applications (Zheng *et al.*, 2021). Regenerating and reusing adsorbents in PGE recovery processes is crucial for optimizing resource efficiency and sustainability. By using adsorbents multiple times, industries can considerably reduce expenses associated with acquiring new materials, thus improving the overall economic sustainability of their operations. This approach also helps decrease waste generation, supporting environmental sustainability objectives.

4.7.3 Economic and environmental aspects of desorption processes

The economic benefits of desorption processes are significantly influenced by the efficiency and purity of the PGEs recovered. Achieving high recovery efficiencies often surpassing 95% can substantially reduce waste and enhance the profitability of PGE operations (Pianowska *et al.*, 2023). In addition, utilizing advanced extractants and improved extraction techniques can reduce operational expenses while maintaining high product quality.

Although solvent extraction is a highly effective technique, it' is crucial to address its potential environmental effects, including solvent toxicity and waste production. Effective management of solvent waste is vital to mitigate pollution, posing a considerable challenge for industry. Moreover, the exploration of greener alternatives, such as deep eutectic solvents, aims to minimize the environmental impact linked to conventional solvents used in desorption processes (Pianowska *et al.*, 2023).

4.8 ENVIRONMENTAL AND ECONOMIC CONSIDERATIONS

Issues associated with the severe environmental and economic drawbacks hinder the recovery of PGEs through adsorption-based approaches. Studies on associated environmental aspects, life-cycle assessment (LCA), and the techno-economic feasibility of PGE recovery processes through adsorption-based approaches that could define levels of sustainability and feasibility at industrial scales, are still in their infancy.

4.8.1 Environmental impacts of adsorbent use and disposal

Adsorptive removal techniques hold much promise as a means of replacing traditional approaches such as pyrometallurgy and hydrometallurgy for PGE recovery, although this again comes with considerations about environmental ramifications. The most crucial among these is the disposal of spent adsorbents, especially those that could be hazardous post-PGE recovery. For example, synthetic adsorbents, for example activated carbon or metal-organic frameworks, entail a substantial environmental footprint with regard to energy-intensive production and difficulties in disposal (Satyam & Patra, 2024).

An eco-friendlier option is that of green adsorbents like those derived from algae or agricultural biomass. Unlike synthetic adsorbents, their natural counterparts are biodegradable and do not harm the environment. They are often recyclable or compostable, which makes them more sustainable (Liu *et al.*, 2024). Additionally, the environmental impacts of the adsorbents themselves and the overall process of adsorption need to be evaluated (Velempini *et al.*, 2023).

4.8.2 LCA and sustainability of adsorption-based PGE recovery processes

LCA provides an assessment of the environmental sustainability of adsorption-based PGE recovery processes. LCAs commonly include raw material extraction, production of the adsorbent, use of the operational system, and end-of-life considerations. Studies demonstrated that the recovery systems by adsorption are more environmentally friendly than traditional procedures if they are not only integrated with effective regeneration techniques but also with reusability of the adsorbent (Kulkarni & Kaware, 2014).

For example, magnetic nanoparticles have great potential for regeneration and reuse, thus making them more environmentally friendly than single-use adsorbents (Rasheed, 2022). Biosorbents, and many other green adsorbents from renewable sources, typically possess lower life-cycle impacts because of the lesser energy demands required for production and biodegradation (Osman *et al.*, 2023). Regeneration of adsorbents is also a critical aspect of making the overall process sustainable. Adsorbents that maintain their adsorption capacity upon repeated cycles of regeneration, such as magnetic nanoparticles and some types of MOFs, can reduce the environmental costs of new adsorbent production every time an adsorbate can be recovered (Kamali, 2024). Green adsorbents also remove incidental problems since they make the process less dependent on synthetic materials and chemicals with higher environmental footprints

4.8.3 Economic viability for industrial-scale implementation

The feasibility of adsorption-based recovery methods for PGEs depends on a few factors, including the cost of making the adsorbent, regeneration capacity and scalability. The production costs for most adsorbent materials, apart from a handful of nanostructures like MOFs, are still prohibitive for adoption on a large scale (GadelHak *et al.*, 2023). This offers potential development in the utilization of green adsorbents that may make the recovery of PGE economically viable. Biosorbents are produced cheaper and readily available. However, these materials must demonstrate high selectivity and efficacy to recover PGEs from complex waste streams to compete on the basis of performance in addition to being cost-effective compared to synthetic adsorbents (Tofan *et al.*, 2022).

The introduction of adsorption technology in industrial systems offers a significant cost-saving potential, as modular adsorption units require minimal structural changes and low initial investment. These systems are applied across various sectors, from wastewater treatment to urban mining and e-waste recovery. However, commercialization depends on advances in adsorbent regeneration and optimization for specific waste streams. To ensure long-term viability, new technologies must focus on cost reduction, while ongoing research will improve adsorbent regeneration and scaling, making adsorption key to sustainable PGE recovery (Jani, 2022; Yousef *et al.*, 2020).

4.9 CONCLUSION AND FUTURE DIRECTIONS

In the backdrop of recent breakthroughs occurring in the field of material science and nanotechnology, PGE recovery through adsorptive methods holds a promising future. These advancements have made adsorptive techniques effective and environmental-friendly solutions for recovering PGEs from various waste streams. This chapter overviewed the mechanisms, along with kinetics and thermodynamics, of PGEs adsorption. Conventional and emerging adsorbents for PGEs are discussed. There are challenges to be addressed such as adsorbent regeneration, selectivity in complex waste streams, and scalability. Cost-effective, novel and hybrid adsorbents for selective PGE recovery need to be explored. Until now, the studies on adsorption of PGEs are mostly confined to batch assays. This implies the need for pilot-scale continuous studies with real effluents (and not only with synthetic PGE supplemented aqueous solutions). This could result in further development of kinetic models, LCA and techno-economic evaluation on adsorption-based PGE recovery. Thus, continued research and further developments are paramount to integrate adsorption-based PGE recovery techniques into industrial systems.

REFERENCES

Abovskaya N. V., Simanova S. A., Boichinova E. S. and Fedorov M. K. (2006). Sorption recovery of sulfate complexes of iridium on hydrated zirconium dioxide. *Russian Journal of Applied Chemistry*, **79**, 563–567, https://doi.org/10.1134/S1070427206040112

Aktas S. and Morcali M. H. (2011). Platinum recovery from dilute platinum solutions using activated carbon. *Transactions of Nonferrous Metals Society of China*, **21**(11), 2554–2558, https://doi.org/10.1016/S1003-6326(11)61091-1

Aljamali N. M., Khdur R. and Alfatlawi I. O. (2021). Physical and chemical adsorption and its applications. *International Journal of Thermodynamics and Chemical Kinetics*, **7**(2), 1–8.

Asere T. G., Mincke S., Folens K., Bussche F. V., Lapeire L., Verbeken K., ... and Stevens C. V. (2019). Dialdehyde carboxymethyl cellulose cross-linked chitosan for the recovery of palladium and platinum from aqueous solution. *Reactive and Functional Polymers*, **141**, 145–154, https://doi.org/10.1016/j.reactfunctpolym.2019.05.008

Buchspies B., Thormann L., Mbohwa C. and Kaltschmitt M. (2017). PGE production in Southern Africa, part II: environmental aspects. *Minerals*, **7**(11), 225, https://doi.org/10.3390/min7110225

Chandrasekaran K. and Karunasagar D. (2016). Dispersive liquid–liquid microextraction for simultaneous preconcentration of platinum group elements (Pd, Os, Ir, and Pt) and selected transition elements (Ag, Cd, Ta, and Re) at parts per trillion levels in water and their determination by inductively coupled plasma-mass spectrometry. *Journal of Analytical Atomic Spectrometry*, **31**(5), 1131–1140, https://doi.org/10.1039/C6JA00035E

Chen M., Li S., Jin C., Shao M., Huang Z. and Xie X. (2021). Removal of metal-cyanide complexes and recovery of Pt (II) and Pd (II) from wastewater using an alkali-tolerant metal-organic resin. *Journal of Hazardous Materials*, **406**, 124315, https://doi.org/10.1016/j.jhazmat.2020.124315

Chidunchi I., Kulikov M., Safarov R. and Kopishev E. (2024). Extraction of platinum group metals from catalytic converters. *Heliyon*, **10**(3), e25283, https://doi.org/10.1016/j.heliyon.2024.e25283

Cilek E. C. and Uysal K. (2018). Froth stabilisation using nanoparticles in mineral flotation. *Physicochemical Problems of Mineral Processing*, **54**, 878–889, https://doi.org/10.5277/ppmp1889

Compagnone M., González-Cortés J. J., Pilar Yeste M., Cantero D. and Ramírez M. (2023). Sustainable recovery of platinum group metals from spent automotive three-way catalysts through a biogenic thiosulfate-copper-ammonia system. *Molecules*, **28**(24), 8078, https://doi.org/10.3390/molecules28248078

Conroy E. (2017). Rhodium recovery: separation of a precious metal for reuse in catalysis of sunovion chemical process. Doctoral dissertation, Worcester Polytechnic Institute.

Cui J. and Zhang L. (2008). Metallurgical recovery of metals from electronic waste: a review. *Journal of Hazardous Materials*, **158**(2–3), 228–256, https://doi.org/10.1016/j.jhazmat.2008.02.001

Dada A. O., Ojediran J. O., Okunola A. A., Dada F. E., Lawal A. I., Olalekan A. P. and Dada O. (2019). Modeling of biosorption of Pb (II) and Zn (II) ions onto PaMRH: Langmuir, Freundlich, Temkin, Dubinin-Raduskevich, Jovanovic, Flory-Huggins, Fowler-Guggenheim and Kiselev comparative isotherm studies. *International Journal of Mechanical Engineering and Technology (IJMET)*, **10**(2), 1048–1058.

de Oliveira Demarco J., Cadore J. S., Veit H. M., Madalosso H. B., Tanabe E. H. and Bertuol D. A. (2020). Leaching of platinum group metals from spent automotive catalysts using organic acids. *Minerals Engineering*, **159**, 106634, https://doi.org/10.1016/j.mineng.2020.106634

Döker S., Malcı S., Doğan M. and Salih B. (2005). New poly (N-(hydroxymethyl) methacrylamide-1-allyl-2-thiourea) hydrogels prepared by radiation-induced polymerisation: selective adsorption, recovery and pre-concentration of Pt (II) and Pd (II). *Analytica Chimica Acta*, **553**(1–2), 73–82, https://doi.org/10.1016/j.aca.2005.08.022

Ek K. H., Morrison G. M. and Rauch S. (2004). Environmental routes for platinum group elements to biological materials – a review. *Science of the Total Environment*, **334**, 21–38, https://doi.org/10.1016/j.scitotenv.2004.04.027

Fu F. and Wang Q. (2011). Removal of heavy metal ions from wastewaters: a review. *Journal of Environmental Management*, **92**(3), 407–418, https://doi.org/10.1016/j.jenvman.2010.11.011

Furukawa H., Cordova K. E., O'Keeffe M. and Yaghi O. M. (2013). The chemistry and applications of metal-organic frameworks. *Science*, **341**(6149), 1230444, https://doi.org/10.1126/science.1230444

GadelHak Y., El-Azazy M., Shibl M. F. and Mahmoud R. K. (2023). Cost estimation of synthesis and utilization of nano-adsorbents on the laboratory and industrial scales: a detailed review. *Science of the Total Environment*, **875**, 162629, https://doi.org/10.1016/j.scitotenv.2023.162629

Goc K., Benke G., Kluczka J., Pianowska K., Malarz J. and Leszczyńska-Sejda K. (2024). Application of ion-exchange dynamic conditions in the recovery of precious metals from refining waste. *Scientific Reports*, **14**(1), 15026, https://doi.org/10.1038/s41598-024-66086-x

Godlewska-Żyłkiewicz B., Sawicka S. and Karpińska J. (2019). Removal of platinum and palladium from wastewater by means of biosorption on fungi *Aspergillus* sp. and yeast *Saccharomyces* sp. *Water*, **11**(7), 1522, https://doi.org/10.3390/w11071522

Grilli M. L., Slobozeanu A. E., Larosa C., Paneva D., Yakoumis I. and Cherkezova-Zheleva Z. (2023). Platinum group metals: green recovery from spent auto-catalysts and reuse in new catalysts – a review. *Crystals*, **13**(4), 550, https://doi.org/10.3390/cryst13040550

Hosseinzadeh M. and Petersen J. (2024). Platinum (IV), palladium (II), and rhodium (III) recovery from mixed acidic chloride solutions using chelating ion exchange resin Puromet MTS9600. *Solvent Extraction and Ion Exchange*, **42**(6–7), 609–635, https://doi.org/10.1080/07366299.2024.2446472

Iqbal M. N. and Ashraf A. (2018). Environmental pollution: heavy metals removal from water sources. *International Journal of Alternative Fuels and Energy*, **2**(1), 14–15.

Jani Y. (2022). Adsorption: a cost-effective wastewater treatment technology for removal of conventional and emerging organic contaminants. In: M. Nasr, and A. M. Negm, (Eds.), Cost-Efficient Wastewater Treatment Technologies: Engineered Systems (1st ed., pp. 17–33). Springer International Publishing, Cham, Switzerland.

Jha M. K., Joshi S., Sharma R. K., Kim A. A., Pant B., Park M. and Pant H. R. (2021). Surface modified activated carbons: sustainable bio-based materials for environmental remediation. *Nanomaterials*, **11**(11), 3140, https://doi.org/10.3390/nano11113140

Kalam S., Abu-Khamsin S. A., Kamal M. S. and Patil S. (2021). Surfactant adsorption isotherms: a review. *ACS Omega*, **6**(48), 32342–32348, https://doi.org/10.1021/acsomega.1c04661

Kamali A. (2024). Thermal Regeneration of Graphene-Encapsulated Magnetic Nanoparticles as an Efficient Adsorbent for Dye Removal from Wastewater. *Available at SSRN 4877962*.

Kang T., Park Y. and Yi J. (2004). Highly selective adsorption of Pt^{2+} and Pd^{2+} using thiol-functionalized mesoporous silica. *Industrial & Engineering Chemistry Research*, **43**(6), 1478–1484, https://doi.org/10.1021/ie030590k

Karim S. and Ting Y. P. (2021). Recycling pathways for platinum group metals from spent automotive catalyst: a review on conventional approaches and bio-processes. *Resources, Conservation and Recycling*, **170**, 105588, https://doi.org/10.1016/j.resconrec.2021.105588

Kramer J., Scholten A., Driessen W. L. and Reedijk J. (2001). Recovery of rhodium-containing catalysts by silica-based chelating ion exchangers containing N and S donor atoms. *Inorganica Chimica Acta*, **315**(2), 183–190, https://doi.org/10.1016/S0020-1693(01)00356-5

Kramer J., Erkelens J. A., García A. R., Driessen W. L. and Reedijk J. (2002). Remarkably fast and selective recovery of a rhodium-containing catalyst with silica-based (poly) amine ion exchangers. *New Journal of Chemistry*, **26**(7), 822–826, https://doi.org/10.1039/b200141a

Kramer J., Driessen W. L., Koch K. R. and Reedijk J. (2005). Highly selective and efficient recovery of Pd, Pt, and Rh from precious metal-containing industrial effluents with silica-based (poly)amine ion exchangers. *Separation Science and Technology*, **39**(1), 63–75, https://doi.org/10.1081/SS-120027401

Kulkarni S. and Kaware J. (2014). Regeneration and recovery in adsorption – a review. *International Journal of Innovative Science, Engineering and Technology*, **1**(8), 61–64.

Liu Y., Biswas B., Hassan M. and Naidu R. (2024). Green adsorbents for environmental remediation: synthesis methods, ecotoxicity, and reusability prospects. *Processes*, **12**(6), 1195, https://doi.org/10.3390/pr12061195

MacDonald L., Zhang D. and Karamalidis A. (2024). Platinum group metals: key solid phase adsorption technologies for separation from primary and secondary sources. *Resources, Conservation and Recycling*, **205**, 107590, https://doi.org/10.1016/j.resconrec.2024.107590

Macheyeki A. S., Li X., Kafumu D. P. and Yuan F. (2020). Elements of exploration geochemistry. In: A.S. Macheyaki, X. Li, D.P. Kafumu and F. Yuan (Eds), Applied Geochemistry: Advances in Mineral Exploration Techniques, pp. 2–43. Elsevier, Amsterdam, The Netherlands.

Maponya T. C., Makgopa K., Somo T. R., Tshwane D. M. and Modibane K. D. (2023). Highly adsorptive removal of palladium and platinum ions from wastewater using novel ethylenediamine-glutaraldehyde-grafted metal organic framework. *Environmental Nanotechnology, Monitoring & Management*, **20**, 100805, https://doi.org/10.1016/j.enmm.2023.100805

Mavhungu A., Mbaya R. K. K. and Moropeng M. L. (2013). Recovery of platinum and palladium ions from aqueous solution using grape stalk waste. *International Journal of Chemical Engineering and Applications*, **4**(6), 354, https://doi.org/10.7763/IJCEA.2013.V4.324

McNeice J. and Mahandra H. (2023). Application of nanomaterials for precious metals recovery. In: U. Shanker, C. M. Hussain, and M. Rani, Handbook of Green and Sustainable Nanotechnology: Fundamentals, Developments and Applications, Springer International Publishing, Cham, pp. 2501–2532.

Minoda A., Ueda S., Miyashita S. I., Ogura T., Natori S., Sun J. and Takahashi Y. (2023). Reversible adsorption of iridium in lyophilized cells of the unicellular red alga *Galdieria sulphuraria*. *RSC Advances*, **13**(21), 14217–14223, https://doi.org/10.1039/D3RA01249B

Mishra S. P. (2014). Adsorption–desorption of heavy metal ions. *Current Science*, **107**, 601–612.

Mokhodoeva O., Shkinev V., Maksimova V., Dzhenloda R. and Spivakov B. (2020). Recovery of platinum group metals using magnetic nanoparticles modified with ionic liquids. *Separation and Purification Technology*, **248**, 117049, https://doi.org/10.1016/j.seppur.2020.117049

Mosai A. K., Chimuka L., Cukrowska E. M., Kotzé I. A. and Tutu H. (2019). The recovery of Platinum (IV) from aqueous solutions by hydrazine-functionalised zeolite. *Minerals Engineering*, **131**, 304–312, https://doi.org/10.1016/j.mineng.2018.11.028

Mosai A. K., Chimuka L., Cukrowska E. M., Kotzé I. A. and Tutu H. (2020). Removal of platinum (IV) from aqueous solutions with yeast-functionalised bentonite. *Chemosphere*, **239**, 124768, https://doi.org/10.1016/j.chemosphere.2019.124768

Mphahlele M. R., Mosai A. K., Tutu H. and Kotzé I. A. (2024). Advancements in sustainable platinum and palladium recovery: unveiling superior adsorption efficiency and selectivity with a novel silica-anchored acylthiourea adsorbent. *RSC Advances*, **14**(11), 7507–7516, https://doi.org/10.1039/D3RA08169A

Musah M., Azeh Y., Mathew J. T., Umar M. T., Abdulhamid Z. and Muhammad A. I. (2022). Adsorption kinetics and isotherm models: a review. *CaJoST*, **4**(1), 20–26, https://doi.org/10.4314/cajost.v4i1.3

Nakhjiri A. T., Sanaeepur H., Amooghin A. E. and Shirazi M. M. A. (2022). Recovery of precious metals from industrial wastewater towards resource recovery and environmental sustainability: a critical review. *Desalination*, **527**, 115510, https://doi.org/10.1016/j.desal.2021.115510

Nouri A., and Wen C. (2021). Noble metal alloys for load-bearing implant applications. In C. Wen (Ed.), *Structural Biomaterials: Properties, Characteristics and Selection* (pp. 127–156). Woodhead Publishing, Duxford, United Kingdom.

Osman A. I., El-Monaem E. M. A., Elgarahy A. M., Aniagor C. O., Hosny M., Farghali M., … and Eltaweil A. S. (2023). Methods to prepare biosorbents and magnetic sorbents for water treatment: a review. *Environmental Chemistry Letters*, **21**(4), 2337–2398, https://doi.org/10.1007/s10311-023-01603-4

Petrov G., Zotova I., Nikitina T. and Fokina S. (2021). Sorption recovery of platinum metals from production solutions of sulfate-chloride leaching of chromite wastes. *Metals*, **11**(4), 569, https://doi.org/10.3390/met11040569

Pianowska K., Kluczka J., Benke G., Goc K., Malarz J., Ochmański M. and Leszczyńska-Sejda K. (2023). Solvent extraction as a method of recovery and separation of platinum group metals. *Materials*, **16**(13), 4681, https://doi.org/10.3390/ma16134681

Prasetyo E. and Anderson C. (2020). Platinum group elements recovery from used catalytic converters by acidic fusion and leaching. *Metals*, **10**(4), 485, https://doi.org/10.3390/met10040485

Rasheed T. (2022). Magnetic nanomaterials: greener and sustainable alternatives for the adsorption of hazardous environmental contaminants. *Journal of Cleaner Production*, **362**, 132338, https://doi.org/10.1016/j.jclepro.2022.132338

Saitoh N., Nomura T. and Konishi Y. (2017). Biotechnological recovery of platinum group metals from leachates of spent automotive catalysts. In: H. Kim, S. Alam, N. R. Neelameggham, H. Oosterhof, T. Ouchi and X. Guan, Rare Metal Technology 2017, Springer International Publishing, Cham, Switzerland, pp. 129–135.

Satyam S. and Patra S. (2024). Innovations and challenges in adsorption-based wastewater remediation: a comprehensive review. *Heliyon*, **10**(9), e29573.

Schulte R. F. (2019). 'Platinum Group Metals' *2019 Minerals Yearbook*.

Sefako R., SekgarametsoK. and Sibanda V. (2017). Potential processing routes for recovery of platinum group metals from Southern African oxidized PGM ores: a review. *Journal of Sustainable Metallurgy*, **3**(4), 797–807, https://doi.org/10.1007/s40831-017-0146-0

Sharma S., Kumar A. S. K. and Rajesh N. (2017). A perspective on diverse adsorbent materials to recover precious palladium and the way forward. *RSC Advances*, **7**(82), 52133–52142, https://doi.org/10.1039/C7RA10153H

Siwela L. B., Hardwick E. K., Falconer J. G., Mathibela M. E. and Rolfe W. (2022). Recovering Platinum Group Metals from acidic and high chloride effluent streams using ion exchange resins.

Sronsri C., Panitantum N., Sittipol W., Kongpop U. and Kerdphol P. (2022). Optimization of selective gold recovery from electronic wastes through hydrometallurgy and adsorption. *Process Safety and Environmental Protection*, **163**, 659–668, https://doi.org/10.1016/j.psep.2022.06.001

Suoranta T. (2016). Advanced analytical methods for platinum group elements: applications in the research of catalyst materials, recycling and environmental issues.

Taylor S. (2012). Recovery of platinum group metals from waste sources. Doctoral dissertation, University of Birmingham.

Treber J. V., González P., Gil R., Robledo S. and Almeida C. (2024). Platinum group metals removal using expanded vermiculite. *Chemical Engineering Communications*, **211**(7), 1015–1029, https://doi.org/10.1080/00986445.2024.2324428

Trieu A. (2018). Design of hybrid nano-composite adsorbent for recovery of Pd and Au from electronic wastewater. PhD dissertation, Université de Lyon.

Tofan L., Bojoaga C. N. and Paduraru C. (2022). Biosorption for the recovery and analysis of rare earth elements and platinum group metals from real samples. A review. *Environmental Chemistry Letters*, **20**(2), 1225–1248, https://doi.org/10.1007/s10311-021-01330-8

Uddin M. K. (2017). A review on the adsorption of heavy metals by clay minerals, with special focus on the past decade. *Chemical Engineering Journal*, **308**, 438–462, https://doi.org/10.1016/j.cej.2016.09.029

United States Department of Energy. (2022). Platinum Group Metal Catalysts U.S. Department of Energy (DOE) Response to Executive Order 14017, 'America's Supply Chains' 2022, Washington DC, United States.

Velempini T., Ahamed M. E. H. and Pillay K. (2023). Heavy-metal spent adsorbents reuse in catalytic, energy and forensic applications – a new approach in reducing secondary pollution associated with adsorption. *Results in Chemistry*, **5**, 100901, https://doi.org/10.1016/j.rechem.2023.100901

Wang J. and Guo X. (2023). Adsorption kinetics and isotherm models of heavy metals by various adsorbents: an overview. *Critical Reviews in Environmental Science and Technology*, **53**(21), 1837–1865, https://doi.org/10.1080/10643389.2023.2221157

Weiss F. T., Leuzinger M., Zurbrügg C. and Eggen H. I. (2016). Chemical Pollution in Low- and Middle-Income Countries. Eawag. Swiss Federal Institute of Aquatic Science and Technology, Uberlandstrasse 133, 8600 Dübendorf, Switzerland.

Woods R. (1996). Chemisorption of thiols on metals and metal sulfides. *Modern Aspects of Electrochemistry*, **29**, 401–453.

Wu K., Ning S., Yin X., Xu S., Zhong Y., Li Z., Chen L., Hamza M. F., Fujita T. and Wei Y. (2024). Precise stepwise recovery of platinum group metals from high-level liquid wastes based on SDB polymer-modified SiO_2. *Dalton Transactions*, **53**(4), 1586–1598, https://doi.org/10.1039/D3DT03469K

Yakoumis I., Panou M., Moschovi A. M. and Panias D. (2021). Recovery of platinum group metals from spent automotive catalysts: a review. *Cleaner Engineering and Technology*, **3**, 100112, https://doi.org/10.1016/j.clet.2021.100112

Yousef R., Qiblawey H. and El-Naas M. H. (2020). Adsorption as a process for produced water treatment: a review. *Processes*, **8**(12), 1657, https://doi.org/10.3390/pr8121657

Zhang D., MacDonald L., Raj P. and Karamalidis A. K. (2024). Thiol-functionalized cellulose adsorbents for highly selective separation of palladium over platinum in acidic aqueous solutions. *Chemical Engineering Journal*, **494**, 152948, https://doi.org/10.1016/j.cej.2024.152948

Zheng, H., Ding, Y., Wen, Q., Liu, B. and Zhang, S. (2021). Separation and purification of platinum group metals from aqueous solution: Recent developments and industrial applications. *Resources, Conservation and Recycling*, **167**, 105417. https://doi.org/10.1016/j.resconrec.2021.105417

Zheng Z. and Greedan J. E. (2003). Rare earth elements and materials. In: R. A. Meyers, *Encyclopedia of Physical Science and Technology*, (3rd ed., p. 1–22). Academic Press, Cambridge, United States.

Zhou L., Xu J., Liang X. and Liu Z. (2010). Adsorption of platinum (IV) and palladium (II) from aqueous solution by magnetic cross-linking chitosan nanoparticles modified with ethylenediamine. *Journal of Hazardous Materials*, **182**(1–3), 518–524, https://doi.org/10.1016/j.jhazmat.2010.06.062

Zientek M. L., Loferski P. J., Parks H. L., Schulte R. F. and Seal R. R. II (2017). Platinum-Group Elements (No. 1802-N). US Geological Survey, Reston, VA, United States.

doi: 10.2166/9781789065084_0095

Chapter 5

Microbial biorecovery of platinum group metals using biosorption and bioreduction

Christopher Egan-Morriss and Jinxin Xie**

Department of Earth and Environmental Sciences, University of Manchester, Oxford Road, Manchester M13 9PL, UK

**Corresponding authors: christopher.eganmorriss@manchester.ac.uk; Jinxin.xie@manchester.ac.uk*

ABSTRACT

This chapter reviews the literature on the biorecovery of platinum group metals (PGMs) using biosorption and bioreduction by microorganisms. It covers studies on biorecovery from synthetic monometallic systems, mixed metal solutions, and real industrial effluents, and discusses recycling of the recovered metals. Comparison of the two approaches suggests that bioreduction has more potential than biosorption for industrial PGM biorecovery, though far more research on real industrial effluents is necessary. Overall, microbial biorecovery presents a promising biotechnology to reduce the environmental impact of PGM usage and support sustainable management of these critical resources within a circular economy.

Keywords: Platinum group metals, biorecovery, biosorption, bioreduction

5.1 INTRODUCTION

5.1.1 Application of PGMs

The platinum group metals (PGMs) are of crucial importance across many technology sectors, such as catalysts in the automotive industry and industrial chemical production, petroleum refining, electronics, fuel cells, biomedical devices, photovoltaics, superalloys, and jewellery (Karim & Ting, 2021; Taninouchi & Okabe, 2023). The PGMs are among the rarest and most expensive metals in the world and their global demand far exceeds annual primary production, underscoring the critical need for effective recovery and recycling technologies (Karim & Ting, 2021). The high technical recyclability of PGMs means that over 95% recovery can be achieved once PGM-containing scrap reaches a state-of-the-art refining facility (Hagelüken, 2012). However, economic, and environmental imperatives drive the need to recover these metals from low-concentration industrial waste streams as well. Importantly, the unsustainable nature of the traditional linear economy – characterized by the flow of extraction, use, and disposal – has led to the production of anthropogenic urban waste, such as waste electrical and electronic equipment and spent automotive catalysts, with higher precious metal content by weight than many primary ore deposits (Prichard *et al.*, 2016). The concept of 'urban mining' emphasizes the recovery and recycling of precious metals waste, and the need to shift towards a circular economy focused on the sustainable management of materials over their entire lifecycle (Cossu, 2013).

5.1.2 Microbial biorecovery techniques

Microbial biorecovery offers a potential biotechnological route to recycling precious metals in a circular economy without the use of high-temperatures and harmful chemicals by leveraging the intrinsic capabilities of microorganisms to bind, accumulate, and transform metal ions from solution into recoverable forms. This chapter evaluates the two main strategies, biosorption and bioreduction, applied to the microbial biorecovery of PGMs.

5.1.2.1 Biosorption

Microorganisms are natural biosorbents of metal species due to their high surface-to-volume ratios and abundance of potential physisorption and chemisorption sites (Beveridge, 1989). Biosorption involves the adsorption of metal ions onto biomaterials via ionic interactions (ion exchange, electrostatic forces) or covalent interactions (complexation/chelation) (Naja & Volesky, 2011). The biochemical makeup of microbial cells and their extracellular polymeric substances (EPS) can vary significantly and heavily influences the biosorption process. Metals are typically classified as either 'hard acids' (e.g. Al^{3+}, Mg^{2+}, Ca^{2+}), 'soft acids' (e.g. Pd^{2+}, Pt^{4+}, Au^{3+}), or 'borderline' (e.g. Fe^{3+}, Cu^{2+}, Ni^{2+}) (Pearson, 1968). In terms of biosorption, hard acids will typically favour bonding to functional group atoms in the order $O > N > S$, whereas soft acids will prefer $S > N > O$ (Nieboer & Richardson, 1980).

Microbial biosorption is considered a complex process that is influenced by various factors such as the type of microbial biomass used, the chemical, stereo-chemical, and coordination characteristics of the metals, and the solution chemistry, such as pH and competing ions (Tsezos *et al.*, 2006). Microbial biosorption typically occurs independently of metabolic activity, offering advantages such as the use of non-viable biomass and reusability of biosorbents, but can also occur via metabolically dependent processes such as bioprecipitation, biosequestration by metal-binding proteins, or active transport into the cell (Gadd, 2009). Significant challenges include poor metal selectivity, rapid saturation, and the need for significant modification of the biomass to enhance its performance. It is important to note that metal biosorption recovery technologies have been researched for decades without any significant industrial exploitation (Gadd, 2009).

5.1.2.2 Bioreduction

Microorganisms can facilitate the bioreduction of metal species from solution to less oxidized states as precipitates that are readily incorporated into the biomass matrix for easy recovery. The enzymatic bioreduction of metal ions is coupled to the oxidation of an electron donor resulting in the transfer of electrons from an organic or inorganic substrate to the metal. Bioreduction is particularly advantageous for recovering precious metals (PGMs, Au, and Ag) resulting in the formation of metallic nanoparticles supported on microbial biomass (De Corte *et al.*, 2011b; Konishi *et al.*, 2007; Law *et al.*, 2008; Lloyd *et al.*, 1998), which can be highly valuable as catalysts in industrial processes and environmental remediation (Egan-Morriss *et al.*, 2022). Additionally, bioreduction from mixed metal solutions can result in the co-deposition of precious metals on cells as bimetallic nanoparticles (Kimber *et al.*, 2021), potentially enhancing the selectivity and efficiency of PGM recovery while improving the catalytic properties of the nanoparticles formed.

The most utilized electron donor for metal biorecovery is hydrogen, which is oxidized by hydrogenase enzymes to produce two protons and two electrons (Sickerman & Hu, 2019). However, a key drawback of bioreduction systems is that these enzymes can become denatured or deactivated under harsh conditions, such as highly acidic pH or in the presence of toxic pollutants or metals. To circumvent this, a combined approach of metal biosorption followed by non-biological hydrogen-driven reduction can be employed, which lessens the dependency on purely metabolic bioreduction processes. Another key challenge lies in using large quantities of viable biomass for PGM bioreduction without incurring prohibitively high costs, making waste biomass an important resource.

5.1.3 Challenges of biorecovery from real industrial effluents

The application of microbial biorecovery technologies to real-world industrial effluents presents significant challenges. These effluents are inherently complex, containing a diverse range of PGM concentrations alongside high levels of contaminant metals, organic and inorganic pollutants, and anions. Microbial systems must therefore possess a high metal binding capacity and strong selectivity for PGMs from these complex mixtures. Additionally, effluents from the same source exhibit considerable variability over time, necessitating a robust biorecovery system that can effectively adapt to fluctuations in waste composition.

Microbial biorecovery of PGMs faces substantial obstacles, including scalability, process control, and managing the intricate nature of waste streams. While research has shown promising results in synthetic and simulated environments, translating these successes to real industrial waste scenarios has progressed slowly. Effective recovery from such effluents demands a thorough understanding of all waste stream components, a complex task especially when dealing with commercially sensitive materials. Moreover, much of the existing research has been conducted under batch conditions, and there is a pressing need to shift focus towards continuous flow systems to achieve industrial relevance.

5.2 BIORECOVERY FROM SINGLE COMPONENT SYNTHETIC PGM SOLUTIONS

To date, research on microbial PGM recovery has largely centred on monometallic synthetic systems. While these are far from accurate representations of real waste conditions, they are important for understanding the fundamentals of microbe–metal interactions, which will be discussed here for the PGMs.

5.2.1 Palladium

5.2.1.1 Biosorption

The biochemical makeup of different microbial cell walls profoundly affects metal binding. The optimal pH for Pd(II) biosorption in various bacteria was observed at pH 3–4 (Cui *et al.*, 2017; Park *et al.*, 2010; Tan *et al.*, 2017; Xu *et al.*, 2018; Yong *et al.*, 2002a), whereas pH 1.6–3 was optimal for the yeast *S. cerevisiae* (Godlewska-Zyłkiewicz, 2003; Godlewska-Żyłkiewicz *et al.*, 2019), and pH 1.6–2 for the alga *C. vulgaris* (Dziwulska *et al.*, 2004; Godlewska-Zyłkiewicz, 2003). In contrast, optimal Pd(II) biosorption by the fungus *Aspergillus* sp. was largely pH independent, occurring at pH 4–11, likely due to complexation with amine groups within the fungal cell wall (Godlewska-Żyłkiewicz *et al.*, 2019).

Although the mechanism and efficiency of biosorption varies across microbial species, Pd(II) biosorption is widely reported to occur via amine, amide, carboxyl, hydroxyl and phosphoryl groups (Chen & Chen, 2022; Cui *et al.*, 2017; Lin, 2002; Tan *et al.*, 2017; Xu *et al.*, 2018). These groups can become protonated, or deprotonated with changing pH, affecting their ionic interactions with Pd(II) complexes. The S, O, and N atoms of cell functional groups can act as ligands that coordinate to Pd(II). For example, Fahmy *et al.* (2006) identified the carboxylates of aspartate and glutamate as coordination sites for Pd(II) binding in strains of the bacterium *Bacillus sphaericus* (Fahmy *et al.*, 2006), and Adams *et al.* (2021) reported the coordination of Pd(II) to either N or S containing groups in *G. sulphuraria* at highly acidic pH, which was largely dependent on the Cl^- concentration (Adams *et al.*, 2021). Additionally, *E. coli* functionalized with a polymer rich in amine groups possessed a significantly increased Pd(II) biosorption capacity of 88.3 mg/g, compared to 56.2 mg/g for the raw biomass (Park *et al.*, 2010). Engineering overexpression of a metal-binding peptide containing sulfhydryl sites in *E. coli* significantly improved Pd(II) biosorption compared to native cells, from 143.77 to 200 mg Pd/g biomass (Tan *et al.*, 2018).

The microbial biosorption of Pd(II) also depends significantly on the metal speciation. Remoudaki *et al.* (1999) observed that $[Pd(NH_3)_4]^{2+}$ and $[PdCl_4]^{2-}$ displayed optimal biosorption on heavy metal-resistant bacteria under different conditions (Remoudaki *et al.*, 1999). Low pH was optimal for $[PdCl_4]^{2-}$ biosorption and was highly dependent on the Cl^- concentration. At high Cl^- concentration the only species present was $PdCl_4^{2-}$ and could only be biosorbed by protonated functional groups, whereas at low Cl^- concentration hydrolysis of $[PdCl_4]^{2-}$ occurred, producing $[PdCl_3(H_2O)]^-$ and polynuclear hydroxocomplexes, which can be removed through both biosorption and precipitation (Remoudaki *et al.*, 1999; Troitskii *et al.*, 1993, 1995). In contrast, optimal $[Pd(NH_3)_4]^{2+}$ biosorption was recorded at pH 8 due to the proclivity of deprotonated carboxyl (COO^-) groups to coordinate with Pd(II) via ligand substitution. Similarly, Yong *et al.* (2002a, 2002b) found $Pd(NH_3)_4^{2+}$ was recovered far less via biosorption at pH <7 than anionic Pd complexes (Yong *et al.*, 2002a).

5.2.1.2 Bioreduction

The bioreduction of Pd(II) to Pd(0) is a well-studied biosynthesis route for the recovery of Pd and synthesis of bio-Pd nanoparticles across different microbial species employing various electron donors and specific terminal reductase enzymes (Egan-Morriss *et al.*, 2022). Hydrogen and formate are commonly utilized as electron donors under anaerobic conditions. The [NiFe]-hydrogenase localized in the periplasm of many bacteria is crucial for Pd(II) hydrogen-driven bioreduction, resulting in Pd(0) particles also localized within the periplasm (Dundas *et al.*, 2018; Mikheenko *et al.*, 2008). Gene knockout studies have demonstrated expression of hydrogenase enzymes can modify both the location and the efficiency of Pd(II) reduction. Formate-driven bioreduction was shown to result in Pd(0) deposition in the periplasm in bacteria, the site of hydrogenase and formate dehydrogenase enzymes (Deplanche *et al.*, 2010; Hou *et al.*, 2017; Yang *et al.*, 2020); although formate-driven Pd(II) may also result in the formation of intracellular or extracellular Pd(0) nanoparticles (Figure 5.1) (Foulkes *et al.*, 2016; Omajali *et al.*, 2015). Dissimilatory metal-reducing bacteria (DMRB) can utilize a diverse range of electron donors for Pd(II) bioreduction, for example *Shewanella oneidensis* can use lactate, pyruvate, ethanol, hydrogen, and formate (Cheng *et al.*, 2021; De Windt *et al.*, 2005; Ng *et al.*, 2013; Yang *et al.*, 2020). DMRB have been shown to bioreduce Pd(II) at their outer membrane, and extracellularly, via extracellular electron transfer processes through outer membrane c-type cytochromes, which can be enhanced in the presence of electron shuttles, such as riboflavin (Dundas *et al.*, 2018; Voegtlin *et al.*, 2022; Wu *et al.*, 2011; Yates *et al.*, 2013). Supplying the DMRB *Geobacter sulfurreducens* with different electron donors, and initial Pd(II) concentrations, controlled the deposition location of Pd nanoparticles within the cells, in either the periplasm, outer membrane, or extracellular polymeric substances, enabling control over the catalytic activity of the nanoparticles (Egan-Morriss *et al.*, 2024).

Solution chemistry plays an important role in Pd(II) bioreduction, for example, anionic Pd complexes from Na_2PdCl_4 solutions were recovered to a greater extent than $Pd(NH_3)_4^{2+}$ at pH 7 via bioreduction by *G. sulfurreducens* supplied with acetate as electron donor (Egan-Morriss *et al.*, 2023). Anionic Pd(II) complexes were more prone to hydrolysis at pH 7, resulting in greater recovery and smaller, more dispersed nanoparticles, compared to the more stable $Pd(NH_3)_4^{2+}$ complex, which formed large aggregates of nanoparticles upon bioreduction.

Acidic pH is not favourable for Pd(II) bioreduction by neutrophilic microbes due to the denaturing of terminal reductase enzymes. $Pd(NH_3)_4^{2+}$ was poorly

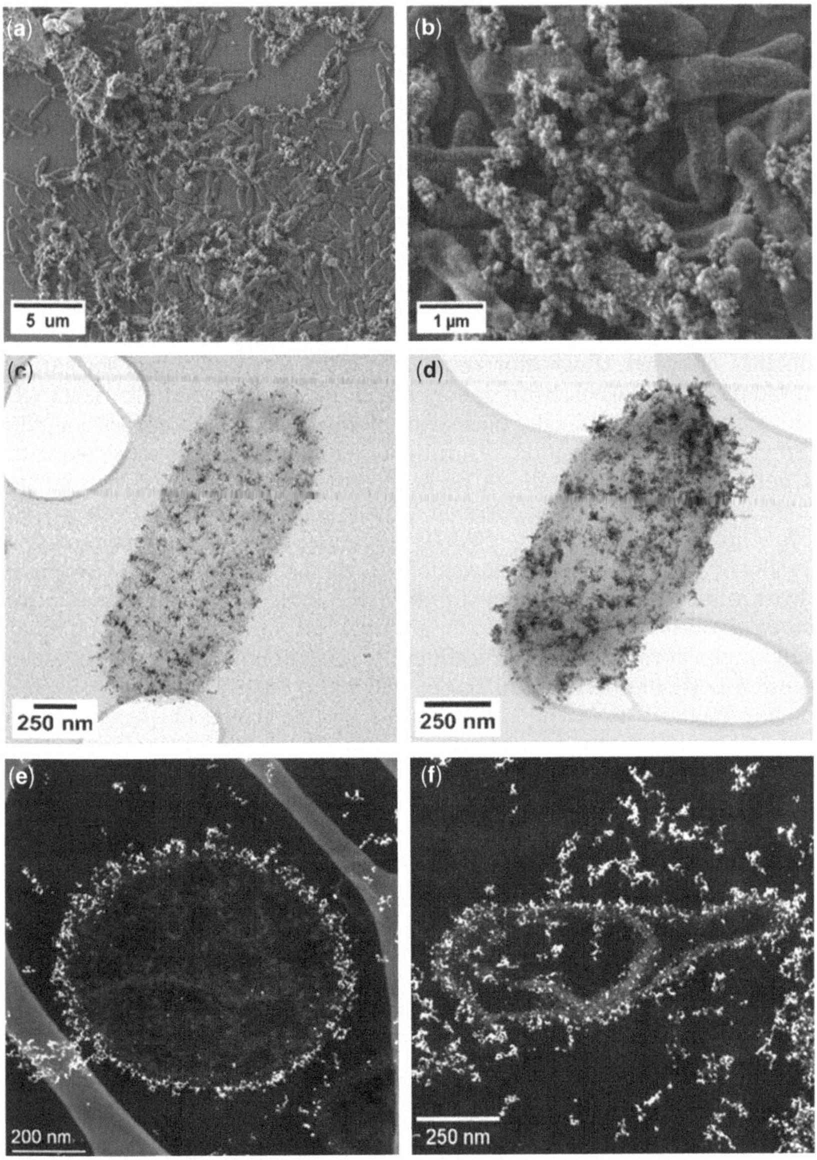

Figure 5.1 Scanning electron microscope (SEM) images of bio-Pd nanoparticles supported on the bacterium *Shewanella oneidensis* (a, b). Transmission electron microscope (TEM) images of drop-cast bio-Pd nanoparticles on the bacterium *Geobacter sulfurreducens* synthesised at (c) low Pd load and (d) high Pd load. TEM images of ultrathin sections of bio-Pd showing intracellular and extracellular nanoparticles synthesised at high Pd load using (e) hydrogen or (f) formate as electron donor.

bioreduced at pH <3 by *D. desulfuricans* with both hydrogen or formate as electron donor (Yong *et al.*, 2002b). Similarly, the formate-driven bioreduction of $Pd(NH_3)_4^{2+}$ was optimal at pH 6 for *Citrobacter* sp. bacteria, but fell significantly at pH 2 (Matsena & Chirwa, 2021). Acidophilic metal-reducing microbes may be a useful alternative, for example, *Acidocella aromatica* and *Acidiphilium cryptum* completely reduced Pd(II) at pH 2.5 within a few hours using formate as electron donor (Okibe *et al.*, 2017). Similarly, the acidophilic archaeon *Sulfolobus tokodaii* fully reduced Pd(II) at pH 2 using formate as electron donor under micro-aerobic conditions within 10 h (Kitjanukit *et al.*, 2019).

5.2.2 Platinum

5.2.2.1 Biosorption

Similarly to Pd, the microbial biosorption of Pt largely depends on the pH, metal speciation, and the species of microorganism used. Optimal biosorption of Pt(IV) occurred at pH 2–3 on the bacteria *B. subtilis, S. oneidensis* (Tanaka & Watanabe, 2015), and *D. desulfuricans* (De Vargas *et al.*, 2004), pH 1.8–3.5 for the yeast *S. cerevisiae,* pH 2.5 for the red alga *G. sulphuraria* (Ju *et al.*, 2016), and pH 1.6–2.2 for the alga *C. vulgaris* (Godlewska-Zyłkiewicz, 2003); primarily through electrostatic interaction between anionic Pt complexes and protonated cell functional groups. In contrast, near complete biosorption was observed at pH 2–11 on the fungi *Aspergillus* sp, suggesting a complexation mechanism by cell wall amine groups (Godlewska-Żyłkiewicz *et al.*, 2019). The microbial cell surface functional groups most commonly associated with Pt biosorption are carboxyl, amine, and phosphoryl groups (Campbell *et al.*, 2018; Pavlova *et al.*, 2018; Tan *et al.*, 2019; Tanaka & Watanabe, 2015; Wang *et al.*, 2017). Sulfur containing groups are also important, for example, Tan *et al.* (2019) showed that increasing sulfhydryl sites on *E. coli* significantly improved Pt(IV) biosorption compared to native cells, from 173.22 to 239.92 mg Pt/g biomass (Tan *et al.*, 2019). *G. sulphuraria* recovered 94% of 10 mg/L Pt(IV) from 2 M HCl (pH −0.301), depositing Pt(IV) intracellularly, resulting in reduction to Pt(II) and the formation of inner-sphere complexes with S in thiol residues of lyophilized cells (Miyashita *et al.*, 2021).

Important factors in metal biosorption are metal concentration and speciation. The maximum Pt biosorption capacity of microbes has widely been shown to increase with increasing initial Pt concentration, although cells rapidly become saturated with Pt after a few minutes, independent of the metal concentration (De Vargas *et al.*, 2004; Mack *et al.*, 2008; Tan *et al.*, 2019). Campbell *et al.* (2018) showed that the metal-resistant bacterium *Cupriavidus metallidurans* CH34 possessed higher biosorption capacity for Pt(II) over Pt(IV), as the higher oxidation state of Pt(IV) was more toxic to cells likely due to inducing greater oxidative stress (Campbell *et al.*, 2018). Additionally, increasing the Cl^- concentration reduced biosorption of Pt(II) and Pt(IV) complexes, as Cl^- ions compete with anionic Pt complexes for protonated cell surface functional groups (De Vargas *et al.*, 2004; Tanaka & Watanabe, 2015).

5.2.2.2 Bioreduction

The microbial bioreduction of Pt(IV) and Pt(II) to Pt(0) has been shown to occur via enzyme-mediated processes at circumneutral pH. *Shewanella algae* supplied with lactate as electron donor deposited periplasmic Pt(0) from the bioreduction of Pt(IV) (Konishi *et al.*, 2007). A sulfate-reducing bacterial consortium bioreduced Pt(IV) utilizing hydrogen, lactate, or ethanol as electron donors (Rashamuse & Whiteley, 2007). Hydrogen-driven bioreduction produced the highest metal recovery, and hydrogenases were implicated as the terminal reductase. A wastewater treatment anaerobic sludge utilized hydrogen, formate, or ethanol as electron donor for bioreduction to Pt(0); the fastest bioreduction rate was achieved with formate (Simon-Pascual *et al.*, 2019). Hydrogen-driven Pt(II) bioreduction, performed using *E. coli* at pH 2.3, resulted in spherical Pt(0) nanoparticles, which were used as a catalyst (Attard *et al.*, 2012; Bennett *et al.*, 2012). Filamentous cyanobacteria were shown to gradually reduce Pt(IV), without an electron donor, to organometallic Pt(II) complexes upon cell death from metal toxicity, which upon heating formed crystalline Pt(0) (Lengke *et al.*, 2006). The bioreduction of Pt(IV) to Pt(0) was performed at pH 2.5 using acidophilic extremophilic bacteria, *Acidocella aromatica* and *Acidiphilium crytpum*, with formate as electron donor (Matsumoto *et al.*, 2021).

Metal speciation dictates the bioavailability of Pt complexes to microbial bioreduction. Maes *et al.* (2017) investigated the bioreduction of five Pt complexes, including $Pt(II)Cl_4^{2-}$, $Pt(IV)Cl_6^{2-}$, $Pt(II)(NH_3)_4Cl_2$, and the pharmaceutical Pt(II) complexes cisplatin and carboplatin, for hydrogen-driven bioreduction from dilute aqueous solutions at pH 2 utilizing five different bacterial species (Maes *et al.*, 2017). Under the experiment conditions, low pH, the type of platinum complexes and their reduction products significantly affect the viability of cultures, carboplatin is less toxic than other Pt complexes and *C. metallidurans* showing greater tolerance. All five bacterial species fully recovered $Pt(II)Cl_4^{2-}$ and $Pt(IV)Cl_6^{2-}$, whereas only *S. oneidensis* and *C. metallidurans* completely recovered cisplatin. Less than 10% recovery was observed for carboplatin, and no recovery of $Pt(II)(NH_3)_4Cl_2$ was observed for any bacteria. In a subsequent study, the authors demonstrated that four different strains of isolated halophilic bacteria precipitated Pt(0) both intra- and extracellularly when supplied with hydrogen at pH 2 (Maes *et al.*, 2016). All bacteria achieved the complete recovery of Pt(II) and Pt(IV) as Pt(0), however, the bioreduction rate of $Pt(II)Cl_4^{2-}$ was significantly faster than that of $Pt(IV)Cl_6^{2-}$. Similarly, Simon-Pascual *et al.* (2019) observed faster rates of bioreduction for Pt(II) than Pt(IV) (Simon-Pascual *et al.*, 2019).

5.2.3 Rhodium

The microbial biorecovery of Rh(III) was first demonstrated by Ngwenya and Whiteley (2006) using a resting sulfate-reducing bacteria (SRB) consortium. In synthetic Rh(III) solutions, supplying hydrogen as electron donor was the most effective Rh(III) recovery strategy, achieving a maximum uptake capacity of 66 mg Rh/g biomass, precipitating Rh likely via bioreduction to Rh(0). A mildly acidic pH 3–5 was optimal for Rh(III) biorecovery, suggesting preferential binding via electrostatic interactions between anionic Rh(III) complexes and protonated cell functional groups, and the extent of Rh(III) removal increased with increasing initial Rh(III) concentrations. Tamaoki *et al.* (2013) performed Rh(III) bioreduction at pH 7 from synthetic buffered solutions using the marine iron(III)-reducing bacterium *Shewanella algae*, supplied with lactate as electron donor (Tamaoki *et al.*, 2013). *S. algae* recovered 90% Rh from a 0.1 mM solution, and 45% Rh from a 1 mM solution. Rh was biomineralized as nanoparticles, which XANES analysis revealed were comprised of 48% Rh(0) and 52% Rh(III). Without lactate, *S. algae* only recovered 12% Rh from the 1 mM solution via biosorption of Rh(III), showing that bioreduction was necessary to maximize Rh biorecovery. Manipura & Burgess (2008) observed that the biosorption efficiency of *S. cerevisiae* approximately doubled to 51.1% after denitrification; demonstrating that competing ions hindered Rh(III) recovery, either through preferentially occupying cell binding sites or complexing Rh(III) (Manipura & Burgess, 2008).

5.2.4 Iridium and osmium

Few studies have reported on the biorecovery of iridium or osmium. Ruiz *et al.* (2003) found that chitosan beads functionalized with polyethyleneimine (PEI) exhibited high biosorption capacities of 650 mg Os/g and 340 mg Ir/g, primarily due to the high amine content of PEI (Ruiz *et al.*, 2003). The biosorption capacities for both metals increased with higher initial metal concentrations. Over 90% of Os was biosorbed within the first 6 hours, and was challenging to desorb indicating strong affinity, whereas Ir demonstrated weaker interactions, evidenced by its release from the biosorbent after 12 h. Minoda *et al.* (2023) reported the biorecovery of Ir from HCl solutions (pH 0–1) using lyophilized *G. sulphuraria*, which efficiently recovered 87% of Ir within 3 min from a 0.2 M HCl solution containing 1800 mg/L Ir, reaching a biosorption capacity of 52 mg Ir/g (dry weight) (Minoda & Burgess, 2023). Biosorption of Ir was shown to occur via outer-sphere complexation of Ir(IV)Cl62- primarily by amine groups. However, the alga was consistently outperformed by a commercial ion-exchange resin (DIAION™ SA-10A).

5.3 BIORECOVERY FROM MULTI-COMPONENT SYNTHETIC PGM SOLUTIONS

This section discusses PGM biorecovery via both biosorption and bioreduction from synthetic PGM-containing solutions with multiple components, such as co-contaminant metals and anions. The complexity of microbial metal biorecovery increases significantly when multiple components are introduced, and studies of these systems are essential to understand how microbes interact with several metals at once, relating to real industrial effluents.

5.3.1 Biosorption

In multi-metal solutions, metals with similar properties will compete for the same binding sites, due to characteristics such as ionic radius and electronegativity.

Tsezos *et al.* (1996) observed significant ionic competition effects for metals belonging to the same class. The presence of a hard acid, such as yttrium, had no effect on the biosorption of a soft acid, such as silver, whereas a soft acid, such as Pd, had a significantly negative effect on the biosorption of another soft acid, such as gold (Tsezos *et al.*, 1996). Mack *et al.* (2011) compared the effect of different contaminant metals in synthetic solutions on Pt biosorption at pH 1.5 by immobilized and functionalized *S. cerevisiae*, which was suppressed by up to roughly 80% by Pd, 65% by Au, and 60% by Rh, showing that biosorption was non-specific between precious metal chloro-complexes and protonated surface groups (Mack *et al.*, 2011). Heavy transition metals significantly suppressed Pt biosorption in the order Co > Cu > Ni > Fe. Interestingly, base metals (Na, K, Ca, Mg) were not biosorbed but did suppress Pt biosorption, indicating that interactions between anionic Pt complexes and cationic base metals reduced the binding affinity of Pt, possibly due to complexation or hydrolysis of Pt complexes. In addition, increasing the anion concentration of Cl^-, HNO_3^-, or SO_4^{2-} greatly suppressed Pt biosorption, which is in agreement with studies on Pd(II) and Pt(IV) biosorption, as anions compete with PGM chloro-complexes for protonated binding sites (De Vargas *et al.*, 2004; Mack *et al.*, 2011, Yong *et al.*, 2002a).

Several studies have assessed microbial biosorbents for the biorecovery of Pd(II) and Pt(IV) from bicomponent solutions. In all cases, Pd(II) and Pt(IV) compete for the same biosorption sites and biosorbents display greater affinity for Pd(II), resulting in the displacement of Pt(IV) and a far greater Pd biorecovery (Chassary *et al.*, 2005; Wang *et al.*, 2017; Xu *et al.*, 2017). Comparing waste *E. coli* biomass and ion exchange resins, Kim *et al.* (2015) showed that primary amine sites selectively bound Pd(II) over Pt(IV), whereas quaternary amines did not (Kim *et al.*, 2015). Microbial cells possess a diverse array of functional groups resulting in non-specific metal binding, whereas ion exchange resins may be specifically engineered with highly selective metal-binding functional groups (Gadd, 2009).

The red alga, *G. sulphuraria*, demonstrated promising biosorption selectivity for Pd from an extremely acidic 4 M aqua regia solution containing 10 mg/L each of Pd, Cu, Fe, Ni, Sn, and Zn, achieving near-complete Pd recovery while recovering negligible amounts of the other metals (Adams *et al.*, 2024). However, selectivity for Ir was less strong from a synthetic 0.2 M HCl solution containing 20 mg/L each of Ir, Rh, Zn, Mn, Fe, Cu, and Ni, resulting in 83% recovery of Ir along with 64% co-recovery of Fe (Minoda *et al.*, 2023).

5.3.2 Bioreduction

A key advantage to metal recovery methods using bioreduction is that metals such as Pd and Pt do not have to compete for the same binding sites, in contrast to biosorption, as reduced PGMs can act as nucleation sites for further autocatalytic PGM reduction, leading to the growth of bi- or even multi-metallic nanoparticles in which PGMs may form alloyed phases (Kusada *et al.*, 2020). Bimetallic nanoparticles may be comprised of an alloy of the two metals, or two separate populations of nanoparticles of each metal may be formed in the same location on the cell. The study of microbial bimetallic nanoparticles is a relatively new field and Pd-based bimetallic particles mostly have been studied, such as PdAu and PdAg systems (De Corte *et al.*, 2011a; Kimber *et al.*, 2021; Tran *et al.*, 2012). The metal-reducing bacterium *S. oneidensis* was shown to biosynthesize bimetallic PdPt nanoparticles from synthetic solutions at pH 7, and alloying of Pd and Pt was proposed (Tuo *et al.*, 2017; Xu *et al.*, 2019).

PGM-containing effluents are usually highly acidic mixed metal solutions, which are a poor medium to perform microbial metal bioreduction due to the deactivation of bacterial terminal reductase enzymes at low pH, and inhibitory effects of contaminant metals, anions, and organics (Yong *et al.*, 2002a). The Macaskie group utilized a two-step approach to synthesize PGM bimetallic nanoparticles from acidic mixed metal synthetic solutions (Murray *et al.*, 2017, 2018). First, small amounts of bioreduced Pd nanoparticles are 'pre-seeded' onto bacteria using hydrogen as electron donor. An initial biosorption step is used in the pre-seeding step that promotes favourable biosorption of anionic Pd(II) complexes onto protonated cell surface groups. The pre-seeded bacteria are then supplied with hydrogen and challenged with acidic mixed metal PGM solutions resulting in the chemically mediated reduction of PGMs onto the Pd nanoparticle scaffolds. For example, Murray *et al.* (2018) pre-seeded *Escherichia coli* with either Pd(0) or Pt(0) nanoparticles at different initial loadings and then challenged the cells with a synthetic solution of 0.34 mM Pd(II) and 0.42 mM Pt(IV) in nitric acid. Under all conditions, *E. coli* bioreduced and recovered more Pt than Pd, in contrast to biosorption results, and a higher initial loading of pre-seeded nanoparticles resulted in a greater total weight of metal recovered. Furthermore, these bimetallic PdPt nanoparticles were highly active catalysts for the reduction of Cr(VI) and the hydrogenation of soybean oil, with a 2 wt% Pd preloading resulting in the best catalytic performance (Murray *et al.*, 2018). Using the same method, bimetallic PdPt nanoparticles were synthesized from a synthetic solution at pH 2 using *D. desulfuricans* and *B. benzeovorans*, which outperformed a commercial catalyst for the upgrading of heavy oil (Omajali *et al.*, 2017).

Bimetallic PdRu nanoparticles were synthesized by *E. coli*, *D. desulfuricans*, and *B. benzeovorans*, via the reduction of Pd(II) and Ru(III) using hydrogen on pre-seeded microbial Pd(0) nanoparticles (Gomez-Bolivar *et al.*, 2019; Mikheenko *et al.*, 2019; Omajali *et al.*, 2019). Across these studies, near-complete metal removal was observed from synthetic acidic solutions; Pd and Ru were not alloyed; Ru(III) was not reduced and was present as Ru(III), Ru(IV), and Ru(IV) species; Pd was present as Pd(0) metal and Pd(II) oxide; and Ru was mostly deposited at the cell surface, whereas Pd tended to be deposited intracellularly. A consortium of acidophilic sulfidogens was also used to synthesize microbial PdRu bimetallic nanoparticles resulting in the production of Pd-S and Ru-S compounds, which outperformed *D. desulfuricans* PdRu nanoparticles for the hydrogenation synthesis of drop in fuels (Mikheenko *et al.*, 2019). Overall, these results highlight the potential to generate mixed metal catalysts from waste solutions.

5.4 BIORECOVERY OF PGMs FROM REAL INDUSTRIAL EFFLUENTS

The two primary waste types for microbial PGM biorecovery are acidic leachates and industrial wastewaters, which are both highly acidic (generally pH 0–2) complex mixtures of multiple metals with many inorganic and organic components all at extremely variable concentrations. Acidic leachates are generated from various solid materials, such as spent automotive catalytic converters and waste electronic and electrical equipment, typically via the use of hydrochloric acid. Industrial wastewaters are effluents discharged from a wide variety of industrial activities with a wide variety of solution chemistries. Microbial biotechnologies must be able to achieve high efficiency and selectivity for PGM biorecovery from these effluents, which usually contain much higher concentrations of base and heavy metals than PGMs. This section discusses the performance of both biosorption and bioreduction processes.

5.4.1 Biosorption from single PGM containing effluents

5.4.1.1 Biosorption from industrial effluents

Several studies have investigated the biosorption of a single PGM from mixed metal industrial effluents. Biomass can be functionalized to increase the concentration of specific functional groups, such as amines, to preferentially bind PGMs over base and heavy metal contaminants. Mack *et al.* (2011) utilized *S. cerevisiae* immobilized with PEI and cross-linked with glutaraldehyde to biorecover Pt(IV) from an industrial PGM refinery wastewater (Mack *et al.*, 2011). The wastewater contained 18 mg/L Pt at pH 1.2 and many contaminants (Table 5.1). However, Pt biosorption was limited to around 40% from the industrial effluent, compared to >90% Pt biosorption from synthetic solutions. Biosorption of $Pt(IV)Cl_6^{2-}$ was negatively affected by competition for binding sites with high concentrations of anions, such as Cl^- and SO_4^{2-}, and the presence of base metals. A series of studies by Yun and colleagues (Kwak *et al.*, 2013; Song *et al.*, 2013; Won *et al.*, 2014b; Kim *et al.*, 2016; Song *et al.*, 2018) used microbial biosorption with powdered waste *Corynebacterium glutamicum* bacteria from industrial fermentation, functionalized with PEI and cross-linked with glutaraldehyde, to recover Ru from acetic acid industrial effluents from the Cativa™ process (Kim *et al.*, 2016; Kwak *et al.*, 2013; Won *et al.*, 2014b, 2015); producing an initial significantly higher biosorption capacity than raw biomass of 47.1 mg/g (Song *et al.*, 2013). This biosorbent was combined with chitosan fibres and applied to an effluent containing 1822.9 mg/L Ru in a continuous packed column system. The PEI-modified bacterial fibres achieved a Ru biosorption capacity of 38.24 mg/g and a breakthrough time of 2.23 h, whereas a commercial ion-exchange resin (Lewatit MonoPlus M600) achieved a Ru biosorption capacity of 1.67 mg/g and breakthrough time of 0.22 h (Kwak *et al.*, 2013). However, the microbial biosorbent showed poor selectivity for Ru over contaminant metals; after biosorption from mixed metal effluent the biosorbent contained by rough wt%: >40% I, >20% Ni, >10% Fe, >10% K, 5% Ru, and small amounts of Mo, Cr, and W (Won *et al.*, 2015). In addition, the chitosan fibres experienced significant swelling due to low stability under acidic conditions, limiting their use in large-scale continuous flow processes (Song *et al.*, 2018). Waste *E. coli* biomass from industrial fermentation

Table 5.1 Microbial biosorption studies reporting on the biorecovery of PGMs from real mixed metal industrial effluents.

Industrial Waste	Working Conditions	PGMs Concentration	Contaminants Concentration	Microbial Biorecovery System	PGM Recovery Performance	Reference
Industrial PGM refinery wastewater	pH 1.2 5 g/L biosorbent 24 h	Pt: 18 mg/L	Cl^-: 27.5 g/L SO_4^{2-}: 5.7 g/L NO_3^- 660 mg/L Cu: 0.3 mg/L Ni: 0.6 mg/L Na: 2.1 g/L Zn: 4.4 mg/L	*Saccharomyces cerevisiae* immobilized with PEI, cross-linked with glutaraldehyde, treated with hot 3% KOH	Pt: 40%	Mack *et al.* (2011)
Industrial wastewater	pH 0.56 10 g/L biosorbent 3 h	Pd: 24.51 mg/L Pt: 3.21 mg/L	Al: 10 350 mg/L Fe: 133.20 mg/L	*Escherichia coli* expressing EC20 metal binding protein	**First**: Pt: 97.5% Pd:100% **Second**: Pt: 20% Pd: >85% **Third**: Pt: 0% Pd: >50%	Tan *et al.* (2020)
Dilute PGM refinery process solution	pH 1.8 (adjusted to pH 3) Flow through column system; varied flow rate and bed depth 3 h	Pd: 0.248 mg/L Pt: 1.88 mg/L Ir: 8.25 mg/L Ru: 0.937 mg/L Rh: 0.065 mg/L	Na: 45 100 mg/L Cl: 3497 mg/L NH_3: 79.1 mg/L SO_4^{2-} : 10 947 mg/L Se: 416 mg/L Te: 1.24 mg/L Fe: 1.17 mg/L Ni: 2.08 mg/L Cu: 0.275 mg/L Co: 0.026 mg/L Zn: 0.077 mg/L Au: 0.015 mg/L Ag: 0.004 mg/L	*Saccharomyces cerevisiae* ethanol treated and immobilized in Plaster of Paris beads	Max. breakthrough capacity (mg/g) Pd: 0.056 Pt: 0.058 Ir: 0.258 Ru: 0.012 Rh: n/a Cl: 109.535 SO_4^{2-}: 411.46 Na: 282.5 Te: 0.046 Se: 5.212 Ni: 0.065	Oke (2017)

(*Continued*)

Table 5.1 Microbial biosorption studies reporting on the biorecovery of PGMs from real mixed metal industrial effluents. (*Continued*)

Industrial Waste	Working Conditions	PGMs Concentration	Contaminants Concentration	Microbial Biorecovery System	PGM Recovery Performance	Reference
Industrial wastewater	'Extremely acidic' 0–60 g/L biosorbent 4 h	Pt: 1554.9 mg/L Rh: 201.2 mg/L Ir: 10.9 mg/L	As: 812.0 mg/L Ca: 130.0 mg/L Na: 78.8 mg/L K: 52.7 mg/L Bi: 17.8 mg/L Mg: 16.8 mg/L Ag: 13.9 mg/L Nb: 10.9 mg/L Sb: 5.2 mg/L	*Escherichia coli* immobilized with PEI, cross-linked with glutaraldehyde then freeze-dried	Pt: 108.8 mg/g 98.7% Pt recovery of biosorbed Pt via incineration	Won *et al.* (2010)
Electrolytic metal refinery anode slime	pH 3.5 for Pd(II) recovery pH 2 for Pt(IV) recovery 2 h	Pd: 7 μg/g Pt: 6 μg/g	Ag: 40% Pb: 30% Sb: 1–3% Se: 1–2% Cu: 1–2% Au: 200 μg/g	*Aspergillus* species	Pd: 97.9% Pt: 74.8%	Godlewska-Żyłkiewicz *et al.* (2019)
Aqua regia wastewater (10-fold dilution)	pH 0.25 20 mg/mL biosorbent 16 h	Pd: 12 mg/L Pt: 4 mg/L	Cu: 480 mg/L Au: 53 mg/L Fe: 57 mg/L Ni: 46 mg/L Sn: 5 mg/L Zn: 11 mg/L	Lyophilized *Galdieria sulphararia*	Pd: 95% Pt: 19%	Ju *et al.* (2016)

(*Continued*)

Table 5.1 Microbial biosorption studies reporting on the biorecovery of PGMs from real mixed metal industrial effluents. (*Continued*)

Industrial Waste	Working Conditions	PGMs Concentration	Contaminants Concentration	Microbial Biorecovery System	PGM Recovery Performance	Reference
Ru electroplating effluent	Ru_A: pH 1.8 Ru_B: pH 8.7 65°C	Ru_A: 3250 mg/L |Ru_B: 1800 mg/L	Ru_A Cu: 29 mg/L Zn: 88 mg/L Ni: 980 mg/L Fe: 688 mg/L Ru_B Cu: 9.5 mg/L Zn: 30 mg/L Ni: 29 mg/L Fe: 32 mg/L	*Rhodopseudomonas palustris* strains 42OL, AV33, AV32a, SC0, and CGA009	Maximum Ru biosorption capacity 40 mg/g Ru_A: 72% recovery Ru_B: 42% recovery	Colica *et al.* (2012)
Rh electroplating effluent	pH 1.2 0.25 g/L biosorbent 150 rpm 30°C 2 h	Rh: 48.54 mg/L	Na: ~85 mg/L Fe: ~15 mg/L K: ~8 mg/L Mg: ~5 mg/L Ca: <5 mg/L Mo: <5 mg/L	Waste *Pichia pastoris* *Pichia pastoris* with GS-R Rh-binding peptide	48.49% Rh biosorption 65.86% Rh selectivity	Gao *et al.* (2022)

was combined with PEI and polysulfone fibres to enhance the mechanical strength and chemical stability of the biosorbent, resulting in a maximum Ru biosorption capacity of 121.28 mg/g, higher than conventional ion exchange resins, Lewatit MonoPlus M500 (17.91 mg/g), Amberjet 4200 (31.22 mg/g) and Lewatit MonoPlus TP214 (61.93 mg/g), respectively (Kim *et al.*, 2016). Additionally, the biosorbent treated 4.63 L of 78 mg/L Ru waste effluent in a continuous flow packed column, with a breakthrough time of 192 h, however, metal selectivity of the sorbents was not reported.

5.4.1.2 Biosorption from electroplating wastewater

Electroplating wastewater is a good candidate for microbial biosorbents to biorecover PGMs from as they typically contain higher concentrations of, and fewer types of, metals than other industrial effluents. Commercial dry baker's yeast (*S. cerevisiae*) was used to biosorb Pd(II) from spent electroplating solutions, achieving 75% Pd(II) sorption from a solution containing 15 mg/L Pd and 300 mg/L Cu at pH 0.8, with negligible Cu removal (Saitoh *et al.*, 2020). In contrast, the biosorption of Sn was favoured 10-fold over Pd from a solution containing 90 mg/L Pd and 8000 mg/L Sn at pH 0.8.

Gao *et al.* (2022) studied Rh(III) biosorption using *P. pastoris* from acidic industrial electroplating wastewater containing 48.54 mg/L Rh and roughly double the amount of Na^+, as well as other contaminant metals (Table 5.1) (Gao *et al.*, 2022). The biosorption efficiency of *P. pastoris* decreased significantly in the industrial effluent compared to synthetic solutions, likely due to the high concentrations of competing metal ions such as Na(I), Fe(III), and Mg(II). Engineered *P. pastoris*, containing a Rh(III) binding peptide, achieved a maximum Rh(III) biosorption efficiency of 48.49%, compared to only 27.98% efficiency for waste *P. pastoris* from industrial enzyme production. Engineering *P. pastoris* increased the selectivity of Rh(II) biosorption from 48.17% to 65.86%, as well as increasing Fe(III) biosorption, but reducing Na(I) biosorption.

Colica *et al.* (2012) investigated five different strains of the purple nonsulfur bacterium (PNSB) *Rhodopseudomonas palustris* to biorecover Ru from electroplating industrial effluents. Both acidic (pH 1.8, Ru 3250 mg/L) and alkaline (pH 8.7, |Ru 1800 mg/L) effluents were tested, which contained contaminant metals (Table 5.1). Optimal Ru biosorption capacities were achieved for the acidic (133 mg/g) and alkaline (95 mg/g) effluents, respectively, using different strains of *R. palustris*. An acid pretreatment step of the biomass significantly favoured Ru biorecovery from the alkaline effluent, whereas untreated cells gave the highest Ru biosorption capacity from the acidic effluent. Notably, Ru selectivity was high, and under optimal conditions the biosorption capacity of contaminant metals did not exceed 1 mg/g. Dilution of the effluent 1:10 with water was important as the maximum Ru biosorption capacity decreased approximately 4-fold and 2.5-fold for the undiluted acidic and alkaline effluents, respectively.

5.4.2 Biosorption from mixed metal PGM-containing effluents

The microbial biosorption recovery of PGMs has been performed for a wide range of industrial effluents containing many metals and mixed PGMs. In general, biosorbents display non-specific binding of contaminant metals, and strong competitive binding effects between precious metals (Garcia *et al.*, 2021). Tan *et al.* (2020) reported on the biosorption of Pd(II) and Pt(IV) from industrial wastewater using genetically modified *E. coli* EC20, displaying cysteine-rich and thiol-reactive peptides on the outer membrane (Tan *et al.*, 2020). The effluents contained 24.51 mg/L Pd, 3.21 mg/L Pt and many contaminant metals at pH 0.56 (Table 5.1). The cells achieved excellent biosorption efficiencies of 97.5% and 100% for Pd and Pt, respectively, and were separated and reused in fresh effluent five times. The biosorption efficiency of Pd was >85% in the second cycle and >50% after the third use, whereas the Pt biosorption efficiency dropped to 20% in the second biosorption cycle, and 0% in the third cycle. After five cycles 21.98 mg/L Pd, 0.97 mg/L Pt, 22.96 mg/L Al, and 3.33 mg/L Fe were recovered, demonstrating good biorecovery of Pd, but lacking overall selectivity for PGMs from an effluent with very high concentrations of contaminant metals. Oke (2017) utilized waste brewer's yeast, *S. cerevisiae*, immobilized on plaster of Paris to treat a dilute PGM processing effluent containing 0.248 mg/L Pd, 1.88 mg/L Pt, 8.25 mg/L Ir, 0.937 mg/L Ru, and 0.065 mg/L Rh at a pH of 1.8 (adjusted to pH 3), with many contaminants (Table 5.1) (Oke, 2017). The biosorbent displayed good biosorption of the PGMs, except for Rh, as well as base metals and other trace elements, overall displaying poor PGM selectivity. The biosorbent displayed metal affinities in the order: $Pd^{2+} > SO_4^{2-} > Te^{2+} > Pt^{2+} > Ir^{3+} > Ni^{2+} > Cl^- > Ru^{3+} > Se^{2+} > Na^+$. The effluent was treated in a flow-through packed-bed column with an adsorption time of >180 min to reach bed saturation observed for most metals.

Won *et al.* (2010) utilized waste *E. coli* modified with PEI to biorecover Pt(IV) from dilute aqua regia inductively coupled plasma industrial wastewater, which contained 1554.9 mg/L Pt, 201.2 mg/L Rh, and 10.9 mg/L Ir, along with many contaminant metals (Table 5.1) (Won *et al.*, 2010). A maximum Pt biosorption capacity of 108.8 mg/g was attained after 60 min, although removal of contaminant metals was not reported, however, the biosorbent was incinerated and the resulting ash contained 56.6% Pt, as either Pt(0) or platinum oxide (PtO). Godlewska-Żyłkiewicz *et al.* (2019) biorecovered Pd(II) and Pt(IV) using the fungus *Aspergillus* sp. from electrolytic metal refinery anode slime containing 7 μg/g of Pd and 6 μg/g of Pt, and other mixed metals (Table 5.1). The pH was adjusted to 3.5 for optimal Pd(II) biosorption, and pH 2 for optimal Pt(IV) biosorption, leading to high biorecovery efficiencies of 97.9% for Pd, and 74.8% for Pt, although the biosorption of other metals was not reported (Godlewska-Żyłkiewicz *et al.*, 2019). The red alga *G. sulphararia* was used to recover metals from an aqua regia wastewater (5.6 M acid) containing 120 mg/L Pd^{2+}, 40 mg/L Pt^{4+}, and 530 mg/L Au^{3+}, and many contaminants (Table 5.1) (Ju *et al.*, 2016). Poor recovery efficiency was observed from the raw wastewater, however, a 10-fold dilution enabled the recovery efficiency to reach 95% and 90% for Pd(II) and Au(III), respectively. However, high amounts of Cu were also biosorbed, with only 19% Pt biosorption, highlighting competitive binding between, and poor selectivity for, precious metals.

5.4.3 Bioreduction from mixed metal PGM-containing effluents

As previously discussed, exposing bacteria to highly acidic waste PGM effluents generally results in poor PGM bioreduction due to both deactivation of bacterial terminal reductase enzymes at pH <2 and inhibitory effects from toxic metals (Mabbett *et al.*, 2006; Murray *et al.*, 2007; Yong *et al.*, 2002a). For example, Ngwenya and Whiteley (2006) used resting SRB cells to biorecover Rh from a precious metal refinery industrial effluent containing 210 mg/L of Rh at pH 1.31, along with high concentrations of chlorides, sulfates, and other metals (Table 5.2). The bacteria recovered 64% of Rh after 30 min with hydrogen as electron donor, however, by 60 min all Rh was lost to the effluent, suggesting a toxicity response by the cells.

E. coli pre-seeded with Pd(0) nanoparticles was used to biorecover PGMs from waste electronic and spent automotive catalyst leachates adjusted to pH 1.5, which contained low concentrations of Pt, Pd, Rh, and very high concentrations of Al and Fe (Table 5.2) (Mabbett *et al.*, 2006). The biorecovered mixed metals were comprised by weight in the order Al > Pd > Fe > Mg > Pt > Cu > Rh > Zn, showing that hydrogen-driven bioreduction lacked PGM specificity. *E. coli* pre-seeded with Pt(0) nanoparticles was used to biorecover 65% of PGMs from a pH 1.6 leachate containing high concentrations of Pt, Pd, Ir, Rh, and Ru (Table 5.2) – although PGM selectivity was not quantified – whereas fresh cells gave negligible recovery (Murray *et al.*, 2007). Pre-seeded *E. coli* with Pd(0) nanoparticles biorecovered PGMs from spent automotive catalyst leachates diluted to pH 2.2 (Murray *et al.*, 2017), with complete removal achieved after 60 h. Silica and alumina were implicated as compounds in the leachate that could inhibit or significantly slow PGM bioreduction. Gauthier *et al.* (2010) utilized either fresh or pre-palladized bacteria to recover PGMs via hydrogen-driven bioreduction from an acidic leachate (pH 1.4) containing Pt, Pd, Rh, Ir, Ru, and many contaminant metals (Table 5.2). Surprisingly, the fresh bacteria performed comparably to pre-palladized bacteria, with both biorecovering 96–100% Pd, 70–74% Pt, and 55–57% Rh from the undiluted leachate – other metals were not quantified, but were not identified using energy-dispersive X-ray spectroscopy.

Previous research found that halophilic bacteria could not biosorb or bioreduce the 11 mg/L of tetraamine Pt-complex ($Pt[NH_3]_4{\cdot}2HCO_3$) present in an industrial process stream (Table 5.2) (Maes *et al.*, 2016). Instead of pre-seeding the bacteria, the authors spiked the solution with $K_2Pt(II)Cl_4$, producing in situ Pt(0) nanoparticle formation, resulting in the reduction and recovery of 95% Pt. After 12 successive batch cycles a total of 0.132 mg Pt/g biomass was biorecovered.

Growing bacteria as an immobilized biofilm may improve the robustness of bacteria in acidic PGM effluents. For example, a biofilm of *D. desulfuricans* grown on a Pd-Ag electrode of an electrobioreactor was used to biorecover PGMs from an automotive catalyst leachate and a processing waste containing Pd, Pt, and Rh at pH 2.5 (Table 5.2) (Yong *et al.*, 2003). The biofilm was supplied with hydrogen as electron donor from an electrolytic cell, and the effluent was continuously flown (1–2 mL/min) at constant current (20 mA), resulting in the effective removal of Pd (85.5–88.4%), Pt (80.8–96.5%), and Rh (73.8–77.9%) from the effluent, outperforming the no cell electrode or dead biofilm controls.

Table 5.2 Microbial bioreduction studies reporting on the biorecovery of PGMs from real mixed metal industrial effluents.

Industrial Waste	Working Conditions	PGMs Concentration	Contaminants Concentration	Microbial Biorecovery System	Biorecovery Performance	Reference
Precious metal refinery industrial effluent	pH 1.31 0.5 g/L cells H_2 as electron donor 1 h	Rh: 210 mg/L	Cl: 38.23 mg/L SO_4^{2-}: 2.62 mg/L Ag, Au, Cu, Ni, Os, Pb and Zn	Sulfate reducing bacteria consortium	64% Rh at 30 min 0% Rh at 60 min	Ngwenya and Whiteley (2006)
Electronic and spent automotive catalyst waste leachates	pH 1.5 3:10 dilution H_2 as electron donor 5 wt% biomass:Pd	SI Pt: 81 mg/L Pd: 29 mg/L Rh:7 mg/L SII Pt: 51 mg/L Pd: 19 mg/L Rh: 5 mg/L	SI Cl: 54 312 mg/L Al: 4050 mg/L |Fe: 1203 mg/L Zn: 102 mg/L Cu: 24 mg/L SII Cl: 27 924 mg/L Al: 3630 mg/L Fe: 741 mg/L Zn: 63 mg/L Cu: 19 mg/L	*Escherichia coli* pre-seeded with bio-Pd(0)	SI Pd: 5.8 wt% Pt: 0.9 wt% Rh: 0 wt% Al: 7.7 wt% Fe: 3 wt% Mg: 2.5 wt% Zn: 0.3 wt% Cu: 0.2 wt% SII Pd: 5.3 wt% Pt: 0.9 wt% Rh: 0.4 wt% Al: 5.7 wt% Fe: 2.2 wt% Mg: 1.5 wt% Zn: 0.1 wt% Cu: 0.4 wt%	Mabbett *et al.* (2006)
Spent automotive catalyst leachate	pH 1.6 1:100 dilution H_2 as electron donor	Pt: 4.1 mM Pd: 6.6 mM Ir: 0.4 mM Rh: 2.2 mM Ru: 1.5 mM	–	*Escherichia coli* pre-seeded with bio-Pt(0)	65% of PGMs (selectivity not studied)	Murray *et al.* (2007)

(Continued)

Table 5.2 Microbial bioreduction studies reporting on the biorecovery of PGMs from real mixed metal industrial effluents. (*Continued*)

Industrial Waste	Working Conditions	PGMs Concentration	Contaminants Concentration	Microbial Biorecovery System	Biorecovery Performance	Reference
Acidic leachate	pH 1.4 Freeze–thaw desalination H_2 as electron donor	Pt: 280 mg/L Pd: 110 mg/L Rh: 93 mg/L Ir: 2 mg/L Ru: 2 mg/L	Ag: 1 mg/L Au: 9 mg/L Pb: 29 mg/L Cu: 5 mg/L Bi, Te, Se, Fe, and Sb	Fresh or pre-palladized *Cupriavidus necator* (DS M428) and *Cupriavidus metallidurans* (DS M2839)	Pd: 96–100% Pt: 70–74% Rh: 55–57%	Gauthier *et al.* (2010)
Refinery process stream	pH 1.65 adjusted to 2.3 H_2 as electron donor 20–72 h Repeated batch cycles	Pt: 11 mg/L	NH_4Cl: 13 000 mg/L	*Halophilic bacteria* (from Artemia cysts)	95% Pt recovery 0.132 mg Pt/g biomass after 12 cycles	Maes *et al.* (2016)
Spent automotive catalyst leachate and processing waste	pH 2.5 Continuous flow electrobioreactor with electrochemical H_2 generation as electron donor	Leachate Pd: 1.4 mM Pt: 1.2 mM Rh: 0.4 mM Process waste Pd: 2.8 mM Pt: 1.9 mM Rh: 0.14 mM	Leachate Cl^-: 8.7 M NO_3^-: 3.9 M	Pre-palladized *Desulfovibrio desulfuricans* biofilm	Leachate Pd: 85.5–86.9% Pt: 80.8% Rh: 73.8% Process waste Pd: 87.6–88.4% Pt: 81.8–96.5% Rh: 75–77.9%	Yong *et al.* (2003)
Diluted spent Pd catalyst leachate	pH 2.5 Formate as electron donor 20 h	Pd: 20 mg/L	Al: 2240 mg/L Cl^-: 187 mM	Acidophilic metal-reducing bacteria, *Acidocella aromatica* and *Acidiphilium cryptum*	Pd: 100% removal from solution No Al removal	Okibe *et al.* (2017)

Acidophilic extremophile microbes are also an interesting alternative for the biorecovery of PGMs from acidic effluents, as their acid tolerance may result in improved enzymatically driven PGM bioreduction compared to neutrophilic bacteria. The acidophilic metal-reducing bacteria *Acidocella aromatica* and *Acidiphilium cryptum* were grown at pH 2.5, in a diluted spent Pd catalyst leachate containing 20 mg/L Pd, 2240 mg/L Al, and 187 mM Cl, adjusted to pH 2.5 (Table 5.2) (Okibe *et al.*, 2017). Providing 100 mM formate as electron donor resulted in the complete removal of Pd(II) from solution, with no Al(III) removal, whereas using hydrogen as electron donor resulted in abiotic precipitation of Pd(0) not attached to the cells.

5.5 RECYCLING OF PGMs AFTER MICROBIAL BIORECOVERY

A key aspect of microbial PGM biorecovery processes is how the precious metals are to be recovered after being concentrated from wastes by microbes, to recycle and reuse them in a circular economy.

5.5.1 Bioreduction: reuse and recycling of bio-PGM nanoparticles

Microbial bioreduction processes have been demonstrated to produce metallic PGM bionanoparticles, which are simple to separate from solution and are highly amenable to integration into industrial PGM refinery processes (Hagelüken, 2012). In addition, recovered bio-PGM nanoparticles could potentially be used as catalysts for industrially important reactions (Egan-Morriss *et al.*, 2022), and could be recycled through the same process streams as industrial heterogeneous nanoparticle catalysts.

5.5.2 Biosorption–desorption process

Ideally, the selective biosorption of PGMs would be followed by the selective desorption of PGMs, using a low-cost, eco-friendly eluent that does not damage the biosorbent, enabling its reuse over several cycles (Won *et al.*, 2014a), however, this has not yet been achieved. Generally, the best performing eluent for Pd and Pt from biosorbents is thiourea, which is then difficult to separate metals from (Chassary *et al.*, 2005; Ilyas *et al.*, 2021; Won *et al.*, 2011). Improved selectivity for the elution of Pd(II) and Au(III) compared to thiourea was achieved with 0.2 M ammonium chloride at pH 11 (Ju *et al.*, 2016). Furthermore, thiourea was ineffective at desorbing Ir and Rh (Minoda *et al.*, 2023; Oke *et al.*, 2014). Molar concentrations of HCl (1–4 M) were efficient at desorbing Ir from cross-linked chitosan gel beads (Ruiz *et al.*, 2003), and lyophilized *G. sulphuraria*, enabling the desorption and reuse of the algae up to seven times (Minoda *et al.*, 2023). The highest desorption efficiency of Rh(III) from *P. pastoris* was 34.49%, achieved using 2 M HNO_3 assisted with ultrasonic waves (Gao *et al.*, 2022). The elution of Ru from lyophilized PNSB was performed using HNO_3 and H_2O_2 at 150°C for 72 h (Colica *et al.*, 2012).

5.5.3 Biosorption-incineration process

A potential alternative to the conventional biosorption–desorption process could be a biosorption–incineration process. Biosorption–incineration process is economically viable if the microbial biosorbent is cheap to produce, for example from waste sources (Won *et al.*, 2014a). During the incineration process the biosorbent is thermally decomposed and oxidized leaving behind an ash residue with significantly higher precious metal content (Song *et al.*, 2018). During this process, the biosorbed metal ions can be reduced to their metallic form (Won *et al.*, 2010, 2013). However, both incineration temperature and metal loading are important factors to enhance metal recovery efficiency and purity (Park *et al.*, 2010; Won *et al.*, 2011, 2013). Won *et al.* (2015) demonstrated a biosorption–incineration–leaching–smelting sequential process (using a $NaClO/KMnO_4$/aqua regia leaching process) that increased the purity of biosorbed Ru from 5.4 wt% in the microbial biosorbent after biosorption to 99.75 wt% of metallic Ru in the residual ash. The concentration of metallic PGMs in small volumes of ash prevents the need for subsequent recovery and reduction steps (Won *et al.*, 2014a), limiting the use of hazardous chemicals for further desorption steps, which pose risks of secondary pollution and require extensive waste management. The pyrolysis of PGM-loaded microbial biosorbents can also lead to the synthesis of supported metallic nanoparticle catalysts (Ng *et al.*, 2019; Zhang *et al.*, 2019). However, the oxidation of organic constituents from the incinerated biosorbents may release hazardous by-products and emissions, especially considering the wide range of possible hazardous contaminants present in industrial effluent waste streams, necessitating detailed environmental safety investigations.

5.6 CONCLUSIONS

This chapter highlights the potential of microbial biorecovery processes to effectively recover PGMs from industrial waste streams, offering a promising alternative or complementary process to conventional recovery methods within a circular economy. Despite decades of research, biosorption has not been applied industrially due to complexities such as pH, metal speciation, competing metals and organics, combined with the significant variability of real waste streams. Microbial biosorbents possess low specificity for precious metals compared to contaminant metals with similar properties, and competitive binding effects between the PGMs further reduce the biosorption efficiency. Overall, considerable innovation is required to enhance the specificity and efficiency of PGM microbial biosorbents. In comparison, the microbial bioreduction of PGMs from industrial wastes, though less developed, shows significant promise. Bioreduction leverages microbial metabolism to reduce soluble metal ions as insoluble and recoverable metal nanoparticles. The co-bioreduction of PGMs, and potential metal alloy formation, could enhance the biorecovery efficiency and selectivity. However, bioreduction depends on the stability and activity of microbial cells and their reductase enzymes, which are challenged by the highly acidic and toxic conditions of industrial effluents. Pre-seeding bioreduction steps and the use of acidophilic or genetically engineered strains may improve efficiency and specificity.

The transition from laboratory to industrial application is challenging due to the complexity and variability of industrial effluents. Developing robust, adaptable flow-through microbial metal recovery systems is crucial. Life-cycle assessments are needed to evaluate the environmental impact and economic feasibility of microbial biorecovery processes, including the sustainability of biomass sources, by-product fate, and overall cost–benefit compared to traditional methods. Quantifying the efficiency and selectivity of metal removal from complex effluents and understanding metal–microbial interactions will be essential for enhancing PGM recovery.

ACKNOWLEDGEMENTS

This work was supported by UK Research and Innovation (UKRI) through Engineering and Physical Sciences Research Council (EPSRC) funding from grant EP/W524347/1, and Biotechnology and Biological Sciences Research Council (BBSRC) funding from grant BB/Y008448/1.

REFERENCES

Adams E., Maeda K., Kato T. and Tokoro C. (2021). Mechanism of gold and palladium adsorption on thermoacidophilic red alga *Galdieria sulphuraria*. *Algal Research*, **60**, 102549, https://doi.org/10.1016/j.algal.2021.102549

Adams E., Maeda K., Kamemoto Y., Hirai K. and Apdila E. T. (2024). Contribution to a sustainable society: biosorption of precious metals using the microalga *Galdieria*. *International Journal of Molecular Sciences*, **25**, 704, https://doi.org/10.3390/ijms25020704

Attard G., Casadesús M., Macaskie L. E. and Deplanche K. (2012). Biosynthesis of platinum nanoparticles by *Escherichia coli* MC4100: Can such nanoparticles exhibit intrinsic surface enantioselectivity?

Bennett J. A., Attard G. A., Deplanche K., Casadesus M., Huxter S. E., Macaskie L. E. and Wood J. (2012). Improving selectivity in 2-butyne-1,4-diol hydrogenation using biogenic Pt catalysts. *ACS Catalysis*, **2**, 504–511, https://doi.org/10.1021/cs200572z

Beveridge T. J. (1989). Role of cellular design in bacterial metal accumulation and mineralization. *Annual Review of Microbiology*, **43**, 147–171, https://doi.org/10.1146/annurev.mi.43.100189.001051

Campbell G., Maclean L., Reith F., Brewe D., Gordon R. A. and Southam G. (2018). Immobilisation of platinum by *Cupriavidus metallidurans*. *Minerals*, **8**, 10, https://doi.org/10.3390/min8010010

Chassary P., Vincent T., Sanchez Marcano J., Macaskie L. E. and Guibal E. (2005). Palladium and platinum recovery from bicomponent mixtures using chitosan derivatives. *Hydrometallurgy*, **76**, 131–147, https://doi.org/10.1016/j.hydromet.2004.10.004

Chen Y. and Chen Y. (2022). Difference in toxicity of Pd (II) and mechanism of action before and after reduction by *Bacillus wiedmannii* MSM. *Environmental Science and Pollution Research*, **29**, 1824–1835, https://doi.org/10.1007/s11356-021-15736-y

Cheng Y.-Y., Wang W.-J., Ding S.-T., Zhang M.-X., Tang A.-G., Zhang L., Li D.-B., Li B.-B., Deng G.-Z. and Wu C. (2021). Pyruvate accelerates palladium reduction by regulating catabolism and the electron transfer pathway in *Shewanella oneidensis*. *Applied and Environmental Microbiology*, **87**, e02716–20.

Colica G., Caparrotta S. and de Philippis R. (2012). Selective biosorption and recovery of ruthenium from industrial effluents with *Rhodopseudomonas palustris* strains. *Applied Microbiology and Biotechnology*, **95**, 381–387, https://doi.org/10.1007/s00253-012-4053-9

Cossu R. (2013). The urban mining concept. *Waste Management*, **33**, 497–498, https://doi.org/10.1016/j.wasman.2013.01.010

Cui J., Zhu N., Kang N., Ha C., Shi C. and Wu P. (2017). Biorecovery mechanism of palladium as nanoparticles by *Enterococcus faecalis*: from biosorption to bioreduction. *Chemical Engineering Journal*, **328**, 1051–1057, https://doi.org/10.1016/j.cej.2017.07.124

De Corte S., Hennebel T., Fitts J. P., Sabbe T., Bliznuk V., Verschuere S., van der Lelie D., Verstraete W. and Boon N. (2011a). Biosupported bimetallic Pd-Au nanocatalysts for dechlorination of environmental contaminants. *Environmental Science & Technology*, **45**, 8506–8513, https://doi.org/10.1021/es2019324

De Corte S., Hennebel T., Verschuere S., Cuvelier C., Verstraete W. and Boon N. (2011b). Gold nanoparticle formation using *Shewanella oneidensis*: a fast biosorption and slow reduction process. *Journal of Chemical Technology & Biotechnology*, **86**, 547–553, https://doi.org/10.1002/jctb.2549

Deplanche K., Caldelari I., Mikheenko I. P., Sargent F. and Macaskie L. E. (2010). Involvement of hydrogenases in the formation of highly catalytic Pd(0) nanoparticles by bioreduction of Pd(II) using *Escherichia coli* mutant strains. *Microbiology*, **156**, 2630–2640, https://doi.org/10.1099/mic.0.036681-0

De Vargas I., Macaskie L. E. and Guibal E. (2004). Biosorption of palladium and platinum by sulfate reducing bacteria. *Journal of Chemical Technology and Biotechnology*, **79**, 49–56, https://doi.org/10.1002/jctb.928

De Windt W., Aelterman P. and Verstraete W. (2005). Bioreductive deposition of palladium (0) nanoparticles on *Shewanella oneidensis* with catalytic activity towards reductive dechlorination of polychlorinated biphenyls. *Environmental Microbiology*, **7**, 314–325, https://doi.org/10.1111/j.1462-2920.2005.00696.x

Dundas C. M., Graham A. J., Romanovicz D. K. and Keitz B. K. (2018). Extracellular electron transfer by *Shewanella oneidensis* controls palladium nanoparticle phenotype. *ACS Synthetic Biology*, **7**, 2726–2736, https://doi.org/10.1021/acssynbio.8b00218

Dziwulska U., Bajguz A. and Godlewska-Żyłkiewicz B. (2004). The use of algae *Chlorella vulgaris* immobilized on Cellex-T support for separation/preconcentration of trace amounts of platinum and palladium before GFAAS determination. *Analytical Letters*, **37**, 2189–2203, https://doi.org/10.1081/AL-200026696

Egan-Morriss C., Kimber R. L., Powell N. A. and Lloyd J. R. (2022). Biotechnological synthesis of Pd-based nanoparticle catalysts. *Nanoscale Advances*, **4**, 654–679, https://doi.org/10.1039/D1NA00686J

Egan-Morriss C., Kimber R. L., Powell N. A. and Lloyd J. R. (2023). Impact of solution chemistry on the biotechnological synthesis and properties of palladium nanoparticles. *Johnson Matthey Technology Review*, **67**, 438–448, https://doi.org/10.1595/205651323X16813753335431

Egan-Morriss C., Cheung C. K., Nunn E., Parmeggiani F., Powell N. A., Kimber R. L., Haigh S. J. and Lloyd J. R. (2024). Biosynthesis parameters control the physicochemical and catalytic properties of microbially supported Pd nanoparticles. *Small*, **20**, 2311016, https://doi.org/10.1002/smll.202311016

Fahmy K., Merroun M., Pollmann K., Raff J., Savchuk O., Hennig C. and Selenska-Pobell S. (2006). Secondary structure and Pd(II) coordination in S-layer proteins from *Bacillus sphaericus* studied by infrared and X-ray absorption spectroscopy.

Foulkes J. M., Deplanche K., Sargent F., Macaskie L. E. and Lloyd J. R. (2016). A novel aerobic mechanism for reductive palladium biomineralization and recovery by *Escherichia coli*.

Gadd G. M. (2009). Biosorption: critical review of scientific rationale, environmental importance and significance for pollution treatment. *Journal of Chemical Technology & Biotechnology*, **84**, 13–28, https://doi.org/10.1002/jctb.1999

Gao M., Zhou Y., Yan J., Zhu L., Li Z., Hu X. and Zhan X. (2022). Efficient precious metal Rh(III) adsorption by waste *P. pastoris* and *P. pastoris* surface display from high-density culture. *Journal of Hazardous Materials*, **427**, 128140, https://doi.org/10.1016/j.jhazmat.2021.128140

Garcia A., Deyris P.-A., Adler P., Pelissier F., Dumas T., Legrand Y.-M. and Grison C. (2021). I – ecologically responsible and efficient recycling of Pd from aqueous effluents using biosorption on biomass feedstock. *Journal of Cleaner Production*, **299**, 126895, https://doi.org/10.1016/j.jclepro.2021.126895

Gauthier D., Søbjerg L. S., Jensen K. M., Lindhardt A. T., Bunge M., Finster K., Meyer R. L. and Skrydstrup T. (2010). Environmentally benign recovery and reactivation of palladium from industrial waste by using Gram-negative bacteria. *ChemSusChem*, **3**, 1036–1039, https://doi.org/10.1002/cssc.201000091

Godlewska-Zyłkiewicz B. (2003). Biosorption of platinum and palladium for their separation/ preconcentration prior to graphite furnace atomic absorption spectrometric determination. *Spectrochimica Acta Part B: Atomic Spectroscopy*, **58**, 1531–1540, https://doi.org/10.1016/S0584-8547(03)00076-4

Godlewska-Żyłkiewicz B., Sawicka S. and Karpińska J. (2019). Removal of platinum and palladium from wastewater by means of biosorption on fungi *Aspergillus* sp. and yeast *Saccharomyces* sp. *Water*, **11**, 1522, https://doi.org/10.3390/w11071522

Gomez-Bolivar J., Mikheenko I. P., Orozco R. L., Sharma S., Banerjee D., Walker M., Hand R. A., Merroun M. L. and Macaskie L. E. (2019). Synthesis of Pd/Ru bimetallic nanoparticles by *Escherichia coli* and potential as a catalyst for upgrading 5-hydroxymethyl furfural into liquid fuel precursors. *Frontiers in Microbiology*, **10**, p. 1276, https://doi.org/10.3389/fmicb.2019.01276

Hagelüken B. C. (2012). Recycling the platinum group metals: a European perspective. *Platinum Metals Review*, **56**, 29–35, https://doi.org/10.1595/147106712X611733

Hou Y. N., Zhang B., Yun H., Yang Z. N., Han J. L., Zhou J., Wang A. J. and Cheng H. Y. (2017). Palladized cells as suspension catalyst and electrochemical catalyst for reductively degrading aromatics contaminants: roles of Pd size and distribution. *Water Research*, **125**, 288–297, https://doi.org/10.1016/j.watres.2017.08.055

Ilyas S., Kim H. and Srivastava R. (2021). Recovery of precious metals by solvent extraction. In: Sustainable Urban Mining of Precious Metals. CRC Press, Boca Raton, p. 173–186.

Ju X., Igarashi K., Miyashita S.-I., Mitsuhashi H., Inagaki K., Fujii S.-I., Sawada H., Kuwabara T. and Minoda A. (2016). Effective and selective recovery of gold and palladium ions from metal wastewater using a sulfothermophilic red alga, *Galdieria sulphuraria*. *Bioresource Technology*, **211**, 759–764, https://doi.org/10.1016/j.biortech.2016.01.061

Karim S. and Ting Y.-P. (2021). Recycling pathways for platinum group metals from spent automotive catalyst: A review on conventional approaches and bio-processes. *Resources, Conservation and Recycling*, **170**, 105588, https://doi.org/10.1016/j.resconrec.2021.105588

Kim S., Song M.-H., Wei W. and Yun Y.-S. (2015). Selective biosorption behavior of *Escherichia coli* biomass toward Pd(II) in Pt(IV)–Pd(II) binary solution. *Journal of Hazardous Materials*, **283**, 657–662, https://doi.org/10.1016/j.jhazmat.2014.10.008

Kim S., Choi Y.-E. and Yun Y.-S. (2016). Ruthenium recovery from acetic acid industrial effluent using chemically stable and high-performance polyethylenimine-coated polysulfone-*Escherichia coli* biomass composite fibers. *Journal of Hazardous Materials*, **313**, 29–36, https://doi.org/10.1016/j.jhazmat.2016.03.075

Kimber R. L., Parmeggiani F., Neill T. S., Merroun M. L., Goodlet G., Powell N. A., Turner N. J. and Lloyd J. R. (2021). Biotechnological synthesis of Pd/Ag and Pd/

Au nanoparticles for enhanced Suzuki–Miyaura cross-coupling activity. *Microbial Biotechnology*, **14**, 2435–2447, https://doi.org/10.1111/1751-7915.13762

Kitjanukit S., Sasaki K. and Okibe N. (2019). Production of highly catalytic, archaeal Pd(0) bionanoparticles using *Sulfolobus tokodaii.*

Konishi Y., Ohno K., Saitoh N., Nomura T., Nagamine S., Hishida H., Takahashi Y. and Uruga T. (2007). Bioreductive deposition of platinum nanoparticles on the bacterium *Shewanella algae*. *Journal of Biotechnology*, **128**, 648–653, https://doi.org/10.1016/j.jbiotec.2006.11.014

Kusada K., Wu D. and Kitagawa H. (2020). New aspects of platinum group metal-based solid-solution alloy nanoparticles: binary to high-entropy alloys. *Chemistry – A European Journal*, **26**, 5105–5130, https://doi.org/10.1002/chem.201903928

Kwak I. S., Won S. W., Chung Y. S. and Yun Y.-S. (2013). Ruthenium recovery from acetic acid waste water through sorption with bacterial biosorbent fibers. *Bioresource Technology*, **128**, 30–35, https://doi.org/10.1016/j.biortech.2012.10.146

Law N., Ansari S., Livens Francis R., Renshaw Joanna C. and Lloyd Jonathan R. (2008). Formation of nanoscale elemental silver particles via enzymatic reduction by *Geobacter sulfurreducens*. *Applied and Environmental Microbiology*, **74**, 7090–7093, https://doi.org/10.1128/AEM.01069-08

Lengke M. F., Fleet M. E. and Southam G. (2006). Synthesis of platinum nanoparticles by reaction of filamentous cyanobacteria with platinum(IV)–chlorido complex *Langmuir*, **22**, 7318–7323, https://doi.org/10.1021/la060873s

Lin Z. (2002). Adsorption and reduction of palladium (Pd^{2+}) by *Bacillus licheniformis* R08. *Chinese Science Bulletin*, **47**, 1262, https://doi.org/10.1360/02tb9279

Lloyd J. R., Yong P. and Macaskie L. E. (1998). Enzymatic recovery of elemental palladium by using sulfate-reducing bacteria. *Applied and Environmental Microbiology*, **64**, 4607–4609, https://doi.org/10.1128/AEM.64.11.4607-4609.1998

Mabbett A. N., Sanyahumbi D., Yong P. and Macaskie L. E. (2006). Biorecovered precious metals from industrial wastes: single-step conversion of a mixed metal liquid waste to a bioinorganic catalyst with environmental application. *Environmental Science & Technology*, **40**, 1015–1021, https://doi.org/10.1021/es0509836

Mack C. L., Wilhelmi B., Duncan J. R. and Burgess J. E. (2008). A kinetic study of the recovery of platinum ions from an artificial aqueous solution by immobilized *Saccharomyces cerevisiae* biomass. *Minerals Engineering*, **21**, 31–37, https://doi.org/10.1016/j.mineng.2007.07.001

Mack C. L., Wilhelmi B., Duncan J. R. and Burgess J. E. (2011). Biosorptive recovery of platinum from platinum group metal refining wastewaters by immobilised *Saccharomyces cerevisiae*. *Water Science and Technology*, **63**, 149–155, https://doi.org/10.2166/wst.2011.025

Maes S., Claus M., Verbeken K., Wallaert E., De Smet R., Vanhaecke F., Boon N. and Hennebel T. (2016). Platinum recovery from industrial process streams by halophilic bacteria: influence of salt species and platinum speciation. *Water Research*, **105**, 436–443, https://doi.org/10.1016/j.watres.2016.09.023

Maes S., Props R., Fitts J. P., De Smet R., Vanhaecke F., Boon N. and Hennebel T. (2017). Biological recovery of platinum complexes from diluted aqueous streams by axenic cultures. *PLoS One*, **12**, e0169093.

Manipura A. and Burgess J. E. (2008). Biological nitrogenous compound removal and subsequent rhodium recovery from simulated metal refinery wastewater. *Minerals Engineering*, **21**, 652–659, https://doi.org/10.1016/j.mineng.2007.12.008

Matsena M. T. and Chirwa E. M. N. (2021). Comparative analysis of biological versus chemical synthesis of palladium nanoparticles for catalysis of chromium (VI) reduction. *Scientific Reports*, **11**, 16674, https://doi.org/10.1038/s41598-021-96024-0

Matsumoto T., Phann I. and Okibe N. (2021). Biogenic platinum nanoparticles' production by extremely acidophilic Fe(III)-reducing bacteria. *Minerals*, **11**, 1175, https://doi.org/10.3390/min11111175

Mikheenko I. P., Rousset M., Dementin S. and Macaskie L. E. (2008). Bioaccumulation of palladium by *Desulfovibrio fructosivorans* wild-type and hydrogenase-deficient strains. *Applied and Environmental Microbiology*, **74**, 6144–6146, https://doi.org/10.1128/AEM.02538-07

Mikheenko I. P., Gomez-Bolivar J., Merroun M. L., Macaskie L. E., Sharma S., Walker M., Hand R. A., Grail B. M., Johnson D. B. and Orozco R. L. (2019). Upconversion of cellulosic waste into a potential "drop in fuel" via novel catalyst generated using *Desulfovibrio desulfuricans* and a consortium of acidophilic sulfidogens. *Frontiers in Microbiology*, **10**, p. 970, https://doi.org/10.3389/fmicb.2019.00970

Minoda A., Ueda S., Miyashita S.-I., Ogura T., Natori S., Sun J. and Takahashi Y. (2023). Reversible adsorption of iridium in lyophilized cells of the unicellular red alga *Galdieria sulphuraria*. *RSC Advances*, **13**, 14217–14223, https://doi.org/10.1039/D3RA01249B

Miyashita S.-I., Ogura T., Fujii S.-I., Inagaki K., Takahashi Y. and Minoda A. (2021). Effect of lyophilization on the acid resistance of a unicellular red alga *Galdieria sulphuraria* during platinum recovery. *Journal of Hazardous Materials Advances*, **3**, 100015, https://doi.org/10.1016/j.hazadv.2021.100015

Murray A. J., Mikheenko I. P., Goralska E., Rowson N. A. and Macaskie L. E. (2007). Biorecovery of platinum group metals from secondary sources. *Advanced Materials Research*, **20–21**, 651–654, https://doi.org/10.4028/www.scientific.net/AMR.20-21.651

Murray A. J., Zhu J., Wood J. and Macaskie L. E. (2017). A novel biorefinery: biorecovery of precious metals from spent automotive catalyst leachates into new catalysts effective in metal reduction and in the hydrogenation of 2-pentyne. *Minerals Engineering*, **113**, 102–108, https://doi.org/10.1016/j.mineng.2017.08.011

Murray A. J., Zhu J., Wood J. and Macaskie L. E. (2018). Biorefining of platinum group metals from model waste solutions into catalytically active bimetallic nanoparticles. *Microbial Biotechnology*, **11**, 359–368, https://doi.org/10.1111/1751-7915.13030

Naja G. and Volesky B. (2011). The mechanism of metal cation and anion biosorption. In: Microbial Biosorption of Metals, P. Kotrba, M. Mackova and T. Macek (eds.), Springer Netherlands, Dordrecht, p. 19–58.

Ng C. K., Cai Tan T. K., Song H. and Cao B. (2013). Reductive formation of palladium nanoparticles by *Shewanella oneidensis*: role of outer membrane cytochromes and hydrogenases. *RSC Advances*, **3**, 22498–22503, https://doi.org/10.1039/c3ra44143a

Ng C. K., Karahan H. E., Loo S. C. J., Chen Y. and Cao B. (2019). Biofilm-templated heteroatom-doped carbon–palladium nanocomposite catalyst for hexavalent chromium reduction. *ACS Applied Materials & Interfaces*, **11**, 24018–24026, https://doi.org/10.1021/acsami.9b04095

Ngwenya N. and Whiteley C. G. (2006). Recovery of Rhodium(III) from solutions and industrial wastewaters by a sulfate-reducing bacteria consortium. *Biotechnology Progress*, **22**, 1604–1611, https://doi.org/10.1002/bp060167h

Nieboer E. and Richardson D. H. S. (1980). The replacement of the nondescript term 'heavy metals' by a biologically and chemically significant classification of metal ions. *Environmental Pollution Series B, Chemical and Physical*, **1**, 3–26, https://doi.org/10.1016/0143-148X(80)90017-8

Oke D. (2017). Purification of a dilute platinum group metals process stream using waste yeast biomass immobilized on plaster of Paris. *Journal of the Southern African Institute of Mining and Metallurgy*, **117**, 757–763, https://doi.org/10.17159/2411-9717/2017/v117n8a5

Oke D., Ndlovu S. and Sibanda V. (2014). Removal of platinum group metals from dilute process streams: identification of influential factors using DOE approach. *Journal of Environmental Chemical Engineering*, **2**, 1061–1069, https://doi.org/10.1016/j.jece.2014.04.003

Okibe N., Nakayama D. and Matsumoto T. (2017). Palladium bionanoparticles production from acidic Pd(II) solutions and spent catalyst leachate using acidophilic Fe(III)-reducing bacteria. *Extremophiles*, **21**, 1091–1100, https://doi.org/10.1007/s00792-017-0969-4

Omajali J. B., Mikheenko I. P., Merroun M. L., Wood J. and Macaskie L. E. (2015). Characterization of Intracellular Palladium Nanoparticles Synthesized by Desulfovibrio desulfuricans and Bacillus Benzeovorans. *Journal of Nanoparticle Research*, **17**(6), p. 264, https://doi.org/10.1007/s11051-015-3067-5

Omajali J. B., Hart A., Walker M., Wood J. and Macaskie L. E. (2017). In-situ catalytic upgrading of heavy oil using dispersed bionanoparticles supported on gram-positive and Gram-negative bacteria. *Applied Catalysis B: Environmental*, **203**, 807–819, https://doi.org/10.1016/j.apcatb.2016.10.074

Omajali J. B., Gomez-Bolivar J., Mikheenko I. P., Sharma S., Kayode B., Al-Duri B., Banerjee D., Walker M., Merroun M. L. and Macaskie L. E. (2019). Novel catalytically active Pd/Ru bimetallic nanoparticles synthesized by *Bacillus benzeovorans*. *Scientific Reports*, **9**, 4715, https://doi.org/10.1038/s41598-019-40312-3

Park J., Won S. W., Mao J., Kwak I. S. and Yun Y.-S. (2010). Recovery of Pd(II) from hydrochloric solution using polyallylamine hydrochloride-modified *Escherichia coli* biomass. *Journal of Hazardous Materials*, **181**, 794–800, https://doi.org/10.1016/j.jhazmat.2010.05.083

Pavlova L. M., Radomskaya V. I., Shumilova L. P., Ionov A. M., Ivanov V. V. and Poselyuzhnaya A. V. (2018). Experimental modeling of platinum biomineralization by microscopic fungi isolated from a lignite deposit. *Geochemistry International*, **56**, 450–461, https://doi.org/10.1134/S0016702918050063

Pearson R. G. (1968). Hard and soft acids and bases, HSAB, part 1: fundamental principles. *Journal of Chemical Education*, **45**, 581, https://doi.org/10.1021/ed045p581

Prichard H. M., Wedin F., Sampson J., Jackson M. T. and Fisher P. C. (2016). Precious metals in urban waste. *Water and Environment Journal*, **30**, 151–156, https://doi.org/10.1111/wej.12166

Rashamuse K. J. and Whiteley C. G. (2007). Bioreduction of Pt (IV) from aqueous solution using sulphate-reducing bacteria. *Applied Microbiology and Biotechnology*, **75**, 1429–1435, https://doi.org/10.1007/s00253-007-0963-3

Remoudaki E., Tsezos M., Hatzikioseyian A. and Karakoussis V. (1999). Mechanism of palladium biosorption by microbial biomass. The effects of metal ionic speciation and solution co-ions. In: Process Metallurgy, R. Amils and A. Ballester (eds.), Elsevier, p. 449–462

Ruiz M., Sastre A. and Guibal E. (2003). Osmium and iridium sorption on chitosan derivatives. *Solvent Extraction and Ion Exchange*, **21**, 307–329, https://doi.org/10.1081/SEI-120018952

Saitoh N., Fujimori R., Yoshimura T., Tanaka H., Kondoh A., Nomura T. and Konishi Y. (2020). Microbial recovery of palladium by baker's yeast through bioreductive deposition and biosorption. *Hydrometallurgy*, **196**, 105413, https://doi.org/10.1016/j.hydromet.2020.105413

Sickerman N. S. and Hu Y. (2019). Hydrogenases. In: Metalloproteins: Methods and Protocols, Y. Hu (ed.), Springer New York, New York, NY, p. 65–88.

Simon-Pascual A., Sierra-Alvarez R. and Field J. A. (2019). Platinum(II) reduction to platinum nanoparticles in anaerobic sludge. *Journal of Chemical Technology & Biotechnology*, **94**, 468–474, https://doi.org/10.1002/jctb.5791

Song M.-H., Won S. W. and Yun Y.-S. (2013). Decarboxylated polyethylenimine-modified bacterial biosorbent for Ru biosorption from Ru-bearing acetic acid wastewater. *Chemical Engineering Journal*, **230**, 303–307, https://doi.org/10.1016/j.cej.2013.06.087

Song M.-H., Kim J. A., Wei W., Kim S. and Yun Y.-S. (2018). Polyethylenimine-coated biomass-chitosan composite fibers for recovery of ruthenium from industrial effluents: effects of chitosan molecular weight and drying method. *Hydrometallurgy*, **182**, 114–120, https://doi.org/10.1016/j.hydromet.2018.11.001

Tamaoki K., Saito N., Nomura T. and Konishi Y. (2013). Microbial recovery of rhodium from dilute solutions by the metal ion-reducing bacterium *Shewanella* algae. *Hydrometallurgy*, **139**, 26–29, https://doi.org/10.1016/j.hydromet.2013.06.019

Tan L., Dong H., Liu X., He J., Xu H. and Xie J. (2017). Mechanism of palladium(II) biosorption by *Providencia vermicola*. *RSC Advances*, **7**, p. 7060–7072, https://doi.org/10.1039/C6RA27589C.

Tan L., Xiao Y., Cui H., Xu H., Xu M., Wu H., Dong H., Qiu G., Liu X. and Xie J. (2018). Influence of sulfhydryl sites on palladium (II) adsorption by displaying EC20 on the surface of *Escherichia coli*. *Journal of Chemical Technology & Biotechnology*, **93**, 3569–3581, https://doi.org/10.1002/jctb.5734

Tan L., Cui H., Xiao Y., Xu H., Xu M., Wu H., Dong H., Qiu G., Liu X. and Xie J. (2019). Enhancement of platinum biosorption by surface-displaying EC20 on *Escherichia coli*. *Ecotoxicology and Environmental Safety*, **169**, 103–111, https://doi.org/10.1016/j.ecoenv.2018.10.116

Tan L., Wu H., Cui H., Xu H., Xu M., Xiao Y., Qiu G., Liu X., Dong H. and Xie J. (2020). Selective adsorption of palladium and platinum from secondary wastewater using *Escherichia coli* BL21 and *Providencia vermicola*. *Bioprocess and Biosystems Engineering*, **43**, 1885–1897, https://doi.org/10.1007/s00449-020-02378-6

Tanaka K. and Watanabe N. (2015). Study on the coordination structure of Pt sorbed on bacterial cells using X-ray absorption fine structure spectroscopy. *PLoS ONE*, **10**, e0127417.

Taninouchi Y.-K. and Okabe T. H. (2023). Trends of technological development of platinum group metal recycling: solubilization and physical concentration processes. *Materials Transactions*, **64**, 627–637, https://doi.org/10.2320/matertrans.MT-M2022150

Tran D. T., Jones I. P., Preece J. A., Johnston R. L., Deplanche K. and Macaskie L. E. (2012).Configuration of microbially synthesized Pd-Au nanoparticles studied by STEM-based techniques.

Troitskii S. Y., Fedotov M. A. and Likholobov V. A. (1993). Studies of the compositions of Pd(II) hydrolysis products. *Russian Chemical Bulletin*, **42**, 634–639, https://doi.org/10.1007/BF00703995

Troitskii S. Y., Chuvilin A. L., Kochubei D. I., Novgorodov B. N., Kolomiichuk V. N. and Likholobov V. A. (1995). Structure of polynuclear palladium(ii) hydroxocomplexes formed upon alkaline hydrolysis of palladium(ii) chloride complexes. *Russian Chemical Bulletin*, **44**, 1822–1826, https://doi.org/10.1007/BF00707203

Tsezos M., Remoudaki E. and Angelatou V. (1996). A study of the effects of competing ions on the biosorption of metals. *International Biodeterioration & Biodegradation*, **38**, 19–29, https://doi.org/10.1016/S0964-8305(96)00011-X

Tsezos M., Remoundaki E. and Hatzikioseyian A. (2006). Biosorption-principles and applications for metal immobilization from waste-water streams. Proceedings of EU-Asia workshop on clean production and nanotechnologies, Seoul, pp. 23–33.

Tuo Y., Liu G., Dong B., Yu H., Zhou J., Wang J. and Jin R. (2017). Microbial synthesis of bimetallic PdPt nanoparticles for catalytic reduction of 4-nitrophenol. *Environmental Science and Pollution Research*, **24**, 5249–5258, https://doi.org/10.1007/s11356-016-8276-7

Voegtlin S. P., Barnes R. J., Hubert C. R. J., Larter S. R. and Bryant S. L. (2022). Formation of biologically influenced palladium microstructures by *Desulfovibrio desulfuricans* and *Desulfovibrio ferrophilus* IS5. *New Biotechnology*, **72**, 128–138, https://doi.org/10.1016/j.nbt.2022.11.001

Wang S., Vincent T., Roux J.-C., Faur C. and Guibal E. (2017). Pd(II) and Pt(IV) sorption using alginate and algal-based beads. *Chemical Engineering Journal*, **313**, 567–579, https://doi.org/10.1016/j.cej.2016.12.039

Won S. W., Mao J., Kwak I.-S., Sathishkumar M. and Yun Y.-S. (2010). Platinum recovery from ICP wastewater by a combined method of biosorption and incineration. *Bioresource Technology*, **101**, 1135–1140, https://doi.org/10.1016/j.biortech.2009.09.056

Won S. W., Park J., Mao J. and Yun Y.-S. (2011). Utilization of PEI-modified *Corynebacterium glutamicum* biomass for the recovery of Pd(II) in hydrochloric solution. *Bioresource Technology*, **102**, 3888–3893, https://doi.org/10.1016/j.biortech.2010.11.106

Won S. W., Lim A. and Yun Y.-S. (2013). Recovery of high-purity metallic Pd from Pd(II)-sorbed biosorbents by incineration. *Bioresource Technology*, **137**, 400–403, https://doi.org/10.1016/j.biortech.2013.03.143

Won S. W., Kotte P., Wei W., Lim A. and Yun Y.-S. (2014a). Biosorbents for recovery of precious metals. *Bioresource Technology*, **160**, 203–212, https://doi.org/10.1016/j.biortech.2014.01.121

Won S. W., Kwak I. S. and Yun Y.-S. (2014b). The role of biomass in polyethylenimine-coated chitosan/bacterial biomass composite biosorbent fiber for removal of Ru from acetic acid waste solution. *Bioresource Technology*, **160**, 93–97, https://doi.org/10.1016/j.biortech.2014.01.082

Won S. W., Kwak I. S., Mao J. and Yun Y.-S. (2015). Biosorption–incineration–leaching–smelting sequential process for Ru recovery from Ru-bearing acetic acid waste solution. *Industrial & Engineering Chemistry Research*, **54**, 7148–7153, https://doi.org/10.1021/acs.iecr.5b01111

Wu X., Zhao F., Rahunen N., Varcoe J. R., Avignone-Rossa C., Thumser A. E. and Slade R. C. T. (2011). A role for microbial palladium nanoparticles in extracellular electron transfer. *Angewandte Chemie International Edition*, **50**, 427–430, https://doi.org/10.1002/anie.201002951

Xu H., Tan L., Dong H., He J., Liu X., Qiu G., He Q. and Xie J. (2017). Competitive biosorption behavior of Pt(iv) and Pd(ii) by *Providencia vermicola*. *RSC Advances*, **7**, 32229–32235, https://doi.org/10.1039/C7RA02786A

Xu H., Tan L., Cui H., Xu M., Xiao Y., Wu H., Dong H., Liu X., Qiu G. and Xie J. (2018). Characterization of Pd(II) biosorption in aqueous solution by *Shewanella oneidensis* MR-1. *Journal of Molecular Liquids*, **255**, 333–340, https://doi.org/10.1016/j.molliq.2018.01.168

Xu H., Xiao Y., Xu M., Cui H., Tan L., Feng N., Liu X., Qiu G., Dong H. and Xie J. (2019). Microbial synthesis of Pd-Pt alloy nanoparticles using *Shewanella oneidensis* MR-1 with enhanced catalytic activity for nitrophenol and azo dyes reduction. *Nanotechnology*, **30**(6) 065607, https://doi.org/10.1088/1361-6528/aaf2a6

Yang Z.-N., Hou Y.-N., Zhang B., Cheng H.-Y., Yong Y.-C., Liu W.-Z., Han J.-L., Liu S.-J. and Wang A.-J. (2020). Insights into palladium nanoparticles produced by *Shewanella oneidensis* MR-1: roles of NADH dehydrogenases and hydrogenases. *Environmental Research*, **191**, 110196, https://doi.org/10.1016/j.envres.2020.110196

Yates M. D., Cusick R. D. and Logan B. E. (2013). Extracellular palladium nanoparticle production using *Geobacter sulfurreducens*.

Yong P., Rowson N. A., Farr J. P. G., Harris I. R. and Macaskie L. E. (2002a). Bioaccumulation of palladium by *Desulfovibrio desulfuricans*. *Chemical Technology and Biotechnology*, **77**, 593–601, https://doi.org/10.1002/jctb.606

Yong P., Rowson N. A., Farr J. P. G., Harris I. R. and Macaskie L. E. (2002b). Bioreduction and biocrystallization of palladium by *Desulfovibrio desulfuricans* NCIMB 8307. *Biotechnology and Bioengineering*, **80**, 369–379, https://doi.org/10.1002/bit.10369

Yong P., Rowson N. A., Farr J. P. G., Harris L. R. and Macaskie L. E. (2003). A novel electrobiotechnology for the recovery of precious metals from spent automotive catalysts. *Environmental Technology*, **24**, 289–297, https://doi.org/10.1080/09593330309385561

Zhang S., Deng X., Chen A., Zhou H., Xie Z., Liang Y. and Zeng F. (2019). Facile synthesis of Pd supported on *Shewanella* as an efficient catalyst for oxygen reduction reaction. *International Journal of Hydrogen Energy*, **44**, 21759–21768, https://doi.org/10.1016/j.ijhydene.2019.06.141

doi: 10.2166/9781789065084_0127

Chapter 6

Phytoremediation and phytomining of palladium and other platinum group elements

Aminu Abdullahi[1,2]*, *Tarangini Korumilli*[1], *K. Jagajjanani Rao*[3], *Selva Kumar T*[3] *and Jeba Sweetly Dharmadhas*[4]

[1]*Centre for Biomaterials & Environmental Biotechnology, Department of Biotechnology, Vel Tech Rangarajan Dr. Sagunthala R&D Institute of Science and Technology, Chennai 600062, India*
[2]*Department of Biotechnology, Modibbo Adama University, Yola PMB 2078, Adamawa State, Nigeria*
[3]*Centre for Interface & Nanomaterials, Department of Biotechnology, Vel Tech Rangarajan Dr. Sagunthala R&D Institute of Science and Technology, Chennai 600062, India*
[4]*Department of Biochemistry and Biotechnology, Avinashilingam Institute for Home Science and Higher Education for Women, Coimbatore, Tamil Nadu, 641108, India*
**Corresponding author: aminuabdubawa71@gmail.com*

ABSTRACT

Increased human activities have led to a growing demand for palladium and platinum. However, since platinum group metals (PGMs) are non-biodegradable, they accumulate in the environment, posing a significant risk of contaminating the food chain. It is crucial to address the problem of contamination by PGMs in various environments, while simultaneously striving to recover these valuable metals in a financially feasible and environmentally sustainable manner. Accordingly, the successful remediation of soil and water is a global concern and an immediate necessity. Phytoremediation and phytomining, as sustainable green technologies, have demonstrated significant potential in addressing complex issues and promoting innovative approaches to the remediation and recovery of heavy metals. The potential implementation of novel hyperaccumulators in the fields of phytoremediation and phytomining is investigated. Molecular biology approaches are used to enhance our understanding of the intricate processes involved in metal uptake, translocation, storage, and tolerance in plants. This chapter discusses different approaches for PGM remediation using plants. It also highlights the basic mechanisms involved in uptake of PGMs by plants as well as pros and cons associated with this eco-friendly and cost-effective technology.

Keywords: Palladium, platinum group metals, phytoremediation, phytomining

6.1 INTRODUCTION

6.1.1 Importance of platinum group metals

Platinum and palladium have significant importance in the sphere of the global market. The platinum group metals (PGMs) include platinum (Pt), palladium (Pd), rhodium (Rh), osmium (Os), iridium (Ir), and ruthenium (Ru). They are extensively used in advanced technological applications, enhancing the quality of life, due to their exceptional characteristics, such as high electrical conductivity, catalytic capabilities and excellent resistance to corrosion and oxidation (Yakoumis *et al.*, 2021). The demand for these valuable group metals (PGMs) is expected to remain high in the future, primarily due to the increasing number of vehicles and growing environmental pressures. According to Johnson Matthey's report, the gross demand for platinum (Pt), palladium (Pd), and rhodium (Rh) in automobile catalytic converters alone stood at 97.2, 216.8, and 24.9 tons, respectively, in 2022. These quantities accounted for over 44% of the overall platinum demand, 72% of the overall palladium demand, and 78% of the overall rhodium demand (Hughes *et al.*, 2021). Furthermore, the recycled amounts of platinum, palladium, and rhodium from these catalysts amounted to 39.7, 57.9, and 8.7 tons, respectively (Murray *et al.*, 2017). Of these, platinum and palladium hold the utmost significance in the commercial sphere due to their extensive application in catalytic converters within the automotive industry, serving as an effective measure to counteract the emission of toxic substances from vehicles (Masinire *et al.*, 2022).

PGMs are exceedingly scarce in the Earth's crust, with both palladium and platinum exhibiting average concentrations of approximately 5 parts per billion (ppb) (Shi, 2016). Despite their limited availability, there has been an observed rise in the contamination by PGMs, both directly and indirectly, as a result of natural and anthropogenic activities (Harumain *et al.*, 2017). The perpetual rise of PGM concentrations in the surroundings poses significant danger to public health and the biosphere due to the toxicological and ecotoxicological hazards associated with these metals (Aribal & Mercado, 2020). They are resistant to decomposition and persist in the natural surroundings. Palladium is the most toxic and its toxicity is accounted for its rapid mobility and high bioavailability.

The conventional physical and chemical approaches to the decontamination of PGMs, including heat treatment, electroremediation, soil replacement, precipitation and chemical leaching, are typically associated with substantial costs and may not be suitable for the remediation of agricultural land. On the other hand, commercial extraction of PGMs is typically carried out from mineral deposits that possess a substantial amount of the desired metals (surpassing the designated threshold) and requires significant financial resources. These ores are confined to limited geographical regions (Chapter 2) and are currently depleting due to the continuous growth of economies, escalating populations, and unorganized industrial expansion (Harumain *et al.*, 2017). The diminishing ore quality limiting the availability of PGM resources and the persistent growth in demand has triggered the development of innovative technologies for the extraction and recovery of PGMs.

Consequently, phytoremediation and phytomining have emerged as a feasible, cost-efficient, and environmental-friendly approach to resolving the contamination or recovery of PGMs from environments. Phytoremediation is an environmental-friendly approach that employs hyper-accumulating plants and the microorganisms present in their rhizosphere to immobilize, translocate, or breakdown contaminants present in soil, water, and the surrounding ecosystem (Yan *et al.*, 2020). It involves the use of plants or their products to restore the contaminated soil, surface water, and groundwater. Phytomining is the process by which bioavailable metal species are absorbed and concentrated from the surrounding environment into the plant biomass through natural processes. It serves as an alternative method for obtaining economically beneficial elements when traditional mining methods are not feasible, such as in cases of low metal concentrations from ore deposits. These elements are then collected in the above-ground parts of the plants, based on the plants' uptake capacity and distribution patterns. After harvesting the plants, the biomass can be used for energy purposes by either burning the phytomass or generating biogas through microbial fermentation.

Figure 6.1 illustrates the synergistic application of phytoremediation and phytomining for the remediation of contaminated environments and the recovery of valuable metals. This chapter discusses phytoremediation and phytomining of PGMs from contaminated environments. It also offers insight into different approaches of phytoremediation and potential challenges. The various mechanisms of phytomining of PGMs have also been elucidated.

6.1.2 Characteristics and application of PGMs

PGMs encompass a collection of metals that possess significant market value and demonstrate limited occurrence within the Earth's crust. They include platinum (Pt), palladium (Pd), rhodium (Rh), osmium (Os), iridium (Ir), and ruthenium (Ru). Historically, PGMs have primarily served as media of exchange and adornments. Nonetheless, owing to technological advancements in the 20th

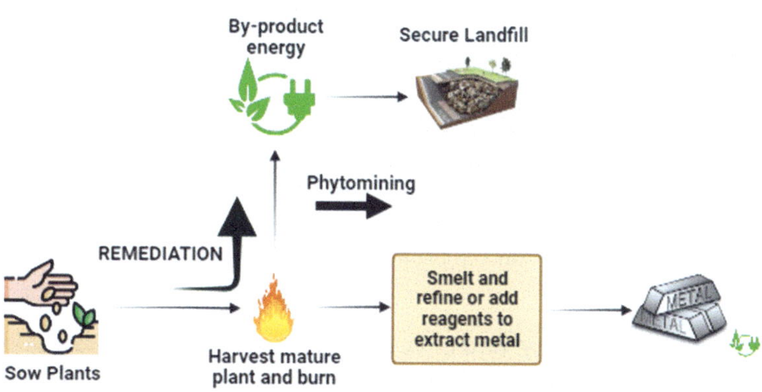

Figure 6.1 Synergistic application of phytoremediation and phytomining.

Figure 6.2 Photographs of the PGM.

and 21st centuries, their characteristics have become indispensable in a variety of contemporary industrial and technological domains, including catalysts, electronics, petroleum refining, and medical applications. Notably, platinum, palladium, and rhodium find themselves irreplaceable in automotive catalytic converters, which are utilized to process exhaust emissions. Conversely, ruthenium, osmium, and iridium do not find application in this context due to their propensity to form volatile oxides (Hughes *et al.*, 2021). PGMs have historically been extracted from ores that contain a relatively high level of the desired component (Murray *et al.*, 2017). With the exception of osmium, these metals appear as silvery grey white elements, often characterized by elevated levels of hardness and brittleness (Figure 6.2). With the exception of osmium and ruthenium, all PGMs are silvery white so they have colourless appearance. The global PGM mineral deposits are estimated to be around 104 000 Mt, with South Africa responsible for more than 60% of the total deposits. Platinum and palladium constitute a substantial proportion, approximately 90%, of the overall deposits.

6.1.3 Occurrence, toxicity, and environmental fate of PGMs

PGMs are rare in the Earth's crust with an average concentration of around 5 ppb for both Pd and Pt (Akinbile *et al.*, 2023). Currently, approximately 71% of the global reserves of PGMs are located within the borders of South Africa. The nation is responsible for the production of roughly 60% of the worldwide supply of three notable precious metals: platinum, palladium, and rhodium (Akinbile *et al.*, 2023). The natural occurrence of these elements in the soil is estimated to range from 0.5 to 4 $\mu g\ kg^{-1}$ (Dinh *et al.*, 2022).

The significance of PGMs primarily arises from their distinctive physical and chemical characteristics, such as notable chemical and thermal resilience, exceptional thermal and electrical conductivity, as well as resistance against corrosion and oxidation. Their ores typically come from extensive PGM deposits (Dinh *et al.*, 2022; Gardea-Torresdey *et al.*, 2005; Yakoumis *et al.*, 2021). The primary sources of PGMs include the natural contribution including the weathering of minerals, erosion, and volcanic eruptions, while the anthropogenic contribution entails mining, agricultural activities, smelting, electroplating, sludge dumping, and industrial effluents. Despite their short supply, the levels of palladium in the environment are on the rise due to a lot of anthropogenic activities. PGMs in general are reported to be highly toxic, allergic and even some reports of damaged DNA due to exposure to PGMs are available (Sadrolvaezin *et al.*, 2023). This is a great risk to public health and a threat to the biosphere as the toxicological and ecotoxicological danger of these metals are yet to be fully explored.

Empirical evidence indicates that automotive catalytic converters are emitting particles of PGMs such as platinum, palladium, and rhodium, which are deposited along road ways. The deposition of these elements results in a progressive accumulation of PGMs within the surrounding ecosystems, thereby provoking concerns regarding the potential environmental consequences and toxicological impacts that may be imposed on various organisms inhabiting these environments (Gagnon *et al.*, 2006). Similar to other metals, increasing soil PGM levels can have a detrimental effect on the plant, leading to reduced growth, impaired photosynthetic activity, and increased oxidative stress (Masinire *et al.*, 2022). They also have the potential to bioaccumulate in the biological tissues of living organisms, thereby presenting significant physiological problems (Linde *et al.*, 2017). Research has linked PGMs to respiratory sensitization, allergic responses, skin dermatitis, asthma, rhinoconjunctivitis, increased lymphocyte proliferation, cytokine release, and a potential risk of cancer (Kalavrouziotis & Koukoulakis, 2009). Also, histological and pathological studies on animal models exposed to PGMs have revealed vacuolization, eosinophilic inclusion bodies in the adrenal glands, and a decrease in the size of kidney glomeruli (Grilli *et al.*, 2023).

The presence of PGMs in the natural environment can be confirmed through the analysis of environmental samples, including roadway drainage and other areas likely to be contaminated with these elements. The mobile fraction of the emitted metals as well as the portion that undergoes conversion to soluble forms, pose a significant threat to both flora and fauna (Krishnasamy *et al.*, 2021). The assessment of PGMs bioavailability and their fate in the environment is hindered by limited data. Despite the growing attention given to this subject, the main obstacles are the low concentrations of these metals in environmental samples and the challenges associated with their detection using contemporary analytical instruments (Chapter 3).

6.2 PHYTOREMEDIATION: A SUSTAINABLE SOLUTION TO PGM POLLUTION

6.2.1 Scope of phytoremediation

Phytoremediation is primarily concerned with the use of plants and their associated soil microorganisms with the aim of eliminating, confining, neutralizing, or breaking down detrimental environmental pollutants (Ali *et al.*, 2013). It is an emerging eco-friendly, cost effective and sustainable technique that can be used for decontamination of heavy metal pollution as well as organic contaminants (Table 6.1). Pollutants originating from human activities can have potential negative impacts on fauna and negative consequences on the well-being of human beings. The presence of even small amounts of contaminants in the ecosystem can pose a hazard due to their potential to accumulate at higher levels in the food chain, a phenomenon known as biomagnification.

Phytoremediation can be used for the decontamination of both organic and inorganic (such as metals) contaminants that are found in low concentrations in the soil. While some metals such as cobalt (Co), copper (Cu), iron (Fe),

Table 6.1 Advantages and limitations of phytoremediation.

Advantages	Limitations
It is a cost-effective sustainable environmental clean-up technology.	May require longer timeframes to achieve desired contamination reduction.
It is an alternative to engineering procedures that are usually more destructive for soil integrity and properties	It has limited application where the bioavailable contaminants are toxic to the desirable plants to be used. Hence cannot address all pollution types.
It takes advantage of the natural plant processes and requires less equipment and labor.	Limited efficiency in high levels of contaminants or soils with mixed contamination.
Trees and smaller plants used in phytoremediation help control soil erosion.	Plants may release chemical vapour into the air (phytovolatization). This might be harmful or allergic to humans or animals.
Applicable for large, shallow contaminated areas, for example agricultural land, wetlands.	Depth of remediation is restricted to the root zone of plants (usually <5 m).
Provides an added benefit by sequestering carbon through plant growth.	Long-term maintenance and monitoring are often required.
Possibility of harvesting biomass for bioenergy or secondary uses, such as metal recovery (phytomining).	Contaminated plant material requires proper treatment and disposal, adding cost and logistical challenges.
Minimal disturbance reduces risk of contaminant migration compared to some conventional technologies.	Some contaminants may volatilize or leach during the process, potentially causing secondary contamination.
Can be applied over longer periods for gradual or continuous remediation.	Not applicable for sites requiring immediate remediation due to slow contaminant uptake and degradation rates.

molybdenum (Mo), manganese (Mn), and zinc (Zn) are critical for plant growth and are classified as essential micronutrients, some metals like platinum (Pt), lead (Pb), and palladium (Pd) are commonly found as contaminants and may be a potential threat to plant growth (Morkunas *et al.*, 2018; Pinho & Ladeiro, 2012). An inherent ability of plants to absorb heavy metals qualifies them as good candidates for the remediation of metal-contaminated environments. Moreover, plants generally manage the pollutants without compromising the soil integrity, thus safeguarding its functionality and productivity (Vangronsveld *et al.*, 2009). Consequently, they have the potential to enhance the fertility of the soil by introducing organic substances.

Traditional approaches considered for environmental decontamination of PGMs, such as precipitation, solvent extraction, and ion exchange, are inadequate, costly and lacks sustainability. Furthermore, conventional methods generate substantial quantities of secondary waste materials. The development of the circular economy concept is presently concentrated not solely on increased PGMs removal, but also on amplified economic and environmental benchmarks by adhering to green principles. Therefore, the associated drawbacks of the conventional methods underscore the significance of cost-effective and environmental-friendly remediation methods for palladium and other PGMs (Diehl & Gagnon, 2007; Tzvetkova *et al.*, 2021).

Phytoremediation is widely accepted as a cost-effective environmental restoration technology and has been applied for the decontamination of palladium and other PGM polluted surroundings (Masinire *et al.*, 2022). It is a viable substitute for engineering methodologies, which typically inflict greater harm to the soil. The desired time frame for phytoremediation to achieve satisfactory levels of pollutants in the environment should not surpass a decade. Particular plant species exhibit superior efficacy in the removal of contaminants compared to others. Moreover, the plants to be employed for phytoremediation must possess the capacity to withstand the presence of various types and concentrations of PGMs, while simultaneously flourishing and surviving within the regional climate.

6.2.2 Mechanisms of phytoremediation

Phytoremediation utilizes the ability of plants to take up pollutants from the environment and accomplish their detoxification using various mechanisms. Certain plants are able to remove or breakdown harmful chemicals from the ground when their roots take in water and nutrients from the contaminated soil, sediment, or groundwater (Masinire *et al.*, 2021). The root system of plants exhibits geotropism, a physiological process that aids in the expansion of roots into the soil matrix. This mechanism also facilitates the significant accumulation of heavy metals, consequently augmenting their bioavailability. As a result, the contaminated soil is effectively remediated, and its fertility is subsequently enhanced (Krishnasamy *et al.*, 2021).

Phytoremediation employs several mechanisms such as rhizo-degradation and phytodegradation for degradation, phytoextraction and rhizofiltration for accumulation, phytovolatilization for dissipation, and hydraulic control and phytostabilization for immobilization of pollutants (Kafle *et al.*, 2022).

Metal uptake by plants generally takes place by specific transporters (channel proteins) or H^+ coupled carrier proteins located on the cell membrane of the root cells. For example, the Fe regulated transporter (IRT1) allows the uptake of Fe. Uptake of other metals also occurs via RT1 transporters, especially if very low concentrations of Fe exist in the soil. Inadvertent uptake of non-essential metals also takes place via other cell membrane transporters. For example, the uptake of As (as AsO_3^{2-}) occurs inadvertently via phosphorous-transporters.

Plants can help to clean up palladium and other PGMs as deep as their roots can reach using natural processes through uptake to (i) store PGMs in the roots, stem, or leaves, (ii) subsequently convert PGMs to less harmful chemical species within the plant or (iii) convert them into volatile compounds, which are released in the air (see section 6.4.4).

6.2.3 Duration of phytoremediation

Phytoremediation requires a significant number of years for the purpose of decontamination of a particular surrounding. The effectiveness of the remediation process depends upon multiple factors. For instance, the duration of the phytoremediation process will be extended in instances where the concentrations of contaminants are high. Furthermore, the magnitude or depth of the contaminated area will also impact the duration of the clean-up. Additionally, the utilization of plant species with long periods of growth will contribute to the duration of the phytoremediation process. Lastly, a brief growth season will prolong the duration required for successful remediation. These factors exhibit variability across different locations. In the event that plants encounter damage due to severe weather conditions, pests, or animals, there may also be a need for their replacement. Consequently, this results in an elongation of the clean-up process.

6.3 PHYTOREMEDIATION-BASED TECHNOLOGIES

Various phytoremediation strategies are available for various environments and types of contaminants (Figure 6.3). These involve different processes such as in situ stabilization, degradation and removal (i.e. volatilization or extraction of contaminant).

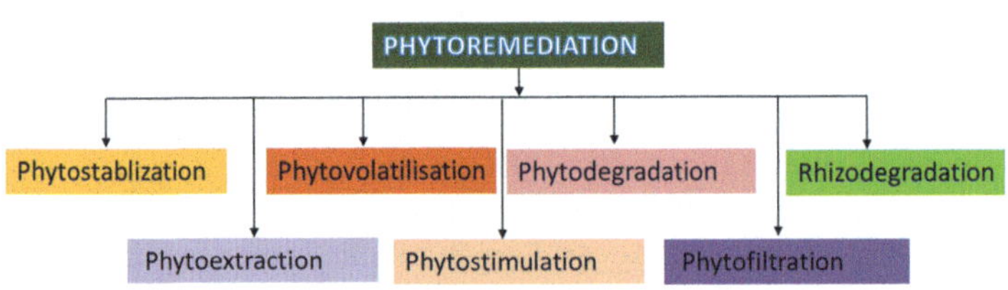

Figure 6.3 Different types of phytoremediation technologies that can be applied to PGM contaminated site.

6.3.1 Phytoextraction

Phytoextraction stands as the primary and highly beneficial approach within the realm of phytoremediation. Its purpose is to eliminate heavy metals, metalloids and other contaminants present in soils, sediments, or water sources (Ali *et al.*, 2013). Phytoextraction harnesses the capacity of plants to sequester pollutants in their aboveground, harvestable biomass. This procedure entails iteratively reaping of the biomass to mitigate the concentration of contaminants in the soil.

Phytoextraction can transpire as an incessant course (employing metal hyperaccumulating plants or rapidly proliferating plants) or as an elicited course (employing chemical agents to enhance the bioaccessibility of metals in the soil). The concept of extracting metals through using plants, commonly known as phytoextraction, was initially evolved around the 1950s, when collections of plant species were found to have significant concentrations of nickel and gold (Akinbile *et al.*, 2023).

Table 6.2 compares phytoextraction with other phytoremediation technologies. Continuous phytoextraction is based on the ability of certain plants to gradually accumulate contaminants (mainly metals) into their biomass. Certain plants can hyper accumulate metals without any toxic effects. These plants are adapted to naturally occurring, metaliferous soils. More than 400 plant species can hyper accumulate various metals. However, most plants can only hyper accumulate one specific metal.

6.3.2 Phytostabilization

Phytostabilization entails the use of plant species that are native to the contaminated soil and with a dense root system are capable of retaining significant amounts of contaminants within the roots, while also having the ability to reproduce autonomously. This form of phytoremediation is called 'hydraulic control'. It reduces the movement of contaminated groundwater towards clean areas offsite. Because tree roots grow deeper, trees such as poplars and willows are used for hydraulic control or to clean up deeper

Table 6.2 Comparing between the different phytoremediation methodologies.

Technique	Action on Pollutant	Type of Contaminant	Nature of Vegetation	Reference
Phytodegradation	Attenuated in situ	Organics	Cover maintained	Ayilara *et al.* (2023)
Phytofiltration	Remove	In organic and metals	Harvested frequently	Khan *et al.* (2022)
Phytostabilization	Retained in situ	Organic and metals	Cover maintained	Kafle *et al.* (2022), Khan *et al.* (2022)
Phytoextraction	Removal	Metals	Harvested frequently	Suman *et al.* (2018)
Phytovolatization	Removal	Organics and metals	Cover maintained	Kafle *et al.* (2022)

soil contamination and contaminated groundwater. Phytostabilization has gained increasing prominence in recent years due to its cost-effectiveness and environment-friendly nature. Additionally, this technique possesses an aesthetically pleasing aspect. It aims to retain contaminants in the soil or water and prevent further dispersal. Contaminants can be stabilized in the roots or within the rhizosphere (Kafle *et al.*, 2022).

Re-vegetation of palladium mine tailings can be a good practice to stabilize further dispersal of PGM contaminants. Mine tailings have been stabilized using commercially available varieties of metal-tolerant grasses such as *Agrostis tetnuis* cv. or *Festuca rubra* cv. merlin. Metal tolerance of plants is generally increased by symbiotic, non-root colonizing, arbuscular mycorrhizal fungi (AMF), through metal sequestration in the AMF hyphae (Dhalaria *et al.*, 2020; Pan *et al.*, 2023). In addition, excretions of the glycoprotein glomulin by AMF that are adapted to metal-contaminated soils have great potential in phytostabilization. Soil microbes can decrease toxic effects of contaminants in the soil. For example, exudates (peptides) from the bacterium *Pseudomonas putida* can decrease cadmium toxicity in plants (Haider *et al.*, 2021).

6.3.3 Phytodegradation

Phytodegradation refers to the breakdown of contaminants either by enzymes released from plant roots or through metabolic processes occurring within the plant tissues (Kafle *et al.*, 2022). Phytodegradation involves the absorption and subsequent uptake of organic pollutants by plant roots, where they are transformed within the plant tissues into less harmful substances. This method has shown particular effectiveness in addressing hydrophobic organic pollutants, like perfluorooctanoic acid and dichlorodiphenyltrichloroethane (DDT). Additionally, plants can alter contaminants to reduce their toxicity or limit their bioavailability. Roots of certain plants release natural chelating agents that can bind with metals in the rhizosphere (Yan *et al.*, 2020). For instance, enzymes on the roots of wetland plants can convert toxic Cr(VI) into the less harmful Cr(III) (Ullah *et al.*, 2023). Poplar trees (*Populus* spp.) have been effectively utilized in the phytodegradation of persistent and toxic organic compounds, such as benzene/toluene/ethylbenzene/xylenes (BTEX), polynuclear aromatic hydrocarbons (PAHs), and long-chain alkanes in petroleum hydrocarbons (Simmer & Schnoor, 2022).

6.3.4 Rhizodegradation

Rhizodegradation is a type of phytodegradation where organic contaminants are broken down into less harmful substances within the rhizosphere through the action of soil microorganisms. This process is facilitated by root exudates, which are organic compounds that nourish soil microbes. To improve the effectiveness of rhizodegradation, specific bacterial inoculants can be introduced to contaminated soils. These inoculants contain bacterial strains with the metabolic capabilities needed to degrade particular contaminants. Encouraging results have been observed when plants are inoculated with genetically modified bacterial strains designed to target specific contaminants (Rebello *et al.*, 2021). Also, populations of soil bacteria can be amplified using the process called biostimulation. This involves, for example, manipulating the nutrient and pH levels of the soil to increase specific bacterial populations (Omokhagbor Adams *et al.*, 2020).

In addition, plants and their associated soil microbes can release chemicals that act as biosurfactants in the soil that increase the uptake of contaminants (Karlapudi *et al.*, 2018). Contaminants can be stabilized in natural and constructed wetlands through a process called phytofiltration. This involves rhizofiltration where metals are precipitated within the rhizosphere. Metal plaque forms typically on the roots of wetland plants through the release of oxygen via the parenchyma of roots (Colmer, 2003). For example, iron oxides precipitate along with other metals into the metal plaque (Guo *et al.*, 2015). Metal plaque on roots acts as a reservoir for active iron (Fe^{2+}), which in turn increases the tolerance of plants to other toxic metals.

6.3.5 Phytovolatization

Phytovolatization involves the uptake of contaminants by plant roots and its conversion to a gaseous state and release into the atmosphere. This process is driven by the evapotranspiration of plants. Plants that have a high evapotranspiration rate are the best option for phytovolatization. Organic and inorganic contaminants, like volatile organic compounds (VOCs) and metals are passively volatilized by plants. For example, hybrid poplar trees have been used to volatilize trichloroethylene (TCE) by converting it to chlorinated acetates and CO_2 (Doucette *et al.*, 2013). Metalloids such as selenium can be volatilized by plants through conversion into dimethylselenide [Se $(CH_3)_2$] (Souza *et al.*, 2000).

Genetic engineering techniques have been used to allow plants to volatilize specific contaminants. For example, the ability of the tulip tree (*Liriodendron tulipifera*) to volatilize methyl-Hg from the soil into the atmosphere as elemental (Hg^0) was improved by incorporating the mercuric ion reductase (merA) gene of *E. coli* that encodes the enzyme (Floreani *et al.*, 2023).

6.3.6 Phytofiltration

Phytofiltration refers to the process of eliminating impurities from polluted surface waters or wastewater through the utilization of plants. This process can be categorized into three methods: rhizofiltration, which involves the use of plant roots, blastofiltration, which employs seedlings, and caulofiltration, which utilizes excised plant shoots. Through phytofiltration, the contaminants are absorbed or adsorbed, thereby minimizing their migration to subterranean water bodies.

6.4 PHYTOREMEDIATION OF PGMs

6.4.1 Role of plants in PGMs uptake and detoxification

There are an insignificant number of plant species reported to take up PGMs from contaminated soil and water. As a result, no detailed mechanism has so far been revealed for the remediation of PGMs. Moreover, among the few species that were reported, these were commercial studies. Hyperaccumulating plants contain more than 1% of a metal in their dry biomass. For example, the hyperaccumulating plant *Berkheya coddii* was found to contain as much as 0.07 (±0.045) mg kg^{-1} dry mass of palladium in the root and aboveground biomass, when grown in PGM-contaminated soil. Liquid culture grown *Arabidopsis* spp. hyperaccumulate and store palladium nanoparticles (Edgar *et al.*, 2021). *Chrysopogon zizanioides* (Vetiver grass) was studied by Masinire *et al.* (2022) for removal and accumulation of Pd, and its tolerance towards elevated Pd concentrations. The grass was cultivated in solutions containing Pd concentrations spanning from 10 to 120 ppm for a 20-day growth period. The Pd removal efficiency varied from 80% at an initial concentration of 10 ppm to 20% at 120 ppm. The deep and dense root systems as well as high biomass yield are characteristics that make it a good choice to be used in the phytoremediation and phytomining of Pd and other PGMs. Recently, a number of plant species has been reported to take up metals and store them as nanoparticles (NPs) in their tissues (Rajput *et al.*, 2020).

The primary constraints on the utilization of hyperaccumulating plants in the process of phytoextraction are the slow growth rate and the limited capacity for biomass production. The efficacy of phytoextraction depends on both the quantity of biomass generated by a given plant and the concentration of pollutants contained within the said biomass. Consequently, crops that exhibit rapid growth and possess the ability to accumulate metallic substances exhibit significant promise for use in phytoextraction. Furthermore, the utilization of crops in the context of phytoextraction can be enhanced through the manipulation of their associated soil microorganisms.

The introduction of plant growth promoting bacteria (PGPB) and *Arbuscular mycorrhizal* fungi (AMF) can serve to augment the production of plant biomass. Low bioavailability of PGMs in the soil is also a limiting factor for their phytoextraction. The bioavailability of metals can be increased by the use of synthetic chelates such as ethylene diaminetetra acetic acid (EDTA) or acidifying chemicals (e.g. $(NH_4)_2SO_4$). The use of synthetic chelates increases the absorption of metals to the root and the translocation of metals from the roots to the foliage. The timing of chelate application is critical, and should ideally take place at the peak of biomass production. Some drawbacks of using synthetic chelates in phytoremediation are the increased risk of metal migration through the soil profile and into the groundwater. However, a possible solution is to treat contaminated soil ex situ in a confined site with an impervious surface. Also, periodic application of low doses of synthetic chelates reduces the risk of metal migration.

6.4.2 Phytomining of PGMs

Since 2011, PGMs have been included in the list of critical raw materials for the European Union (EU) based on their significance as strategic metals for the EU economy and their vulnerability to high supply risk. On a global level, there are only a limited number of locations where PGMs are mined. There exist substantial quantities of PGMs globally, either in small-scale mineralized rock or in mining tailings. However, their exploitation using traditional mining methods is considered economically unviable. Phytomining can be considered a low-cost alternative to extract metals from their low-grade PGM sources.

Phytomining has been suggested as a viable process to recover PGMs from mineralized soil, rock, or from mine wastes. Phytomining of palladium and other PGMs (entails cultivating plants on PGM-rich substrates that possess the capacity to selectively assimilate PGM within their cellular frameworks. In practice, the plants can subsequently be harvested and subjected to regulated pyrolysis to create a substance with stabilized PGM content, from which they can consequently be extracted (Masinire *et al.*, 2022).

The process of phytomining involves the utilization of hyperaccumulating flora in order to extract precious metals from the underlying material. Phytomining has been proposed as a feasible method for the retrieval of PGMs from mineralized soil, rock, or mine wastes (Dinh *et al.*, 2022). Certain plant species possess the ability to accumulate metals to a degree of concentration that surpasses the concentration found in the soil. In cases where most of the metals are available in the soil solution and can be taken up by plants, hyperaccumulating plants have the inherent ability to naturally take up these metals (see section 6.5.3).

6.5 PGM PHYTOMINING AND PHYTOEXTRACTION

6.5.1 Mechanisms of PGM phytomining

Phytomining makes use of plants that possess the capacity to accumulate PGMs from contaminated land, and subsequently retain them in the roots as well as in other parts of the plants, with different distribution than in the soil. The primary reason for the heavy metal intake from the soil can be mainly attributed to the presence of complexing agents, particularly organic acids, which are secreted by the plant into the soil. Literature suggests that platinum group elements, especially palladium, are taken up by the roots from the soil and subsequently bind to biologically active substances that are abundant in sulfur (Savignan *et al.*, 2021). The accumulation process of platinum group elements primarily occurs in the vegetative parts of plants and diminishes in the following sequence: root > stem > leaves (Sobroval *et al.*, 2012).

In the case of hyperaccumulating plants, they actively transport the metals from the roots to the shoots, as the storage capacity within the roots is limited. Metal ions from the same group in the periodic table often display similar characteristics. Therefore, plants may accumulate multiple PGMs, such as platinum and palladium, in an active manner. The content of metals within different parts of the plants can differ considerably, contingent upon the particular plant species and the specific metals involved. The bioaccumulation factor (BF) and translocation factor (TF) are indispensable measures utilized to portray the bioaccumulation characteristics of hyperaccumulators. The bioaccumulation factor is defined as the ratio of the metal concentration in the shoots to that in the soil. This measure serves to elucidate the capacity of plants to absorb the metals from the soils and convey them to their shoots. Conversely, the translocation factor represents the ratio of the shoot/root metal concentration, and a higher translocation factor signifies a greater ability to transport metals from the roots to the shoots. This measure holds significant importance within the phytomining context, as typically only the above-ground parts of the plants are harvested. Hyperaccumulators must have a bioaccumulation factor and translocation factor greater than 1.

6.5.2 Important steps of PGM phytomining

It is possible to extract metals from the harvested biomass in a process termed phytomining (Figure 6.1). Depending on the effectiveness of the mining process, the process of a phytomining operation encompasses multiple conventional stages:

(1) Locate a suitable site containing residual mine materials of PGMs, PGM-contaminated soil, or contaminated water with insufficient levels of the desired metal concentration.
(2) Cultivate a plant species known for its high biomass yield and ability to bioaccumulate significant quantities of the specific PGMs of interest, ideally a hyperaccumulator plant species, while also demonstrating tolerance to other coexisting metals.
(3) If necessary, administer a chelating agent in the proximity to the plant's maturity stage to enhance the bioavailability of the target metal.
(4) Proceed to harvest the plants either at their peak biomass production or when signs of deterioration become apparent.
(5) Harvest the plant biomass and subject it to incineration in order to extract the bio-ore, from which the desired PGM can be reclaimed by means of various techniques such as smelting, freeze-thawing, ashing, leaching, and numerous others.

6.5.3 Factors that affect PGMs phytomining

Several approaches and variables may influence the outcome of the phytomining of PGMs: (i) PGM-rich materials may contain other elements which could inhibit plant growth or metal uptake and (ii) the process of removing Pd or PGMs from the solution and accumulating it in the plant is always influenced by: (a) the initial concentration of a given PGM, (b) the solubility of the metal, and (c) the ability of the plant species to grow in the contaminated soil and produce substantial amounts of biomass.

PGM phytomining is limited by:

(1) The growth rate of plants, solubility, and availability of PGMs in the soil are influenced by various biogeochemical factors such as rhizobiological activity, root exudates, temperature, moisture, pH, and the concentration of competing ions.
(2) The metal needs to establish contact with the root zone of the plant. Consequently, any metals situated beneath the levels of the roots are inevitably rendered inaccessible for mining.
(3) The use of chelators (substances that increase solubility) to enhance the movement of metals may lead to complications if the amount of chemical used exceeds the concentration of metals in the soil solution that can be effectively absorbed by plants.
(4) Most natural metal hyperaccumulators are generally slow growing with a low biomass yield and shallow root system.

(5) Proper storage of harvested biomass will be necessary for phytomining to prevent potential risk pathways, such as introduction into the food chain.
(6) PGMs collected in leaves can be released again to the environment during litter fall, which may eventually contaminate the food chain.

6.5.4 Hyperaccumulation of PGMs

Hyperaccumulation is the ability of a plant to accumulate high levels of heavy metals in its aboveground parts without showing any signs of toxicity, even at concentrations that are 100–1000 times higher than those found in non-accumulating species (Suman *et al.*, 2018). These comprise the establishment of symbiotic relationships with microorganisms in the rhizosphere, the excretion of compounds into the soil and the immobilization of metals, alteration of cell walls, modification in the activation of genes that encode transporters for heavy metals, the chelation and sequestration of heavy metal ions, and the production of heat-shock proteins for regeneration (Skuza *et al.*, 2022). Studies have demonstrated the ability of certain plants to accumulate extraordinary amounts of PGMs in their roots and aboveground parts. Organic acids excreted by hyper accumulators, including acetic, malic, and citric acids, have been identified as having a significant impact on the process of metal absorption from the soil to the roots by means of chelate formation. Table 6.3 shows different plant species reported for the hyperaccumulation of PGMs.

6.5.5 Examples of phytomining PGMs

6.5.5.1 Paladium

Despite the fact that the capacity of various plant species to bioaccumulate PGMs have been elucidated since the 20th century, so far only a few investigations concerning this phenomenon have been documented in recent decades. Fuchs and Rose (1974) are the pioneers in presenting evidence regarding the phytomining of Pd using plants. The evidence was acquired when a concentration of 285 μg kg Pd was identified in the twig ash of *Pinus flexilis* (limber pine). After five years, Kothny (1979) reported a concentration of 400 μg kg^{-1} Pd in the ash of *Quercus chrysolepsis* (oak) that was collected from a sampling site containing 140 μg kg^{-1} Pd (Dinh *et al.*, 2022).

Berkheya coddii was found to be most efficient in accumulating nickel, palladium, and platinum from the soil with mean concentrations of palladium accumulated in the roots and in the leaves ranging between 0.18 (±0.07) dry mass and 0.71 (±0.52) mg kg^{-1} dry mass, respectively (Nemutandani *et al.*, 2006). This was likely the first time to reveal that a plant may have the potential to uptake platinum and palladium; both metals are from the same group of the periodic table as nickel.

Analysis on the accumulation of Pd and its toxic effects on *Lemna minor* (duckweed) were conducted by Jmii and Dewez (2021). The study involved exposing the duckweed to concentrations ranging from 2 to 50 μM. When the concentration of Pd reached 12.5 μM, the accumulation of Pd in the plant reached 63.93 μg/g fresh weight, resulting in a substantial (37.4%) inhibition of

Table 6.3 Plant species used to accumulate palladium and other PGMs.

Element	Plant Species	Substrate	Concentration	Amount Accumulated	Reference
Pd	*Sinus flexilis* (limber pine)	Soil: 3.1 mg kg^{-1} Pd	Pd concentration in the twig ash: Pd: 285 μg kg^{-1}	Natural accumulator	Dinh *et al.* (2022)
Pd	*Quercus chrysolepsis* (oak)	Soil: 140 μg kg^{-1} Pd	Concentration in plant ash: 400 μg kg^{-1} Pd	Natural accumulator	Kothny (1979)
Pt, Pd	*Berkheya coddii* (Asteraceae)	Contaminated soil: 0.04 mg kg^{-1} Pt and 0.07 mg kg^{-1} Pd	In leaf: 0.22 mg kg^{-1} Pt, 0.71 mg kg^{-1} Pd In root: 0.14 mg kg^{-1} Pt and 0.18 mg kg^{-1} Pd	Natural accumulator	Dinh *et al.* (2022)
Pt, Pd, Au	*Berkheya coddii* (Asteraceae)	• Mine tailing: 315 μg kg^{-1} Pd and 61.4 μg kg^{-1} Au • Treatment: KCN (10 g L^{-1})	Concentration in plant: • Pt: 183 μg kg^{-1} • Pd: 7677 μg kg^{-1} • Au: 1580 μg kg^{-}	Induced hyperaccumulator	Suman *et al.* (2018)
Pd, Au	*Cannabis sativa* (hemp)	Gossan (rock): 205.5 μg kg^{-1} Pd, 20.0 μg kg^{-1} Au • Treatment: KCN (8 g L^{-1})	Concentration in aerial biomass. • Pd: mean value 30 336 μg kg^{-1}; highest value 62 420 μg kg^{-1} • Au: mean value	Induced hyperaccumulator	Vicente and Arcas (2005)
Pd	*Salix purpurea* willow (Green Dicks)	• Synthetic ore medium: 50 mg kg^{-1} Pd • Treatment: KCN	Pd in leaf: 820 mg kg^{-1}	Induced accumulation	Dinh *et al.* (2022)
Pd	*Sinapis alba* L. (mustard), *Miscanthus* (silver grass), and Green dicks (willow)	• Synthetic tailings medium: 50 mg kg^{-1} Pd • Treatment: KCN (100 mg kg^{-1})	Palladium in: Aerial plant of mustard: 500 mg kg^{-1} *Miscanthus*: 1500 mg kg^{-1} • Leaf of willow: 800 mg kg^{-1}	Induced accumulation	Dinh *et al.* (2022)

(*Continued*)

Table 6.3 Plant species used to accumulate palladium and other PGMs. (*Continued*)

Element	Plant Species	Substrate	Concentration	Amount Accumulated	Reference
Ag, Pt	Trees (oak, pine, birch, poplar, etc)	Contaminated brown field land	Concentration in plant: • Ag: 11.9 mg kg^{-1} • Pt: 3.06 mg kg^{-1}	Natural accumulation	Dinh *et al.* (2022)
Pt	Grass species accumulation	Soil: vicinity of a motorway	Pt concentration in: • Leaf: 19 42 μg kg^{-1} • Root: 59–137 μg kg^{-1}	Natural	Suman *et al.* (2018), Dinh *et al.* (2022)
Pt, Pd, Rh	*Daucus carota* (wild carrot)	Soil: vicinity of the heavy traffic locations	Mean concentration in plant: • Pt: 14.6 mg kg^{-1} • Pd: 10.2 mg kg^{-1} • Rh: 0.7 mg kg^{-1}	Natural accumulation	Dinh *et al.* (2022c)
Pt, Pd, Rh	*Sinapis alba* (white mustard)	Nutrient solution: 1.0 mg L^{-1} of Pt, Pd, Rh	Concentration of Pt, Pd, Rh respectively in: • Aboveground parts: 95.8; 30.6; 145 mg kg^{-1} • Roots: 5973; 1958; 74.2 mg kg^{-1}	Artificial accumulation	Kowalska *et al.* (2022)
Rh	*Phragmites australis* (common reed)	River side affected by massive urbanization	Rh concentration in plant: 1.11–1.13 mg kg^{-1}	Natural accumulation	Liu *et al.* (2015)
Rh	*Medicago sativa* (alfalfa)	Assarel–Medet copper mine (Bulgaria) High Re levels soil	Leaves: 3985 ± 25 mg kg^{-1}	Natural accumulation	Tzvetkova *et al.* (2021)
Pd	*Lemna minor* (Duckweed)	2–50 μM Pd medium	Roots: 63.93 μg/g Leaves: 255.95 μg/g	Artificial accumulation	Jmii and Dewez (2021)

plant growth. This inhibition was attributed to the presence of oxidative stress in the cytosol and organelles.

Not only terrestrial plants, but also aquatic plants have the capability to bioaccumulate PGMs. Empirical investigations conducted with water hyacinth showed that the concentration of PGMs increases in correspondence to the augmented levels of these metals in the aqueous medium. Similar to other flora, the toxicity of PGMs largely depends upon the metal type. The degree of toxicity rises in the subsequent sequence: Pt(II), Pd(II) > Ru(III) ~ Ru(II) ~ Ir(III) > Pt(IV) ~ Os (IV) >> Rh(III) (Kińska *et al.*, 2019).

6.5.5.2 Rhodium

Research on the existence of rhodium in plants is limited, possibly due to its rarity. Nevertheless, it has been noted that rhodium demonstrates the greatest capacity for dissolution among the PGMs during investigations on soil absorption. This emphasizes the potential of this specific element within the realm of phytomining. The findings derived from the investigation conducted by Tankari Dan-Badjo *et al.* (2008) on various plants, namely watercress, spinach, and nettle, cultivated in the soil obtained from the vicinity of a highway, demonstrate that palladium exhibits the highest degree of bioavailability, subsequently followed by rhodium and platinum. By employing the bioaccumulation coefficient, which represents the ratio between the concentration of an element in plant matter and the concentration of the same element in the soil in close proximity to the plants, it can be concluded that the PGMs may fall in the category of poorly to moderately available to plants, like nickel.

6.6 CONCLUSIONS

The release of PGMs into the environment has the potential to cause unprecedented negative effects on ecosystems. Therefore, in order to make phytoremediation and phytomining of PGMs financially viable, it is necessary to increase their efficiency. The exploitation of phytomining could offer a strong potential for supplementing the supply of primary palladium and other PGMs. From the perspective of resource conservation and environmental protection, the resulting benefits associated with recoverable PGMs would be even greater if the potential limitations of phytoremediation and phytomining were addressed, allowing for large-scale application with lower environmental impacts.

REFERENCES

Akinbile B. J., Makhubela B. C. E. and Ambushe A. A. (2023). Phytomining of valuable metals: status and prospective – a review. *International Journal of Environmental Analytical Chemistry*, **103**(16), 3913–3933.

Ali H., Khan E. and Sajad M. A. (2013). Phytoremediation of heavy metals – concepts and applications. *Chemosphere*, **91**(7), 869–881, https://doi.org/10.1016/j.chemosphere.2013.01.075

Aribal L. and Mercado J. (2020). Phytoremediation of Mangrove species exposed to effluents of mining ultramafic soils.

Ayilara M. S., Adeleke B. S., Adebajo M. T., Akinola S. A., Fayose C. A., Adeyemi U. T., Gbadegesin L. A., Omole R. K., Johnson R. M., Edhemuino M., Ogundolie F. A. and Babalola O. O. (2023). Remediation by enhanced natural attenuation; an environment-friendly remediation approach. *Frontiers in Environmental Science*, **11**, 1182586, https://doi.10.3389/fenvs.2023.1182586

Colmer T. D. (2003). Long-distance transport of gases in plants: a perspective on internal aeration and radial oxygen loss from roots. *Plant, Cell & Environment*, **26**(1), 17–36, https://doi.org/10.1046/j.1365-3040.2003.00846.x

Dhalaria R., Kumar D., Kumar H., Nepovimova E., Kuča K., Torequl Islam M. and Verma R. (2020). Arbuscular mycorrhizal fungi as potential agents in ameliorating heavy metal stress in plants. *Agronomy*, **10**(6), 815, https://doi.org/10.3390/agronomy10060815

Diehl D. B. and Gagnon Z. E. (2007). Interactions between essential nutrients with platinum group metals in submerged aquatic and emergent plants. *Water, Air, and Soil Pollution*, **184**(1–4), 255–267

Dinh T., Dobo Z. and Kovacs H. (2022). Phytomining of noble metals – a review. *Chemosphere*, **286**, 131805, https://doi.org/10.1016/j.chemosphere.2021.131805

Doucette W., Klein H., Chard J., Dupont R., Plaehn W. and Bugbee B. (2013). Volatilization of trichloroethylene from trees and soil: measurement and scaling approaches. *Environmental Science & Technology*, **47**(11), 5813–5820, https://doi.org/10.1021/es304115c

Edgar V.-N., Fabián F.-L., Mario P.-C. J. and Ileana V.-R. (2021). Coupling plant biomass derived from phytoremediation of potential toxic-metal-polluted soils to bioenergy production and high-value by-products: A review. *Applied Sciences*, **11**(7), 2982, https://doi.org/10.3390/app11072982

Floreani F., Pavoni E., Gosar M. and Covelli S. (2023). Evasion of gaseous elemental mercury from forest and urban soils contaminated by historical and modern ore roasting processes (Idrija, Slovenia). *Atmosphere*, **14**, 1036, https://doi.org/10.3390/atmos14061036

Gagnon Z. E., Newkirk C. and Hicks S. (2006). Impact of platinum group metals on the environment: a toxicological, genotoxic and analytical chemistry study. *Journal of Environmental Science and Health. Part A, Toxic/Hazardous Substances & Environmental Engineering*, **41**(3), 397–414.

Gardea-Torresdey J. L., Peralta-Videa J. R., de la Rosa G. and Parsons J. G. (2005). Phytoremediation of heavy metals and study of the metal coordination by X-ray absorption spectroscopy. *Coordination Chemistry Reviews*, **249**(17), 1797–1810, https://doi.org/10.1016/j.ccr.2005.01.001

Grilli M. L., Slobozeanu A. E., Larosa C., Paneva D., Yakoumis I. and Cherkezova-Zheleva Z. (2023). Platinum group metals: green recovery from spent auto-catalysts and reuse in new catalysts: a review. *Crystals*, **13**(4), 550, https://doi.org/10.3390/cryst13040550

Guo Y., Zhu C., Gan L., Ng D. and Xia K. (2015). Effects of exogenous gibberellic acid3 on iron and manganese plaque amounts and iron and manganese uptake in rice. *PLOS ONE*, **10**(2), e0118177, https://doi.org/10.1371/journal.pone.0118177

Haider F. U., Liqun C., Coulter J. A., Cheema S. A., Wu J., Zhang R., Wenjun M. and Farooq M. (2021). Cadmium toxicity in plants: impacts and remediation strategies. *Ecotoxicology and Environmental Safety*, **211**, 111887, https://doi.org/10.1016/j.ecoenv.2020.111887

Harumain Z. A. S., Parker H. L., Muñoz García A., Austin M. J., McElroy C. R., Hunt A. J., Clark J. H., Meech J. A., Anderson C. W. N., Ciacci L., Graedel T. E., Bruce N. C. and Rylott E. L. (2017). Toward financially viable phytoextraction and production of plant-based palladium catalysts. *Environmental Science & Technology*, **51**(5), 2992–3000, https://doi.org/10.1021/acs.est.6b04821

Hughes A. E., Haque N., Northey S. A. and Giddey S. (2021). Platinum group metals: a review of resources, production and usage with a focus on catalysts. *Resources*, **10**(9), 93, https://doi.org/10.3390/resources10090093

Jmii S. and Dewez D. (2021). Toxic responses of palladium accumulation in duckweed (*Lemna minor*): determination of biomarkers. *Environmental Toxicology and Chemistry*, **40**(6), 1630–1638, https://doi.org/10.1002/5011

Kafle A., Timilsina A., Gautam A., Adhikari K., Bhattarai A. and Aryal N. (2022). Phytoremediation: mechanisms, plant selection and enhancement by natural and synthetic agents. *Environmental Advances*, **8**, 100203, https://doi.org/10.1016/j.envadv.2022.100203

Kalavrouziotis I. K. and Koukoulakis P. H. (2009). The environmental impact of the platinum group elements (Pt, Pd, Rh) emitted by the automobile catalyst converters. *Water, Air, and Soil Pollution*, **196**(1), 393–402, https://doi.org/10.1007/s11270-008-9786-9

Karlapudi A. P., Venkateswarulu T. C., Tammineedi J., Kanumuri L., Ravuru B. K., Dirisala V. R. and Kodali V. P. (2018). Role of biosurfactants in bioremediation of oil pollution – a review. *Petroleum*, **4**(3), 241–249, https://doi.org/10.1016/j.petlm.2018.03.007

Khan A. U., Khan A. N., Waris A., Ilyas M. and Zamel D. (2022). Phytoremediation of pollutants from wastewater: a concise review *Open Life Sciences*, **17**(1), 488–496, https://doi.org/10.1515/biol-2022-0056

Kińska K., Bierla K., Krasnodębska-Ostręga B., Godin S., Kowalska J., Preud'Homme H., Lobinski R. and Szpunar J. (2019). A chemical speciation insight into the palladium(ii) uptake and metabolism by *Sinapis alba*. Exposure to Pd induces the synthesis of a Pd–histidine complex. *Metallomics*, **11**(9), 1498–1505, https://doi.org/10.1039/c9mt00126c

Kothny E. L. (1979). Palladium in plant ash. *Plant and Soil*, **53**, 547–550, https://doi.org/10.1007/BF02140726

Kowalska J., Biaduń E., Kińska K., Gniadek M. and Krasnodębska-Ostręga B. (2022). Tracking changes in rhodium nanoparticles in the environment, including their mobility and bioavailability in soil. *Science of the Total Environment*, **806**, 151272, https://doi.org/10.1016/j.scitotenv.2021.151272

Krishnasamy S., Lakshmanan R., Ravichandran M., Krishnasamy S., Lakshmanan R. and Ravichandran M. (2021). Phytoremediation of metal and metalloid pollutants from farmland: an in-situ soil conservation. In: Biodegradation Technology of Organic and Inorganic Pollutants, p. 196. https://doi.org/10.5772/intechopen.111559

Linde S. J. L., Franken A. and du Plessis J. L. (2017). Occupational respiratory exposure to platinum group metals: a review and recommendations. *Chemical Research in Toxicology*, **30**(10), 1778–1790, https://doi.org/10.1021/acs.chemrestox.7b00184

Liu Y., Tian F., Liu C. and Zhang L. (2015). Platinum group elements in the precipitation of the dry region of Xinjiang and factors affecting their deposition to land: the case of Changii City, China. *Atmospheric Pollution Research*, **6**(2), 178–183, https://doi.org/10.5094/APR.2015.021

Masinire F., Adenuga D. O., Tichapondwa S. M. and Chirwa E. M. N. (2021). Phytoremediation of Cr(VI) in wastewater using the vetiver grass (*Chrysopogon zizanioides*). *Minerals Engineering*, **172**, 107141, https://doi.org/10.1016/j.mineng.2021.107141

Masinire F., Adenuga D. O., Tichapondwa S. M. and Chirwa E. M. N. (2022). Palladium phytoremediation and phytomining potential of Vetiver grass (*Chrysopogon zizanioides*). *Chemical Engineering Transactions*, **94**, 1429–1434, https://doi.org/10.3303/CET2294238

Morkunas I., Woźniak A., Mai V. C., Rucińska-Sobkowiak R. and Jeandet P. (2018). The role of heavy metals in plant response to biotic stress. *Molecules: A Journal of Synthetic Chemistry and Natural Product Chemistry*, **23**(9), 2320, https://doi.org/10.3390/molecules23092320

Murray A. J., Zhu J., Wood J. and Macaskie L. E. (2017). Biorefining of platinum group metals from model waste solutions into catalytically active bimetallic nanoparticles. *Microbial Biotechnology*, **11**(2), 359–368, https://doi.org/10.1111/1751-7915.13030

Nemutandani T., Dutertre D., Chimuka L., Cukrowska E. and Tutu H. (2006). The potential of *Berkheya coddii* for phytoextraction of nickel, platinum, and palladium contaminated sites. *Toxicological & Environmental Chemistry*, **88**, 175–185, https://doi.org/10.1080/02772240600585842

Omokhagbor Adams G., Tawari Fufeyin P., Eruke Okoro S. and Ehinomen I. (2020). Bioremediation, biostimulation and bioaugmention: a review. *International Journal of Environmental Bioremediation & Biodegradation*, **3**(1), 28–39, https://doi.org/10.12691/ijebb-3-1-5

Pan J., Cao S., Xu G., Rehman M., Li X., Luo D., Wang C., Fang W., Xiao H., Liao C. and Chen P. (2023). Comprehensive analysis reveals the underlying mechanism of arbuscular mycorrhizal fungi in kenaf cadmium stress alleviation. *Chemosphere*, **314**, 137566, https://doi.org/10.1016/j.chemosphere.2022.137566

Pinho S. and Ladeiro B. (2012). Phytotoxicity by lead as heavy metal focus on oxidative stress. *Journal of Botany*, **2012**, e369572, https://doi.org/10.1155/2012/369572

Rajput V., Minkina T., Mazarji M., Shende S., Sushkova S., Mandzhieva S., Burachevskaya M., Chaplygin V., Singh A. and Jatav H. (2020). Accumulation of nanoparticles in the soil-plant systems and their effects on human health. *Annals of Agricultural Sciences*, **65**(2), 137–143, https://doi.org/10.1016/j.aoas.2020.08.001

Rebello S., Nathan V. K., Sindhu R., Binod P., Awasthi M. K. and Pandey A. (2021). Bioengineered microbes for soil health restoration: present status and future. *Bioengineered*, **12**(2), 12839–12853, https://doi.org/10.1080/21655979.2021.2004645

Sadrolvaezin A., Pezhman A., Zare I., Nasab S. Z., Chamani S., Naghizadeh A. and Mostafavi E. (2023). Systemic allergic contact dermatitis to palladium, platinum, and titanium: mechanisms, clinical manifestations, prevalence, and therapeutic approaches. *MedComm*, **4**(6), e386, https://doi.org/10.1002/mco2.386

Savignan L., Faucher S., Chéry P. and Lespes G. (2021). Platinum group elements contamination in soils: review of the current state. *Chemosphere*, **271**, 129517, https://doi.org/10.1016/j.chemosphere.2020.129517

Shi P. (2016). Geochemical assessment of the bioavailability of platinum group metals for phytomining. https://open.library.ubc.ca/soa/cIRcle/collections/ubctheses/24/items/, https://doi.org/10.14288/1.0300268

Simmer R. A. and Schnoor J. L. (2022). Phytoremediation, bioaugmentation, and the plant microbiome. *Environmental Science & Technology*, **56**(23), 16602–16610, https://doi.org/10.1021/acs.est.2c05970

Skuza L., Szućko-Kociuba I., Filip E. and Bożek I. (2022). Natural molecular mechanisms of plant hyperaccumulation and hypertolerance towards heavy metals. *International Journal of Molecular Sciences*, **23**(16), 9335, https://doi.org/10.3390/ijms23169335

Sobroval P., Zehnalek J., Adam V., Beklova M. and Kizek R., (2012). The effects on soil/water/plant/animal systems by platinum group elements. *Central European Journal of Chemistry*, **10**, 1369–1382, https://doi.org/10.2478/s11532-012-0073-7

Souza M. P., Lytle C. M., Mulholland M. M., Otte M. L. and Terry N. (2000). Selenium assimilation and volatilization from dimethylselenoniopropionate by Indian mustard. *Plant Physiology*, **122**(4), 1281–1288, https://doi.org/10.1104/pp.122.4.1281

Suman J., Uhlik O., Viktorova J. and Macek T. (2018). Phytoextraction of heavy metals: a promising tool for clean-up of polluted environment? *Frontiers in Plant Science*, **9**, 1476, https://doi.org/10.3389/fpls.2018.01476

Tankari Dan-Badjo A., Rychen G. and Ducoulombier C. (2008). Pollution maps of grass contamination by platinum group elements and polycyclic aromatic hydrocarbons from road traffic. *Agronomy for Sustainable Development*, **28**, 457–464, https://doi.org/10.1051/agro:2008032

Tzvetkova C., Novo L. A. B., Atanasova-Vladimirova S. and Vassilev T. (2021). On the uptake of rhenium by plants: accumulation and recovery from plant tissue. *Journal of Cleaner Production*, **328**, 129534, https://doi.org/10.1016/j.jclepro.2021.129534

Ullah S., Liu Q., Wang S., Jan A. U., Sharif H. M. A., Ditta A., Wang G. and Cheng H. (2023). Sources, impacts, factors affecting Cr uptake in plants, and mechanisms behind phytoremediation of Cr-contaminated soils. *Science of the Total Environment*, **899**, 165726, https://doi.org/10.1016/j.scitotenv.2023.165726

Vangronsveld J., Herzig R., Weyens N., Boulet J., Adriaensen K., Ruttens A., Thewys T., Vassilev A., Meers E., Nehnevajova E., van der Lelie D. and Mench M. (2009). Phytoremediation of contaminated soils and groundwater: lessons from the field. *Environmental Science and Pollution Research International*, **16**(7), 765–794, https://doi.org/10.1007/s11356-009-0213-6

Vicente J. and Arcas A. (2005). Aqua palladium complexes: synthesis, properties and applications. *Coordination Chemistry Reviews*, **249**(11), 1135–1154, https://doi.org/10.1016/j.ccr.2004.11.005

Yakoumis I., Panou M., Moschovi A. M. and Panias D. (2021). Recovery of platinum group metals from spent automotive catalysts: a review. *Cleaner Engineering and Technology*, **3**, 100112, https://doi.org/10.1016/j.clet.2021.100112

Yan A., Wang Y., Tan S. N., Mohd Yusof M. L., Ghosh S. and Chen Z. (2020). Phytoremediation: a promising approach for revegetation of heavy metal-polluted land. *Frontiers in Plant Science*, **11**, 359, https://doi.org/10.3389/fpls.2020.00359

Part IV

Recycling of platinum group elements

doi: 10.2166/9781789065084_0151

Chapter 7

Recovery and refinement techniques for circularity of platinum group metals

John Kwame Bediako and Eveliina Repo*

Department of Separation Science, School of Engineering Science, Lappeenranta-Lahti University of Technology (LUT), FI-53850 Lappeenranta, Finland
**Corresponding author: John.Bediako@lut.fi*

ABSTRACT

This chapter explores innovative recovery and refinement techniques aimed at enhancing the circularity of the platinum group metals (PGMs). The chapter begins by highlighting the importance and crucial roles that PGMs play across several sectors and the key challenges that face their supply chain. Next, it veers into circular economy (CE), explaining the principles with regards to PGM recovery and refinement, as well as the economic and environmental benefits. It further provides a comprehensive discussion of state-of-the-art technologies, including mechanical, pyrometallurgical, hydrometallurgical, and biohydrometallurgical processes, and emerging green technologies. The discussion entails extraction of PGMs from diverse sources, such as spent automotive catalysts, electrical and electronic wastes, and industrial by-products. Finally, cases of real-world applications highlighting successful implementations of these techniques by PGM refineries are presented. The integration of cutting-edge techniques, such as adsorption, ion exchange, membrane processes, sustainable solvent extraction, and molecular recognition technology into existing processes help to optimize the PGM recovery while minimizing environmental impact. By presenting a holistic perspective on recovery and refinement strategies, this chapter contributes to the ongoing discourse on sustainable resource management and the transition toward CE, emphasizing the pivotal role of PGMs in fostering technological innovation and environmental responsibility.

Keywords: Platinum group metals, recovery and refinement techniques, circularity, supply chain risks, mechanical separation, pyrometallurgy, (bio)hydrometallurgy

7.1 INTRODUCTION

In the era of sustainable resource management, the spotlight has intensified on the platinum group metals (PGMs), a set of six precious elements, that is platinum (Pt), palladium (Pd), rhodium (Rh), ruthenium (Ru), iridium (Ir), and osmium (Os), which are well-known for their catalytic properties and diverse industrial applications (Xia & Ghahreman, 2023; Yakoumis *et al.*, 2021). As the demand for PGMs continues to rise, driven mainly by their pivotal roles in automotive catalysts, electronics, and fuel cells, the imperative to forge a path toward circularity becomes paramount. The challenge lies not only in securing a reliable supply of these finite resources but also in mitigating the environmental impact associated with traditional mining processes (Granados-Fernández *et al.*, 2021). This chapter seeks to explore the intricate landscape of recovery and refinement techniques tailored to usher PGMs into a circular economy (CE) paradigm. With a steadfast commitment to sustainability, the paradigm call for PGM circularity extends beyond the conventional linear approach, aiming to redefine the life cycle of PGMs through innovative strategies that integrate recovery, recycling, and refining into a cohesive and environmentally conscious framework. The chapter is structured to provide a comprehensive understanding of the current state of PGM circularity, beginning with an understanding of the significance of PGMs in contemporary industries. It will traverse the landscape of PGM recycling through secondary sources, unraveling the associated merits and the looming specter of resource scarcity. Against this backdrop, the chapter will embark on a journey through advanced recovery and refinement methods, exploring cutting-edge technologies that harness urban mining, that is, utilizing waste resources such as electrical and electronic wastes (e-wastes) and end-of-life products as reservoirs of PGMs.

This chapter will examine techniques poised to elevate the purity and quality of reclaimed PGMs, ensuring their seamless integration back into manufacturing processes. From mechanical separation through pyrometallurgical methods to innovative hydrometallurgical approaches, it will navigate the diverse avenues that promise both efficiency and environmental responsibility. As it navigates through the intricate web of PGM recovery and refinement, the overarching goal is to provide a roadmap for stakeholders, ranging from researchers and industry professionals to policymakers and environmental advocates toward a more sustainable and circular future for PGMs. By amalgamating the latest research findings with insights from industry practices, this chapter aspires to catalyze a paradigm shift, where the circularity of PGMs is not merely an aspirational goal but an integral component of responsible resource stewardship.

7.1.1 Challenges in the supply chain of PGMs and need for alternative supply sources

The traditional PGM supply chain is complex, involving multiple stages from extraction to refining and end-use applications. Any disruption or inefficiency in one stage can have cascading effects on the entire supply chain. The supply and demand dynamics of PGMs are influenced by various factors, and understanding the challenges in this domain is critical for developing sustainable strategies.

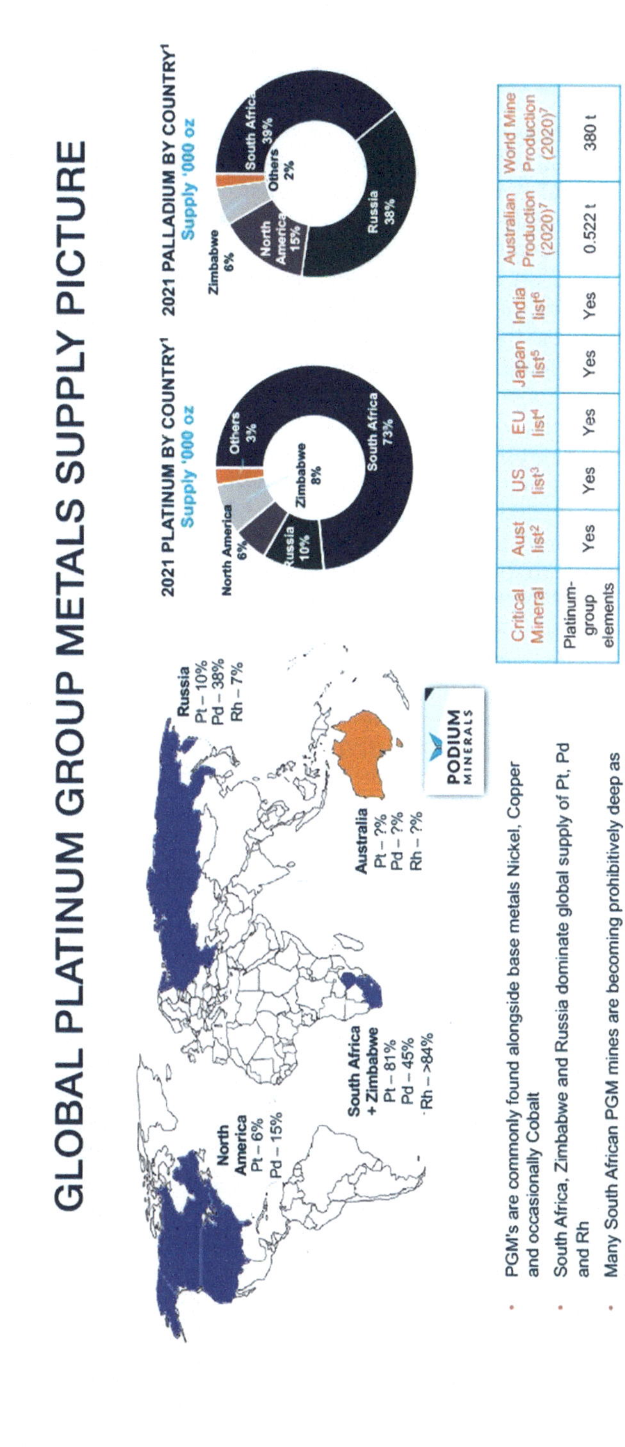

Figure 7.1 Global PGM supply chain depiction (Creamer, 2022).

The PGMs are valued for their unique physical and chemical properties, and they play important roles in various industrial applications. For example, Pt, Pd, and Rh play a pivotal role in catalytic converters, which are essential components for reducing emissions in internal combustion engines (Lillkung & Aromaa, 2012). They are also valuable in the jewellery market and as investment commodities. Being critical components in fuel cells, hydrogen production, sensors, and other green energy technologies is an indication that as the world transitions toward a more sustainable energy landscape, the demand for PGMs in these applications could grow even more significantly. Thus, their scarcity and critical functions make them highly sought after in both industrial and research contexts. However, their supply chain faces several challenges that impact their availability, pricing, and sustainability (Bediako *et al.*, 2023; Løvik *et al.*, 2018).

A pictographic perspective of the global PGM supply chain is depicted in Figure 7.1. In fact, the global reserves of PGMs are approximately 70 000 metric tons, with nearly 90% of it being concentrated in South Africa (Statista.com, Oct 30, 2023; Zhang *et al.*, 2023). Apart from South Africa, other significant contributors include Russia, Zimbabwe, the United States, and Canada. The United States is believed to possess around 900 Mt of Pt metal reserves, although this figure is considerably less compared to the primary reserve holder, South Africa. It is therefore imperative to seek alternative supply sources for PGMs to sustain the many sectors that depend on their supply chain. The alternative sources may include scraps from spent autocatalysts (SACs), waste printed circuit boards (WPCBs), used phone handsets, printers, monitors, and industrial side streams such as spent industrial catalysts (SICs). PGMs can be recovered from these secondary sources through the urban mining process. Urban mining involves the increasing efforts to recover PGMs from end-of-life products, electronic wastes, and industrial by-products through recycling initiatives and it is a key component of circularity of PGMs.

7.1.2 Key challenges facing the PGM supply chain

The PGM supply chain is driven by diverse elements, encompassing economic indicators, political and geopolitical aspects, supply and demand dynamics, investment patterns, as well as environmental and social considerations (Mudd *et al.*, 2018; Xun *et al.*, 2022). Understanding and addressing these challenges is crucial for developing sustainable practices in PGM production, ensuring a stable supply chain, and meeting the increasing demand for these valuable metals in a responsible manner (Xun *et al.*, 2022). The industry's response to these challenges will likely involve a combination of technological innovation, policy adjustments, and a shift toward more circular and sustainable practices. Herein, some causes of PGM supply chain vulnerabilities are discussed.

Geopolitical factors and global pandemics: The supply chain for PGMs faces geopolitical uncertainties. PGMs are considered rare and are geologically scarce, leading to challenges in their extraction and ready availability. Due to the majority of the PGM deposits being concentrated in a few countries (Zhang *et al.*, 2023), geopolitical tensions, trade restrictions, or political instability in these regions can significantly impact the global supply chain (Hurskainen *et al.*, 2025; Xun *et al.*, 2022). This geopolitical vulnerability poses a significant risk to the consistent and reliable supply of PGMs for various industries. Additionally, events like global pandemics or health crises can significantly disrupt the supply chain by impacting mining operations, transportation, and workforce availability, as observed during the COVID-19 pandemic (Gała's *et al.*, 2021; Jowitt, 2020).

Resource scarcity and depletion: PGMs are relatively rare in the Earth's crust (Tang *et al.*, 2023). They are primarily extracted as by-products from nickel and copper mining, and the finite nature of these resources raises concerns about their long-term availability. As market demand for PGMs intensifies, there is a growing risk of resource scarcity and depletion. Sustainable recovery and refinement techniques through recycling, for example from secondary sources such as spent gadgets are essential to mitigate the impact of potential shortages and ensure the longevity of the PGM supply chains (Liu *et al.*, 2024).

Environmental and societal impact of traditional mining: The conventional methods of PGM extraction, such as open-pit and underground mining, often result in direct environmental degradation and significant exposure to multiple fine particles from mining effluent, such as dust, toxic heavy metals, radioactive elements, and chemicals that can cause severe health risks (da Silva-Rêgo *et al.*, 2022). Traditional mining processes can also lead to habitat destruction, soil erosion, and water pollution, especially when done illegally by unregulated small-scale miners. In addition, social issues related to labor conditions and community displacement present challenges for sustainable PGM production. Addressing the environmental impact of PGM mining is crucial for the overall sustainability of the supply chain. CE principles can provide alternative pathways for sourcing PGMs with reduced environmental footprints.

Energy intensity, carbon footprint, and substitution challenges: PGM extraction and refining processes are energy-intensive, contributing to a substantial carbon footprint (Glaister & Mudd, 2010; Yakoumis *et al.*, 2021). As global initiatives emphasize the need for a transition to cleaner energy sources and reduced greenhouse gas emissions, the PGM supply chain confronts the challenge of minimizing its environmental impact. Developing and adopting energy-efficient recovery techniques will be pivotal in aligning the PGM industry with evolving sustainability goals. Besides, while efforts are underway to find or develop alternative materials with properties similar to PGMs, for example catalytic properties and effectiveness (Løvik *et al.*, 2018), finding viable alternatives with similar properties remains a challenge. The EU for instance, is already intensifying efforts by allocating over 30 million euros to substitution or increasing of PGMs in various applications of catalysts in its Horizon 2020 program (Løvik *et al.*, 2018). This quota has significantly increased in the new Horizon Europe program (European Union, 2024). Notwithstanding, further intensification of research and innovation in other jurisdictions may lead to the discovery of more substitutes that can lessen the long-term demand for PGMs.

Technological advancements and process complexity: Rapid advancements in technology may impact the demand for certain PGMs. For example, the rise of electric vehicles could affect the demand for Pd and Rh used in catalytic converters for traditional internal combustion engines. On the other hand, the rise of new technologies, such as fuel cells and green hydrogen production, may increase the demand for PGMs. Therefore, anticipating and addressing the potential surge in demand from emerging technologies is a challenge for the industry (Xun *et al.*, 2022). Besides, the recovery and refinement techniques for PGMs involve complex chemical and metallurgical processes (Chen *et al.*, 2021). The technological challenges associated with these processes include the need for specialized equipment, skilled workforce, and intricate process control. Coupled with the immaturity of most recycling processes, overcoming these complexities is essential for the widespread adoption of circularity practices within the PGM industry.

Economic factors and market volatility: The global trade dynamics and prices of PGMs can be highly volatile due to many factors such as changes in trade policies, imposition of tariffs, geopolitical tensions, and fluctuations in demand from key industries (Burlakovs *et al.*, 2020). For example, economic downturns or financial crises can influence the affordability and accessibility. Export restrictions or trade barriers imposed by producing countries can also affect the supply chain dynamics. For example, Zhang *et al.* (2023) identified how Russia attempted to impose political superiority over neighboring countries by cutting off their natural gas and mineral supply, leveraging on its energy and natural resources advantages to alter the geopolitical landscape. A similar comment can be made regarding the recent trade wars leading to imposition of trade tariffs by the United States and China in particular, which are already reshaping global trade flows and creating significant disruptions in the PGM market of both primary and secondary supply chains. Moreover, the recent military takeovers in some West African countries have witnessed similar fates. Besides, fluctuations in currency exchange rates and economic uncertainties globally can impact both supply and demand. Consequently, this volatility poses challenges for both producers and consumers in planning and decision-making since it influences the cost of production and pricing dynamics. Changes in the market conditions can also impact the economic viability of recovery and refinement activities. Therefore, developing resilient and adaptive strategies to navigate economic uncertainties is important for ensuring the stability of the PGM supply chains.

Scrap recycling constraints: While PGMs are recyclable, the recycling rates are relatively low due to immaturity of the developed processes. In other words, although recycling is a key component of the PGM supply chain, challenges including inadequate collection infrastructure and lack of efficient recycling technologies limit the potential of recycling to meet demand (Chen *et al.*, 2021). Nonetheless, some successes are being chalked; for example, the secondary supplies of Pt and Pd from automobile exhaust catalysts were reported to have increased by 3.7% and 9.91%, respectively in 2019 (Zhang *et al.*, 2023). The challenges here may also be due to insufficient awareness among consumers or a lack of public awareness regarding the importance of recycling as a supplement. Hence, educating the public and consumers in general about the importance of responsible recycling as a sustainable PGM sourcing and circularity would be essential for fostering a collective commitment to mitigating the challenges.

Regulatory and compliance challenges: The growing emphasis on environmental sustainability is impacting the precious metals sector. Stringent environmental regulations and a drive toward more eco-friendly mining practices may limit the supply, resulting in elevated prices (Burlakovs *et al.*, 2020). Other factors may relate to purity demands, labor practices, and ethical sourcing. Compliance with these regulations adds complexity and cost to the recovery and refinement processes. Thus, establishing frameworks for responsible and sustainable PGM sourcing is imperative to meet regulatory standards and build trust within the industry.

Addressing the challenges in the PGM supply chain is pivotal for achieving their circularity. This requires a holistic approach that integrates technological innovation, regulatory frameworks, and stakeholder collaboration to ensure the sustainable recovery and refinement of PGMs, minimizing environmental impact and supporting long-term resource availability.

7.2 CIRCULARITY IN THE CONTEXT OF PGMs

The European Parliament has curated a definition for CE which seeks to prolong the life cycle of products. CE is a model of production and consumption which involves sharing, leasing, reusing, repairing, refurbishing, and recycling existing materials and products as long as possible. The concept is not entirely new; however, it has received global attention in recent years, inspiring environmental advocates, government entities, and businesses. The concept has gained prominence as a sustainable alternative to the traditional linear economic model, that is, take, make, use, and dispose. CE principles prioritize the reduction, reuse, recycling, and recovery of materials, aiming to create closed-loop systems that minimize waste and environmental impact while promoting economic efficiency (Hagelüken & Goldmann, 2022). This section explores the application of CE principles and/or models in the recovery and refinement processes of PGMs, focusing on how these valuable resources can be efficiently and sustainably managed. In the context of PGMs, circularity refers

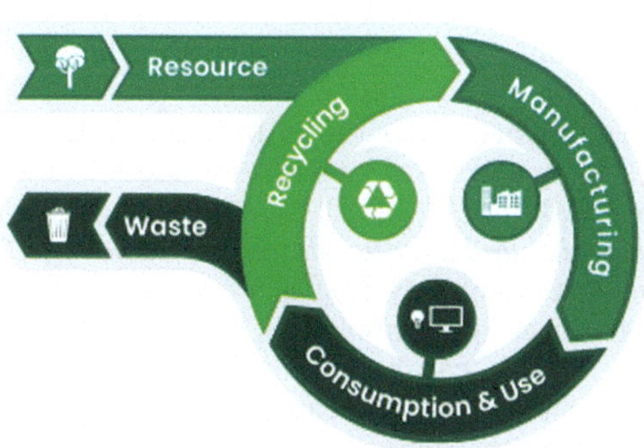

Figure 7.2 CE model: less raw material, less waste, fewer emissions (Source: This image has been designed using assets from Freepik.com).

to the implementation of a sustainable and closed-loop system that maximizes the recovery, recycling, and reuse of PGMs throughout their life cycle. This concept aligns with the general principles of CE. In the CE approach to PGMs, materials are recovered from end-of-life products, industrial processes, and waste/side streams, and then refined and reintegrated into new products or applications. CE practices for PGMs contribute to the conservation of these valuable and often scarce resources, alleviating the environmental impact associated with traditional mining, and fostering a more sustainable and resilient PGM supply chain.

7.2.1 CE principles in PGM recovery and refinement

The CE principles play a crucial role in the recovery and refinement of PGMs. The CE principles broadly include waste management, closed-loop systems, design for recycling, recycling technologies, resource efficiency, and life-cycle assessment (LCA) (Yakoumis *et al.*, 2021). The integration of these principles into PGM recovery and refinement processes represents a critical step toward building a buoyant and eco-friendly supply chain for these precious metals. The CE principles are accompanied by a CE model which promotes use of less raw materials, generation of less wastes and drastically lessening emissions (Figure 7.2).

7.2.2 The CE principles

Waste minimization: Waste minimization is probably the most fundamental principle of CE. It incorporates waste prevention and reduction both in terms of material resource and energy consumption (Mulvaney *et al.*, 2021). The principle persuades resource efficiency and innovation in the production processes of PGMs. Minimizing waste generation during PGM recovery and refinement processes can help to reduce the generation of by-products and waste materials. Advanced separation and recovery techniques such as adsorption, ion exchange and membrane processes are examples of innovative technologies that can be employed to extract and refine PGMs more efficiently, that is to maximize yield, and minimize energy consumption, waste generation and environmental impact (Bediako *et al.*, 2019; Lillkung & Aromaa, 2012; Tanaka *et al.*, 2012). Thus, manufacturers must invest in technologies that minimize waste generation during the extraction and purification of PGMs. It is also important to explore opportunities that utilize by-products generated during the recovery and refining processes, creating additional value, and minimizing environmental perturbations.

Recycling technologies and product design for recycling: This CE principle emphasizes the importance of recycling as a means of extending the life cycle of materials. PGM recycling technologies, such as hydrometallurgical and pyrometallurgical processes, can be optimized to recover PGMs from end-of-life products and industrial waste streams. These methods help to close the material loop and reduce the dependence on primary mining (Reuter & van Schaik, 2016). The principle also advocates for designing products with end-of-life considerations in mind. It promotes the use of PGMs in products that have a longer lifespan, thereby extending the time between the initial production and end of life of the products. For PGM-containing products such as catalytic converters, electronics and jewels for example, manufacturers can adopt design strategies that facilitate easier disassembly and recovery of PGMs at the end of the products' lives. This ensures that valuable materials can be reclaimed more effectively. It is important also to note that materials selection is equally critical in this CE model.

Closed-loop systems: A closed-loop system, also known as a feedback control system, is a type of control system in which the output of the system is monitored and used to adjust or control the input to the system (Sharma, 2011). The primary purpose of a closed-loop system is to maintain or regulate a desired output by continuously comparing it with the reference or desired value and making adjustments as necessary. This feedback loop enables the system to respond to changes and disturbances, ensuring stability and accuracy in its performance. The closed-loop system CE principle thus encourages the continuous recycling and reuse of materials. In PGM refineries for instance, adopting closed-loop processes can ensure that recovered PGMs are reintegrated into the manufacturing processes, reducing the need for new primary resources, and minimizing potential environmental effects (Reuter & van Schaik, 2016; Saidani *et al.*, 2019). In addition, feedback loops between PGM users, recyclers, and refiners may be implemented to optimize the processing, improve recovery rates, and enhance the overall system efficiency.

Life-Cycle Assessment: LCA is a comprehensive methodology used to evaluate the environmental impact of a product, process, or service throughout its entire life cycle (Finnveden & Potting, 2014; Muralikrishna & Manickam, 2017). The life cycle comprises several stages, from raw material extraction and processing to manufacturing, transportation, use, and end-of-life disposal or recycling. It provides a systematic framework to assess the environmental burdens and benefits associated with each stage, allowing for informed decision-making and sustainable practices. In relation to PGMs, conducting comprehensive LCAs helps to evaluate the environmental impact of the PGM recovery and refinement processes. By identifying hotspots and optimizing processes with lower environmental footprints, refineries can align their operations with the general CE goals, striving for sustainability throughout the entire life cycle of PGM-containing products.

While the adoption of CE principles in PGM recovery and refinement presents numerous opportunities for sustainability, it must be acknowledged that there exist challenges. These challenges include the need for standardized recycling practices, collaboration across industries, and technological advancements in the recovery and refinement methods. Therefore, there must be stakeholder engagements and industrial collaborations to foster partnerships among mining companies, recyclers, manufacturers, and policymakers to create a cohesive CE ecosystem for PGMs. There is also the need for consumer awareness creation by educating consumers about the importance of recycling PGM-containing products and the environmental benefits associated with CE practices. Furthermore, to ensure regulatory compliance, stakeholders and policymakers must advocate for and comply with regulations that promote responsible PGM recovery and recycling practices, including incentive programs that support and encourage businesses to adopt CE practices in PGM-related industries.

7.2.3 Economic benefits of adopting CE models in the PGM industry

The PGM industry can experience significant economic boost through the adoption of CE practices. Examples of potential economic benefits are resource efficiency, stable supply chain, job creation and cost savings. The establishment and expansion of recycling facilities for PGM-bearing products can create new job opportunities in the recycling and recovery sector. Implementing circularity in the PGM industry can help to recover the PGMs from end-of-life products, thereby leading to cost savings and reducing the need for primary extraction (Hagelüken & Goldmann, 2022; Mulvaney *et al.*, 2021). Besides, designing products with a focus on longevity and ease of disassembly can enhance the recovery of PGMs, minimizing the loss of these valuable materials during product life cycles. Integrating recycled PGMs into the supply chain can reduce the over reliance on primary mining sources, which may face geopolitical or economic uncertainties. This diversification contributes to more stable and resilient supply chains. CE practices can also lead to the development of new business models, such as product-as-a-service or leasing arrangements, where manufacturers retain ownership of products and are incentivized to recover and recycle PGMs at the end of the product life cycle. Investing in research and development of advanced recycling technologies can create new economic opportunities, foster innovation, and position companies at the forefront of sustainable practices. Embracing CE practices enhances a company's sustainability credentials, which can be a valuable asset in attracting environmentally conscious consumers and investors. This positive reputation can lead to increased customer loyalty and market competitiveness. Finally, proactively adopting CE practices can position PGM industry players to meet evolving environmental regulations and standards. This foresight can prevent costly retrofits and compliance issues in the future.

7.2.4 Environmental benefits of adopting CE models in the PGM industry

Like the economic benefits, CE practices in the PGM industry can lead to several environmental gains including minimization of waste generation, resource conservation, reduced environmental impact, energy savings, reduced carbon footprint, improved air quality, and overall sustainability (Hagelüken & Goldmann, 2022; Saidani *et al.*, 2019). These practices align with broader environmental goals and contribute to the responsible and ethical management of precious mineral resources. Mining operations for PGMs can have significant environmental consequences, including habitat disruption, soil erosion, and water pollution. CE practices in the PGM industry lessen the demand for primary raw materials by promoting the recycling and reuse of PGM-containing products. This, in turn, reduces the need for additional mining activities and helps to ameliorate the corresponding negative environmental effects associated with the conventional mining processes (Yakoumis *et al.*,

2021). Circularity within the PGM industry also emphasizes the importance of extending the lifespan of products and materials bearing PGMs, and the closed-loop systems where materials are continuously circulated through various stages of production, use, and recovery (Saidani *et al.*, 2019). This means designing products with longer lifespans and processes that facilitate easy recovery and recycling of PGMs. By doing so, the need for new PGM extraction from mines is reduced, thus conserving natural resources, minimizing the environmental pollution, and creating a more sustainable and environmental-friendly industry ecosystem. Furthermore, extracting, and refining PGMs from ores is an energy-intensive process and thus recycling PGMs from end-of-life products requires less energy compared to the primary extraction. CE practices that prioritize recycling contribute to energy savings and a lower carbon footprint in the overall production of PGMs. Since PGMs are helpful in reducing harmful emissions from vehicles, encouraging the recycling of used catalytic converters is a huge boost to the ongoing efforts to reduce emissions and contribute to air quality improvement and the fight against climate change.

7.3 OVERVIEW OF PGM RECOVERY AND REFINEMENT TECHNIQUES

Recovery and refinement techniques play a critical role in the retrieval of PGMs from various sources, including primary ores (primary production), industrial wastes, and electronic scraps (secondary production) (Grilli *et al.*, 2023). Efficient and sustainable recovery and refinement are essential to meet the increasing high-quality demands while minimizing environmental effects. High-quality refinement processes contribute to achieving and maintaining certification standards and providing assurance to end-users regarding the quality and reliability of the PGMs, whilst high-purity PGMs command higher market values and open opportunities for trade in global markets. In this section, some common techniques for retrieving PGMs from secondary sources such as automotive wastes, for example SACs, industrial wastes, for example SICs, and electronic wastes, for example WPCBs, are discussed. The conventional methods comprise mechanical, hydrometallurgical and pyrometallurgical processes (Grilli *et al.*, 2023; Hurskainen *et al.*, 2025). However, newer technologies including biohydrometallurgy and modifications or combinations of the existing methods are also explored (Grilli *et al.*, 2023; Sun *et al.*, 2022; Yakoumis *et al.*, 2021).

Irrespective of the source of PGMs and recovery method chosen, a pretreatment step is often required. The whole PGM recovery process can be generally divided into the following stages: pretreatment, enrichment, extraction (leaching), separation and purification, and reduction to metal products (Grilli *et al.*, 2023; Lillkung & Aromaa, 2012; Sun *et al.*, 2022). The secondary production of PGMs can be achieved through these processes due to their mixed matrix composition. For such feedstock, processes that dissolve both the supports and PGMs are necessary. After receiving the PGM-containing wastes, they are first dismantled from the external containers and then put through a pretreatment process involving crushing, grinding, and sieving (homogenization/sampling) to achieve the desired particle size prior to the separation and purification stages.

7.3.1 Mechanical separation processes

The mechanical separation of PGMs involves the extraction and isolation of these precious metals using physical methods. Mechanical separation can be considered as a pretreatment step in most cases of PGM processing (Bizzo *et al.*, 2014). The steps comprise crushing and grading, screening and classification, gravity separation, flotation, magnetic separation, electrostatic separation, Eddy current separation, jigging, and density separation (Bilesan *et al.*, 2021; Cui & Forssberg, 2003). The initial step in mechanical separation involves breaking down the PGM-containing materials into smaller particles through crushing and grinding. This increases the surface area, facilitating subsequent separation processes such as gravity separation or chemical and biological treatment (Duan *et al.*, 2009). Sieving and classification techniques are employed to separate the crushed particles based on size. This can be crucial in achieving a desired particle size for subsequent processing steps. Besides, gravity-based methods which exploit differences in the density of the ground materials, for example centrifugal concentrators, spiral concentrators, and shaking tables are also used for separation.

Alternatively, the ground samples can be suspended in a dense fluid medium such that PGMs with higher densities can be separated from lower-density materials or by jigging (Cui & Forssberg, 2003). Typically, smaller particle size enhances the recovery efficiency, however, according to some sources the efficiency does not improve further if the particle size goes below 100 μm (Baghalha *et al.*, 2009; Lillkung & Aromaa, 2012). Another mechanical separation method is flotation which relies on the selective attachment of air bubbles to crushed PGM-bearing particles, allowing them to rise to the surface and form a froth (Barnwal & Dhawan, 2020; Estrada-Ruiz *et al.*, 2016). This froth can then be collected and further processed to obtain PGM concentrates. Moreover, the magnetic properties of PGMs can be exploited for separation using magnetic fields, for example high-intensity magnetic separators can be effectively used. In addition, an electrostatic separation method can be adopted using the different electrical conductivity properties of PGMs to extract them from other minerals (Cui & Forssberg, 2003). After mechanical separation, PGM concentrates may undergo further processes like smelting and refining to obtain high purities.

7.3.2 Pyrometallurgical processes

Pyrometallurgical processes are the techniques used to concentrate PGMs by physical and chemical conversion at high temperatures (usually in the range of 1500–1650°C), including metal smelting, chlorination volatilization (carbochlorination), and metal vapor treatment (Benson *et al.*, 2000; Grilli *et al.*, 2023; Sun *et al.*, 2022). These processes involve the application of heat to transform raw PGM materials into metal-rich products. Pyrometallurgy has been a traditional and widely employed primary method for metal extraction due to its efficiency and applicability to various types of ores. It plays a significant role in the metallurgical industry, contributing to the production of a wide range of metals essential for various applications. The pyrometallurgical processes are also applied in the recycling of scraps to retrieve PGMs, as has

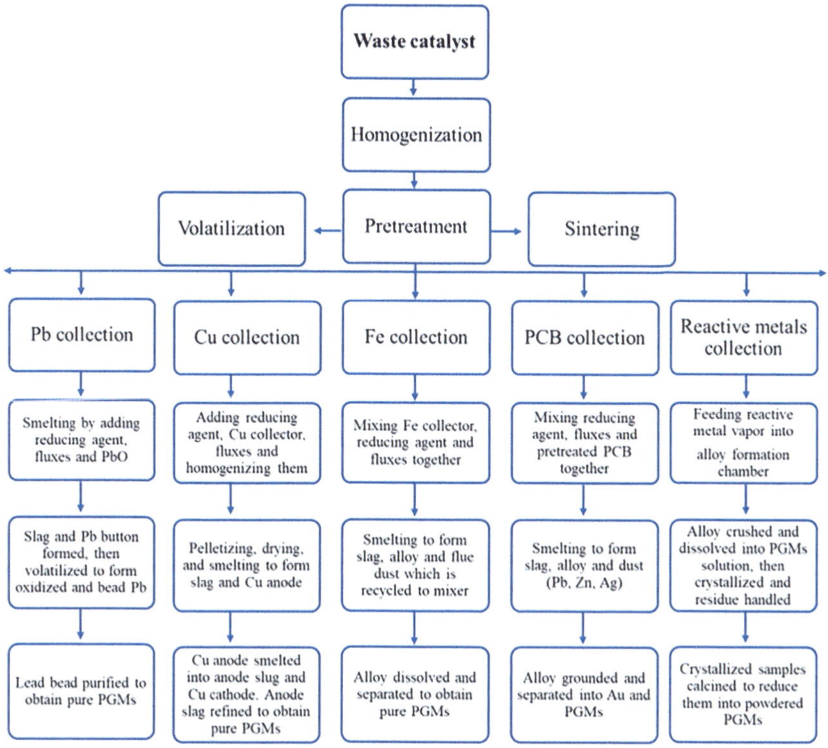

Figure 7.3 Flowsheet illustrating PGMs recovery from spent autocatalysts via pyrometallurgical processes.

been largely used by recycling companies such as Johnson Matthey in the United Kingdom, Sino-Platinum in China, Umicore in Belgium, and Nippon in Japan (Grilli *et al.*, 2023; Sun *et al.*, 2022). The pyrometallurgical processes are energy-intensive due to the high temperatures required for smelting and refining, nonetheless, it remains an important and effective method for large-scale metal production. Some critical aspects of the pyrometallurgical process include pretreatment, smelting, matte formation, conversion and refining, casting, and solidification. Sometimes, the pyrometallurgical process is complemented with hydrometallurgical processes such as solvent extraction, crystallization, and electrochemical refining to obtain high-purity metals (Trinh *et al.*, 2020). Figure 7.3 illustrates a general process flowsheet for the recovery of PGMs from SACs through pyrometallurgical processes.

7.3.2.1 Ore pretreatment and smelting

The first step is preparing the ore (metal-containing waste) for processing, and this may include crushing, grinding, and sometimes beneficiation to concentrate the desired metal contents. The next step is smelting, which is a central stage

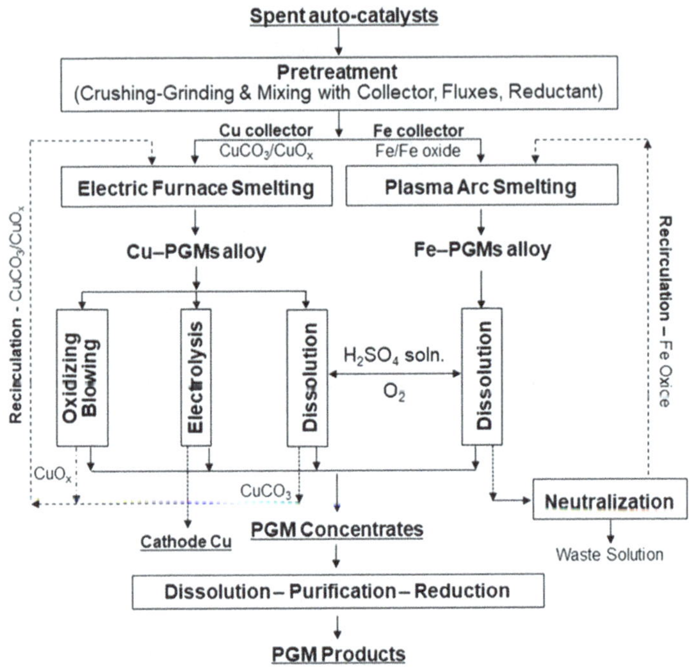

Figure 7.4 Illustration of a smelting process for PGMs recovery from spent autocatalysts. Reproduced from Trinh *et al.* (2020) with permission from Elsevier, © 2020.

in the pyrometallurgical process where the concentrated metal ore is subjected to high temperatures in a furnace, for example electric furnace, plasma arc or induction (Figure 7.4). This heat triggers chemical reactions that result in the reduction of the ore to the metal form. During the smelting, the concentrated ore is mixed with flux, collector and reducing agents so that a PGM-collector alloy is formed (Grilli *et al.*, 2023; Sun *et al.*, 2022). The reducing agents include carbon or carbon monoxide, and the collector substances used are Cu, Pb, Fe, and matte. Furthermore, the flux agents used include cryolite, borax, lime, and soda ash. The most essential parameters in the smelting process are the choice of collector and flux agent, which depend on factors such as the melting points, chemical reactivity, crystal structure, density, and mutual solubility of metals (Benson *et al.*, 2000; Grilli *et al.*, 2023; Sun *et al.*, 2022). The flux agents play the role of dissolving the impurities in the alloy phase to generate a slag that is easily separable from the metallic alloy phase due to density differences (Grilli *et al.*, 2023).

7.3.2.2 Matte collection

In more recent developments, the relatively new Matte collection method which includes Cu, Ni, and Fe matte collection, has been introduced (Sun *et al.*, 2022).

The Matte is a eutectic formed by various heavy non-ferrous metal sulfides with crystal structures and lattice radii similar to those of PGMs, and hence has a good affinity for PGMs. In this method, the smelting process produces an intermediate molten mixture known as matte, which contains the desired metal along with impurities. For instance, Ni and S collectors and Na_2CO_3 (sodium carbonate) and borax fluxes were co-melted with SAC at 1050°C for 30 min, which yielded recovery efficiencies of 90%, 93%, and 88%, for Pt, Pd, and Rh, respectively (You *et al.*, 2016). For WPCBs, roasting may first be done at 600°C for 30–90 min to remove plastics and organic components using a Cu-Sn alloy collector, followed by addition of the fluxes and reducing agents into the roasted mixture and then melting at 1250°C–1400°C for 60 min (Sun *et al.*, 2022). This can also yield a PGM recovery efficiency of over 98%. After smelting, additional steps may be required to convert the matte into a purer form of metal. This can involve processes like converting sulfides to oxides or separating different metals through techniques such as oxidation and reduction reactions. Once the desired metal has been obtained, it is cast into a suitable form for further processing or use. This may involve pouring the molten metal into molds to create ingots or other shapes.

7.3.2.3 Chlorination volatilization or carbochlorination

The chlorination volatilization or carbochlorination method involves the volatilization of PGMs at high temperatures to form gaseous chlorine salts, followed by condensation at lower temperatures to allow separation from the carrier and achieve the desired concentration of PGMs (Grilli *et al.*, 2023; Sun *et al.*, 2022). The process begins with the collection of PGM-bearing materials, such as ores or concentrates containing PGMs. The ore is finely ground to increase its surface area and expose the PGM particles for better reaction with chlorine. The reaction typically involves the formation of volatile chloro-complexes of PGMs, such as chloroplatinate and chloropalladate. The volatile chloro-complexes are separated from the solid residue through gas–solid separation techniques, such as distillation or absorption. This step is critical in capturing the volatile PGM chloro-complexes for further processing. The separated samples are then dissolved in a suitable solvent to form a solution containing PGM chloro-complexes, which may undergo further purification steps to remove impurities. The final step involves the recovery of individual PGMs from the purified solution. This can be achieved through various methods, such as precipitation, solvent extraction, or reduction processes. Using Pt, Pd, and Rh as representatives, the reactions that transform PGMs to chlorine salts can be described by the following equations (Grilli *et al.*, 2023; Sun *et al.*, 2022):

$$Pt + Cl_2 \rightarrow PtCl_2 \tag{7.1}$$

$$Pd + Cl_2 \rightarrow PdCl_2 \tag{7.2}$$

$$\tfrac{2}{3}Rh + Cl_2 \rightarrow \tfrac{2}{3}RhCl_3 \tag{7.3}$$

7.3.2.4 Metal vapor treatment

The metal vapor treatment process also involves the reaction of PGMs with active metal vapors, such as Ca, Mg, and Zn, at high temperatures to form an active metal-PGM alloy, which is more easily dissolvable in acidic solutions, for example aqua regia and HCl compared to the pure PGMs. The Zn vapor is much easier to handle and may react more selectively with PGMs compared to the Ca and Mg vapors (Grilli *et al.*, 2023; Sun *et al.*, 2022). In this process, the molten metal mixtures are subjected to a controlled environment where specific metals are selectively vaporized. This is often achieved by creating a high-temperature and low-pressure environment. The vaporized metal species undergo gas-phase reactions to form volatile metal compounds and are then cooled and condensed back into a solid or liquid state. This phase change allows for separation of the PGMs from the other non-volatile metals and presents the PGMs in a more concentrated and purified form. Additional refining processes may also be applied to achieve the final desired purity levels. It is important to note that specific details of the metal vapor treatment process may vary based on the type of ore, the composition of the metal mixture, and the desired end product.

7.3.2.5 Plasma refining

Plasma refining utilizes plasma technology to extract and refine PGMs. Plasma is the fourth state of matter, consisting of ionized gases with a high temperature (Taylor & Pirzada, 1993). In PGM refinement, plasma is often created by subjecting a gas to a high-energy source, such as radiofrequency or microwaves. This results in the generation of a highly reactive environment capable of breaking chemical bonds. Different plasma methods include plasma arc furnace, DC plasma torch, plasma arc melting furnace, rotary arc furnace, radio frequency (RF) thermal plasma reactor and non-transferred plasma torch (Changming *et al.*, 2018). The plasma arc melting is a key component of the plasma refining process. In this step, the PGM-containing materials are exposed to a high-temperature plasma arc. The intense heat generated in the plasma arc effectively melts the materials, allowing for the separation of the PGMs. The high temperatures achieved in the plasma arc enable selective melting of the PGMs, separating them from the less valuable components (Changming *et al.*, 2018). Since PGMs have unique melting points and properties, the process can be tailored to selectively melt and recover these precious metals. Plasma refining also facilitates gas-phase reactions that aid in the purification of PGMs. Volatile impurities and undesired elements can be driven off as gases, leaving behind purified PGMs. The technique offers advantages in terms of efficiency and reduced environmental impact compared to smelting (Changming *et al.*, 2018). The high temperatures and controlled environment provided by plasma technology allow for the recovery of a higher percentage of PGMs with lower energy consumption and fewer by-products. While plasma refining of PGMs represents an innovative and promising approach, it is important to note that the technology is still evolving, and its widespread adoption depends on factors such as cost-effectiveness, scalability, and continuous technological advancements in the field.

7.3.2.6 Zone refining

Zone refining is commonly used in the purification of PGMs and other precious minerals. Two types of configurations exist, that is the horizontal and vertical orientations. The vertical configuration presents clear benefits of providing improved manipulation of the molten zone and solid–liquid interface when compared with the conventional horizontal configuration (Wen *et al.*, 2023). This ultimately results in a higher efficiency in separating the target metal from the impurities. The zone refining process is based on the principle that impurities tend to concentrate in a molten zone, which is then slowly moved through the material. As the molten zone progresses, impurities are preferentially partitioned into the liquid phase, leading to a higher concentration of impurities in the advancing melt. The final refining efficiency is significantly influenced by the distribution coefficient of each individual impurity. A fundamental principle is that as the deviation of the distribution coefficient (k_x) deviates from unity, impurity x demonstrates a greater inclination to segregate from the metallic substrate (Zhang *et al.*, 2020a). The process begins with a cylindrical rod or ingot of the PGM, often containing various impurities mainly made up of base metals. A narrow molten zone is created along the length of the rod using an induction heater or an electron beam. The molten zone is then moved along the rod at a controlled rate. As the molten zone traverses the rod, impurities migrate into the liquid phase, leaving behind a more purified solid phase. This molten zone is typically moved in one direction, and the process is repeated several times to achieve the desired level of purity. The final product is a rod or ingot with improved purity compared to the initial material. Zone refining is capable of achieving very high purity levels, often exceeding 99.99% (Zhang *et al.*, 2020b; Wen *et al.*, 2023).

7.3.3 Hydrometallurgical processes

Hydrometallurgical processes refer to a set of techniques employed in the extraction and recovery of PGMs and other metals, such as the coinage metals and rare earth elements from aqueous media. Unlike the pyrometallurgical methods that involve high-temperatures and thus high energy consumption, the hydrometallurgical processes rely on chemical reactions in aqueous solutions, typically water-based solutions. This approach is particularly suitable for PGM precursors with low melting points or those that are not easily reduced by conventional smelting methods. Depending on the types of chemicals involved, hydrometallurgy offers advantages such as selective extraction, the ability to process complex waste materials, and eco-friendliness; however, liquid waste management could pose a challenge (Binnemans & Jones, 2023; Zheng *et al.*, 2021). A typical hydrometallurgical process generally involves pretreatment and leaching stages where the PGMs are simultaneously dissolved from their matrices and complexes are formed, followed by separation of the dissolved PGMs from the leach solution and subsequent purification of PGMs (Grilli *et al.*, 2023; Paiva *et al.*, 2022). Factors that influence the leaching efficiency include concentration of the leaching agents, leaching temperature, leaching time, pH of the leachate, and stirring or agitation speed. Examples of hydrometallurgical processes are solvent extraction, solid-phase extraction, chemical precipitation, and ion exchange. To achieve the high purities often desired, combinations of different hydrometallurgical processes or with other methods are usually employed. The following sections discuss various types of hydrometallurgical processes for the extraction, recovery, and refinement of PGMs from various secondary sources.

7.3.3.1 Liquid-phase extraction techniques

Leaching: Leaching as a hydrometallurgical method of PGM recovery is probably the most familiar process. It is the first step after pretreatment and involves PGM dissolution from solid matrices into a liquid phase in an acidic or basic medium. The waste resources, for example SACs and WPCBs, are often crushed and then subjected to a leaching agent that dissolves the target PGMs. Typically, acidic or alkaline leaching agents such as aqua regia or cyanide are commonly used; however, most refineries also leach the PGMs in HCl using chlorine gas. Glasslined or titanium vessels with heating and cooling jackets are often employed (Lillkung & Aromaa, 2012). After leaching, the resulting metal-containing solution is called a leachate. Nowadays, green chemistry principles are being forcefully championed to minimize environmental impact and energy consumption. As a result, sustainable solvents, eco-friendly reagents, and energy-efficient methods are being integrated into the PGM leaching processes. A typical example is the use of ionic liquids (ILs), which are organic salts that remain in liquid state at relatively low temperatures. ILs have shown promise in effectively leaching PGMs from waste streams either as standalones or combinations (Kamisono *et al.*, 2023; Van den Bossche *et al.*, 2021). Another example is deep eutectic solvents (DESs) which are neoteric solvents with interesting advantages such as low cost, low toxicity, high tunability, convenient preparation methods, higher purity, less corrosivity, and easy biodegradability, particularly many of those generated with ChCl (choline chloride) (Jenkin *et al.*, 2016; Yuan *et al.*, 2022). IL- and DES-based leaching generally offer the advantages of improved selectivity, reduced environmental impact, and the ability to operate at lower temperatures. Microwave-assisted extraction (MAE) using ILs and DES is another example of recent leaching innovation with potential for selectivity (Figure 7.3). The MAE utilizes microwave energy to rapidly heat solvents in contact with PGM samples so as to partition them into the solvent. This process was combined with cloud point extraction to recover 91 (±6)% of Pd, 91 (±5)% of Pt, 85 (±6)% of Rh and 66 (±11)% of Ru (Suoranta *et al.*, 2015).

Solvent extraction: Solvent extraction (SE), also known as liquid–liquid extraction, is a PGM separation technique that usually follows leaching, and is used to selectively extract or separate specific target PGMs from solutions containing mixtures of PGMs and other metals (Ilyas *et al.*, 2021; Wei *et al.*, 2016a). A specific organic solvent is introduced to the leachate, and the target metal selectively transfers from the aqueous phase to the organic phase. The solvent extraction process comprises the extraction, scrubbing, and stripping (elution) steps, and three main types of extractants are usually applied. These extractants, distinguished by their mechanisms of extraction, are neutral (e.g. tri-n-butyl phosphate, ethers, and ketones), cationic (e.g. organophosphorus acids like di(2-ethylhexyl) phosphoric acid and β-diketones) and anionic (e.g. amines and quaternary ammonium salts like Aliquat-336) extractants. The mechanisms of extraction include complex formation and solvation, as well as anion and cation exchanges using strong or weak acids and bases (Bernardis *et al.*, 2005; Lillkung & Aromaa, 2012). After extraction, the scrubbing step is required to get rid of co-extracted metals, and the stripping step is used to separate the extracted metal(s) from the organic phase. Commonly used extractants include organophosphorus and amine compounds. Here also, recent trends show the adoption of ILs and DESs for greener extraction of PGMs either as standalones or immobilization in some support matrices (Mokhodoeva *et al.*, 2023; Wei *et al.*, 2016c). A typical SE process employing Aliquat-336 to selectively separate Pt, Pd, and Au is shown in Figure 7.5. SE has the advantage of being faster and showing higher selectivity compared to other methods, however, it requires more dilute solutions, leading to higher volumes of solution to be handled.

Precipitation and cementation: Chemical precipitation involves the selective separation of PGMs from leached solutions by inducing their conversions from dissolved form to solid precipitates. Known for its simplicity, this conventional method enables simultaneous recovery and separation without requiring a separate stripping process. By adjusting factors such as pH and temperature, or by introducing a specific reagent, the desired metal is induced to form solid particles, which can then be separated from the remaining solution via filtration and centrifugation (Xia & Ghahreman, 2023). A similar process, cementation, utilizes Cu, Fe, Al, or Zn powders to reduce PGMs to zero valence. This method was particularly useful for refining Rh up to nearly 100%, however it works best when the leached solution has no impurities (Aktas, 2011). The advantages of precipitation and cementation are that they are relatively simple and easy to control. However, the precipitates usually have small particle sizes such that separating the solid products from the solution poses difficulties. Another disadvantage is the generation of huge volumes of sludge that require secondary treatment.

Distillation: Distillation is a technique that involves heating a metal mixture to vaporize the more volatile ones, which is then condensed back into a liquid form and collected. This process takes advantage of the differences in the boiling points of the various metals. It is particularly effective for separating PGMs with close boiling points, ensuring appreciable purity in the final

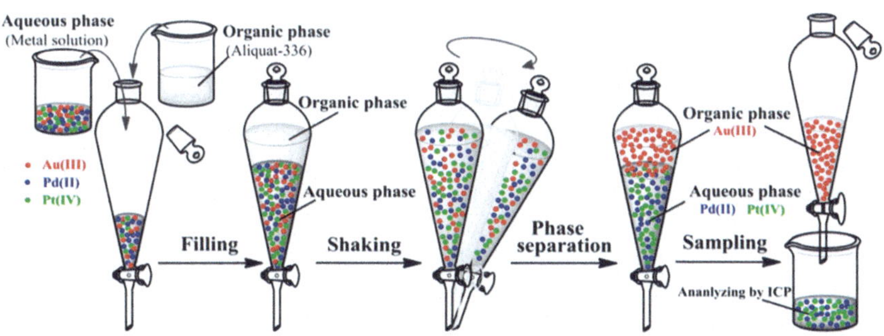

Figure 7.5 Solvent extraction of Pt, Pd and Au using Aliquat-336. Reproduced from Wei *et al.* (2016a) with permission from Elsevier, © 2016.

product. Distillation is often part of a more comprehensive refining process that may include additional steps such as solvent extraction, precipitation, or electrolysis to further purify and separate individual PGMs. A similar process is supercritical fluid (SCF) extraction using solvents such as C_2H_6, CH_3OH, NH_3, CO_2, and H_2O, which transform into the supercritical state above the critical temperature and pressure (Zheng *et al.*, 2021). This extraction method combines the functions and features of SE and distillation. The separation of PGMs through SCF primarily entails the generation of complex anions at the interface of two phases, anion exchange, and the transfer of ion-association complexes. Following interaction of the intricate matrix with SCFs, the PGMs undergo oxidation and complexation at the phase interface. Subsequent ion exchange reactions can take place with the introduction of suitable chelating agents, resulting in products that exhibit increased solubility in supercritical CO_2 (Sc-CO_2). Ultimately, the separation of PGMs is achieved through diffusion and other extraction processes (Zheng *et al.*, 2021).

Crystallization: Crystallization of PGMs involves the formation of ordered structured arrangements of atoms or molecules within the PGM material. The technique exploits the differences in solubility at different temperatures. By controlling the cooling rate of a PGM solution, individual metals can be selectively crystallized and separated from the solution (Kamenetsky *et al.*, 2015; O'Driscoll & González-Jiménez, 2016). The crystallization of PGMs typically occurs through various methods, such as precipitation, zone refining, electrorefining, and crystallization from melt. The PGMs can be crystallized from a molten state by cooling the metal or its alloy, allowing the atoms to arrange themselves into an ordered, crystalline structure (Kramer & van Rosmalen, 2000; Mugavero *et al.*, 2009). This method is commonly used in the production of PGM alloys with specific desired properties. It is also effective for obtaining high-purity PGMs by isolating specific metals through controlled crystallization. The specific crystallization process employed depends on the desired end-use of the PGMs. Crystalline structures are important because they significantly influence the physical, chemical, and mechanical properties of the materials, making them suitable for various applications.

7.3.3.2 Solid-phase extraction techniques

Solid-phase extraction is a widely used technique for the pre-concentration and purification of PGMs from complex matrices. The technique is either used to compliment or replace the liquid-phase extraction methods, particularly SE. It mainly comprises adsorption and ion exchange in batch or chromatography columns, for example batch column chromatography, simulated moving-bed chromatography, continuous annular chromatography, and 'Gatling gun' chromatography (Lillkung & Aromaa, 2012). Solid-phase extraction and SE are based on similar chemical principles. However, they differ in technological aspects: in solid-phase extraction the solid phase plays the role of the organic phase in SE. Thus, the solid-phase extraction process is analogous to SE as it requires both the scrubbing and stripping phases (Bernardis *et al.*, 2005). The first step in solid-phase extraction is choosing a suitable solid-phase material for the PGM extraction. Commonly used precursors include various types of resins, silica-based adsorbents, functionalized polymers, and molecularly imprinted polymers (MIPs). A general solid-phase extraction process using MIPs is shown in Figure 7.6 (Manesiotis *et al.*, 2012). Usually, the choice of the solid phase material depends on the specific PGMs of interest and the characteristics of the mixed matrix. Below, examples of solid-phase extraction techniques are given.

Adsorption technology: Adsorption is generally defined as a process by which dissolved species, e.g., metal ions, adhere to the surface of a solid material. This phenomenon occurs due to attractive forces between the adsorbate and the adsorbent (the material on which adsorption takes place). By utilizing specific adsorbents, target PGMs can be selectively adsorbed and separated from solutions of metal mixtures. The adsorbed PGMs can then be stripped (desorbed) from the adsorbents to obtain concentrated solutions through the use of eluents. It is important to highlight that 'adsorption' must not be mistaken for 'absorption'. Unlike absorption where a substance is taken up and distributed throughout the bulk of another material, adsorption involves a concentration of molecules on the surface of the adsorbent. Adsorption is a relatively low-cost technique because it does not require large amounts of reagents, high-energy inputs, or produces secondary sludge that requires extra treatment as in the case of precipitation and cementation. Moreover, it can achieve high recovery rates and selectivity with specially designed adsorbents at suitable conditions and is feasible even when the concentrations of target metals fall below certain minimum values, for example 100 mg/L (Bediako *et al.*, 2016; Xia & Ghahreman, 2023). Depending on the type of adsorbent and available functional groups, the driving forces for adsorption may include van der Waals forces, electrostatic interactions, and chemical bonding

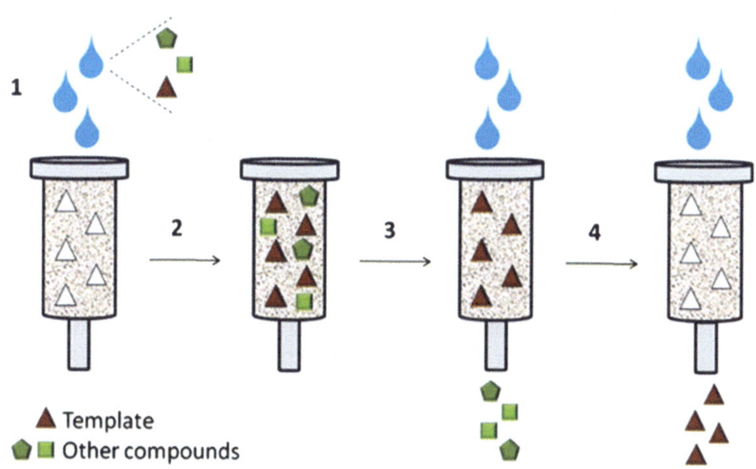

Figure 7.6 Illustration of solid-phase extraction process using molecularly imprinted polymers: 1. loading of sample, 2. washing of column, 3. molecular recognition step, and 4. elution of analytes retained. Reproduced from Manesiotis *et al.* (2012) with permission from Elsevier, © 2012.

between the adsorbate and the adsorbent surface. Adsorption is influenced by factors such as temperature, time, pH, pressure, and the nature of both the adsorbent and adsorbate. With regards to PGMs, efforts are ongoing to develop advanced adsorption materials and processes for their selective recovery from complex solutions (Bediako *et al.*, 2023). These materials, often based on macromaterials, nanomaterials or functionalized surfaces, can selectively capture PGMs from multimetal mixtures, allowing for efficient separation and subsequent refinement.

Ion exchange: Ion exchange involves the replacement of dissolved PGM ions with other ions present on a solid material called ion exchange resin. The ion exchange resins are typically polymers having specific functional groups, such as imidazole, thiourea, mercaptan, sulfide, triisobutylphosphine, or dithiozone (Nikoloski & Ang, 2014; Zheng *et al.*, 2021). The PGM-containing solution comes into contact with the ion exchange resin or similar material with specific affinity for the target metal ions, resulting in the adsorption or attachment of the metal ions onto the solid phase. Each ion exchange resin has a specific number of mobile ion sites, deciding the maximum quantity of exchanges per unit of resin. The different types of ion exchange resins include chelating resins, impregnated resins and impregnated resins with combined fluidized bed (Lillkung & Aromaa, 2012; Tanaka *et al.*, 2012). The advantages of ion exchange are that it enables deployment of uncomplicated equipment, facilitates easy phase separation, and mitigates issues such as reagent losses and phase disengagement, in contrast to SE. Nonetheless, it is important to note that the exchange technology exhibits lower mass transfer rates, necessitates relatively larger equipment, and demands a lengthier process time.

Membrane technology: Membrane separation utilizes membranes with selective permeability to separate PGMs from other components in a solution based on properties such as size, charge, and affinity. Membrane separation is a comparatively better and promising alternative to SE owing to the fact that it consumes less extractant. There exist different membrane types that can be broadly classified into liquid or non-liquid categories.

Liquid membranes employ an organic phase for the retrieval of target metals, whereas non-liquid membranes fix the extractant onto a solid matrix (Xia & Ghahreman, 2023). The permeation process of liquid membranes seamlessly integrates extraction and stripping stages within a singular operation, offering a viable substitute for traditional SE methods. The fundamental concept underlying a liquid membrane process involves minimizing the organic phase to form a film that acts as a barrier between two aqueous phases. These phases can be viewed as analogous to the feed and strip solutions employed in the conventional extraction process. The integration of extraction and stripping stages leads to the ongoing transfer of metal ions through the membrane. In a liquid membrane system, the kinetics dictate the transport of metal ions. The process can be employed for both separation and purification purposes and offers potential for energy-efficient recovery and refinement of PGMs.

The most common form of non-liquid membranes is the polymer inclusion membrane (PIM), which comprises a carrier as the extractant, a polymer base as the matrix and a plasticizer, and exhibits much improved stability (Xia &

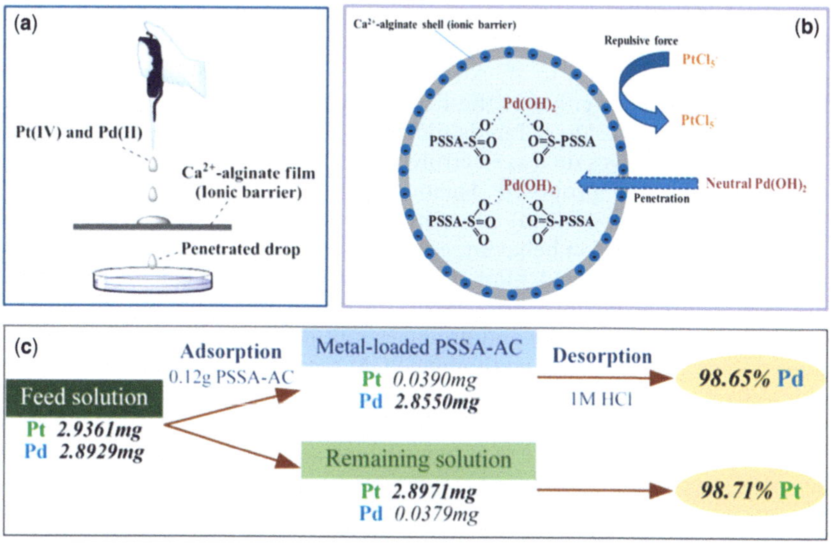

Figure 7.7 Illustration of Pt(IV) and Pd(II) separation using an ionic barrier membrane. (a) PGM penetration through a Ca^{2+}-alginate barrier, (b) depiction of the selective separation mechanism and (c) PGM recovery via selective adsorption–elution routes. Reproduced from Wei *et al.* (2016b) with permission from Elsevier, (c) 2016.

Ghahreman, 2023). Figure 7.7 shows an example of a membrane operation for selectively separating Pt(IV) and Pd(II). In this system, poly(styrenesulfonic acid) was impregnated into a Ca^{2+}-alginate membrane. At pH >3, a neutral $Pd(OH)_2$ readily penetrates the anionic Ca^{2+}-alginate barrier and is adsorbed through chelation reaction with the $-SO_3^-$ (sulfonate) group, whilst the negatively charged $PtCl_5^-$ is rejected by the anionic barrier (Wei *et al.*, 2016b).

Electrochemical method: Electrochemical techniques including electrowinning (electrochemical dissolution) and electrorefining (electrodeposition) are being investigated for their potential in PGM recovery. These methods offer a precise and controlled way to selectively extract PGMs from various matrices, including SACs, SICs, and WPCBs. In the electrowinning step, an electric current is applied to a PGM-containing solution, causing the PGM ions to be reduced and deposited onto an electrode. This electrode becomes coated with a layer of pure PGM. The deposited PGMs may undergo further purification through electrorefining. In this process, the metal-coated electrode is immersed in an electrolyte, and a current is applied. This leads to the dissolution of impurities and the redeposition of the PGMs onto a pure electrode. Electrochemical refining is predominantly useful in the separation of PGMs and Cu as largely applied in many refineries (Sinisalo & Lundström, 2018). For example, the finely ground matte undergoes a series of leaching stages in progressively more oxidizing environments. This process results in a residue containing PGMs along with solutions rich in Ni and Cu. Copper is reclaimed in the form of an electrowon cathode, while Ni is obtained either as a metal powder through hydrogen reduction or as an electrowon cathode. At the Stillwater refinery, a comprehensive Ni-Cu solution is generated and then sent to an external Ni refinery. The residue may undergo additional treatment to enhance the PGM content before it finally moves on to the PGM refinery (Sinisalo & Lundström, 2018).

Recently, alternatives such as electrolytic enrichment of PGMs from Fe-PGMs alloy have also been proposed. Factors considered in this process include effects of potential, base metal concentration, complexing agent, temperature and pH on the dissolution behavior, as well as the effects of C and P impurities contained in the dissolved alloy (Wen *et al.*, 2020). The electrochemical method is relatively cheaper because of less power requirement. The process exclusively utilizes electricity and produces minimal wastewater, making it an environmentally sustainable method for PGM recovery and refinement.

Molecular recognition technology: Recent breakthroughs in separation processes have led to the emergence of the molecular recognition technology (MRT) with a unique attribute of extremely high selectivity in metal separation. The MRT technique employs pre-designed metal-selective ligands, and particularly aims to selectively capture PGMs from complex matrices with high precision and efficiency (Izatt *et al.*, 2016). Molecular imprinting, a technique mimicking the natural recognition processes in biological systems, involves creating polymer matrices with specific binding sites that match the shape and size of PGM ions (Lin *et al.*, 2015). The technique capitalizes on the specific interactions between target PGMs and designed molecular structures. Functionalized polymers offer a versatile platform, where polymer matrices are modified with specific functional groups to enhance PGM binding. The process applies 'lock and key' or 'host guest' chemistry as a basis for its selectivity (Lillkung & Aromaa, 2012). Several molecular recognition strategies have been explored, including the use of selective ligands, functionalized polymers, and molecular imprints (Lin *et al.*, 2015).

Ligand-based approaches involve designing tailored molecules that exhibit a high affinity for PGM ions. The commercial ligand, SuperLig®, has been applied to PGM purification by many recycling companies including Impala Platinum, Tanaka Kikinzoku Kogyo and SepraMet Ltd. (Lillkung & Aromaa, 2012; Zheng *et al.*, 2021). The resin has been successful in selectively separating Ir from Rh and base metals (Figure 7.8), as well as separating Rh from Pd, Pt, Ru, Ir, Ag, Au, and base metals (Figure 7.9) (Izatt *et al.*, 2016; Zheng *et al.*, 2021). The advantages of MRT for PGM recovery lie in its selectivity, efficiency, and potential for target-specific recycling of valuable metals. Despite being promising, the technique faces challenges including optimizing the selectivity and scalability of the process.

7.3.4 Biohydrometallurgical processes

Biohydrometallurgy is a unified term for bioleaching, biomining, biorecovery, and phytomining. It is a comparatively newer emerging technology that integrates hydrometallurgy and electrometallurgy and is considered a sustainable and clean technology for PGM extraction and recovery (Grilli

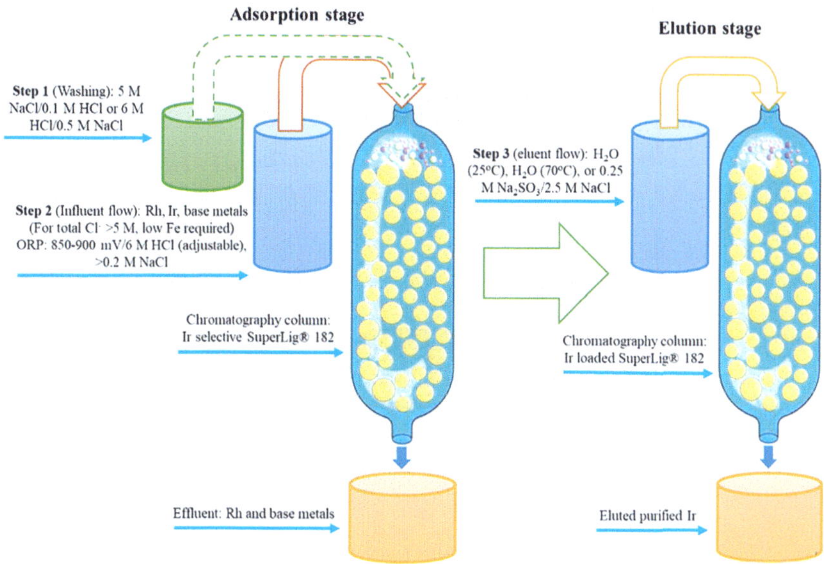

Figure 7.8 Selective separation of Ir from Rh and base metals. ORP = oxidation–reduction potential.

et al., 2023). The process makes use of microorganisms such as bacteria, fungi, archaea, or their combinations, as well as enzymes to transform PGMs into soluble metal ions (Abdalla *et al.*, 2023). These microorganisms have the ability to break down complex precursors and solubilize the PGMs, making them accessible for leaching, recovery, and refinement. This is achieved via proton generation, that is, formation of organic or inorganic acids, by oxidation and reduction reactions or by complexing agents (Hu *et al.*, 2020; Pourhossein *et al.*, 2022).

The advantages of biohydrometallurgy are the use of green solvents, potential selectivity, lower energy requirement, lower carbon footprint, and limited toxic gas emission (Isildar *et al.*, 2019). Nonetheless, controlling key parameters such as leaching temperature, leaching rate, oxygen, and pH can be very perplexing due to the complexity of the solid PGM matrices (Grilli *et al.*, 2023; Yu *et al.*, 2020). Furthermore, biohydrometallurgy is not an end process in itself as post-leaching separation and purification steps such as adsorption and ion exchange are required to refine the PGMs. More distinctively, bioleaching, biomining or biorecovery involve the use of microorganisms or biomolecules to recover PGMs from solutions or solid matrices; however, when certain plants are used, the process is termed as phytomining (Akinbile *et al.*, 2023). Though still in the early stages of development, biohydrometallurgy shows positive prospects as it offers environmental benignity and cost-effective alternatives that reduce the ecological footprint of traditional mining activities.

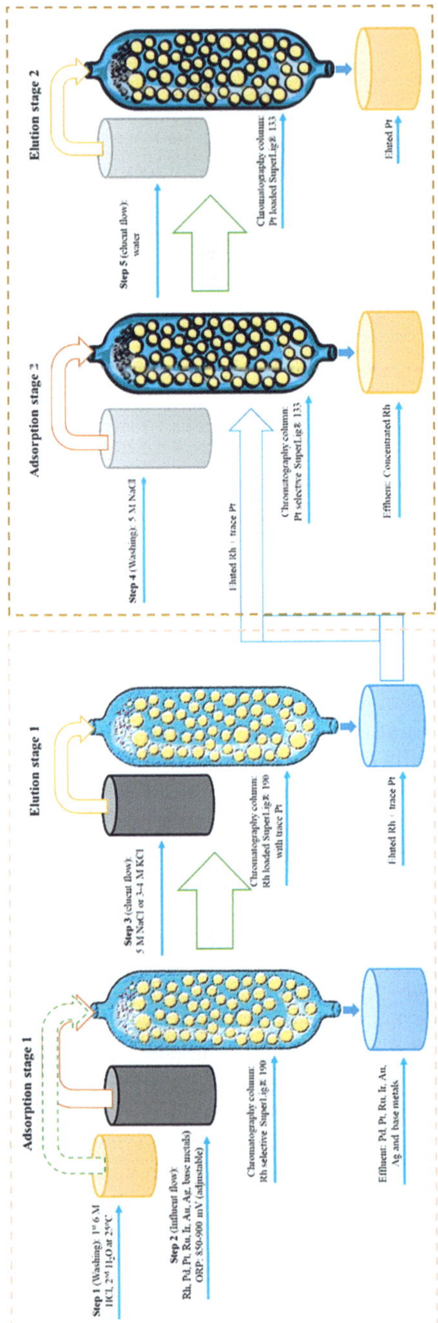

Figure 7.9 Selective separation of Rh from Pd, Pt, Ru, Ir, Ag, Au, and base metals. ORP = oxidation–reduction potential.

7.4 CASES OF RECYCLING AND CE IMPLEMENTATION FOR PGMs

Many famous PGM refining companies have been implementing comprehensive recycling programs that incorporate CE principles to recover PGMs from end-of-life products, such as SACs, SICs, and WPCBs (Zheng *et al.*, 2021). These companies, including Johnson Matthey, Anglo American Platinum, Umicore, Impala, BASF, and Krastsvelmet refineries, have been exploring CE practices to enhance sustainability and reduce environmental impact of these waste materials (Yakoumis *et al.*, 2021). These are achieved through recycling and recovery programs such as urban mining initiatives and closed-loop supply chains. The main techniques mostly employed by these companies for the production of PGMs are the conventional pyrometallurgy, precipitation, solvent extraction, and ion exchange processes (Lillkung & Aromaa, 2012; Zheng *et al.*, 2021). A few of the refinery companies are hereby discussed, and a summary of the waste types and processes used are presented in Table 7.1.

7.4.1 Johnson Matthey's PGM refinery

Johnson Matthey is the world's largest PGMs refiner and mainly engages in the recovery and refinement of Pt, Pd, and Rh from recycled materials. The company uses advanced technologies, mainly pyrometallurgical and hydrometallurgical processes to extract and purify PGMs with high efficiency. The recycling process involves collecting used resources, disassembling them, and then extracting and refining the PGMs for reuse in new applications, for example catalytic converters. In other words, there is first pretreatment, followed by smelting in a furnace at 1200°C for 12 h to separate the PGMs from the non-metallic components. The recovered metals are then passed through further refinement processes, while the non-metallic materials are transferred into the slag after smelting (Xia & Ghahreman, 2023). Johnson Matthey's refinery contributes to a more sustainable PGM supply chain and provides high-quality recycled PGMs for various applications.

7.4.2 Anglo American Platinum's precious metals refinery

Anglo American Platinum, another major player in the PGM industry, has also been implementing CE practices at its Precious Metals Refinery (PMR) in Rustenburg, South Africa. The PMR employs advanced recycling and recovery techniques, including SE and selective precipitation to recover PGMs from primary and secondary sources, including end-of-life products and industrial waste. The main features of their processes are similar to those of the Johnson Matthey's refinery (Lillkung & Aromaa, 2012). The feed (concentrate) usually comprises 50–60% PGMs and is primarily composed of leached magnetic concentrate obtained from the magnetic concentrator plant. This feed is dissolved in HCl using Cl_2 gas at 120°C and a pressure of 4 bar, and all PGMs and base metals undergo dissolution at this stage. In this step also, Os undergoes oxidation to tetroxide, and it is subsequently reclaimed from the offgas through the use of KOH. The gravity concentrates obtained from the mill cyclones of the company's ore concentrators are fed directly into the PMR which uses a separate circuit for the leaching. The recovered metals are refined

Table 7.1 Highlights of companies engaged in PGM recycling (Grilli *et al.*, 2023; Zheng *et al.*, 2021).

Company	Waste Type	Process
Johnson Matthey	Spent process catalysts, autocatalysts, fuel cell catalysts, secondary mine residues and jewellery or scrap metal	Pyrometallurgy and advanced hydrometallurgical processing
Umicore	Large variety of complex, PGM-containing materials, for example e-waste, SACs, and SICs	Smelter, copper leaching and electrowinning plant and precious metals refinery
Anglo	End-of-life products, PGM-bearing materials, for example e-waste, SACs, and SICs	Solvent extraction with b-hydroxyl oxime or trinoctylamine, ion exchange, followed by precipitation with diethylene triammine
Impala	Converter matte and other recycled materials, for example e-waste and SICs	Ion exchange with SuperLig®, alkaline leaching and precipitation with $NH_3{\cdot}H_2O$, NH_4Cl or organic ammine
BASF	Mainly spent autocatalysts, for example SACs	Process of smelting and refining
Multimetco	Recycled materials including SICs	Process of smelting and refining
Hensel recycling	Catalytic converters, e-scrap or other materials containing precious metals	Smelting
Lonmin	Primary concentrates and recycled waste materials	Precipitation with CH_3COONH_4, NH_4Cl or as $Rh(NH_3)_5Cl_3$ in the case of Rh
Krastsvetmet	Primary concentrates, material rich in insoluble PGMs or of low grade	Precipitation with NH_4Cl or as $Na_3Rh(NO_2)_6$ in the case of Rh
Acton refinery	Primary concentrates, anode slimes and secondary materials, for example e-wastes and SACs	Smelting and acid leaching
SepraMet Ltd	Low-grade SICs, high-grade materials, and non-hazardous catalyst materials	Hydrometallurgical and MRT process
Tanaka Kikinzoku Kogyo	Catalysts that contain Pt, Rh and base metals	MRT process using SuperLig 190–medium in column

Note: SAC–spent autocatalysts, SIC–spent industrial catalysts, e-waste–electrical and electronic waste, MRT–molecular recognition technology.

to high purity, contributing to closed-loop systems in various industries. This CE approach is crucial for reducing the environmental impact associated with traditional mining, but also enhances the sustainability of the PGM supply chain. The Rustenburg PMR engages in other CE activities such as supporting waste recycling initiatives in the nearby communities.

7.4.3 Umicore's PGM recycling facility

The Umicore Group is a global materials technology and recycling company located at Hoboken near Antwerp (Belgium), and with industrial operations on all continents and serves a global customer base. The company has a state-of-the-art PGM recycling facility which focuses on the recycling of PGMs from spent catalysts, electronic wastes, industrial residues and by-products from other non-ferrous industries (e.g., drosses, mattes, speiss, and anode slimes) (Hagelüken, 2006). Umicore also utilizes a combination of pyrometallurgical and hydrometallurgical processes to recover and refine PGMs efficiently (Xia & Ghahreman, 2023). The company operates one of the largest precious metals recycling facilities in the world having an annual capacity of more than 50 tons PGMs, 100 tons Au, and 2400 tons Ag (Hagelüken, 2006). Its recently granted patent with publication number: US11879164B2, unveils a process for recovering PGMs from catalysts containing mixtures of silicon carbide and PGMs (Power-technology.com, May 6, 2024). The company is thus closing the loop in multiple industries, providing a sustainable source of PGMs for applications in automotive catalysis, electronics, and renewable energy technologies.

7.4.4 Impala Platinum's ion exchange/MRT process

Impala Platinum is another famous refinery that specializes in the circularity of PGM and other precious metals such as Au and Ag. The company focuses on upgrading and modernizing its processes to improve PGM recovery from mined ores and recycled materials. It also incorporates energy-efficient technologies and waste heat recovery systems. It mainly adopts an MRT-combined ion exchange process using SuperLig®, and in some cases precipitation, for example HCl is added to a dissolved concentrate to precipitate $AgCl_2$ (Lillkung & Aromaa, 2012). The feed material to the plant is a converter matte that contains approximately 65% of PGMs, which is initially dissolved using HCl and Cl_2. The dissolved residue is subsequently leached in NH_4OH, and then Pd is recovered in an ion exchange step that uses MRT. Generally, less than 1% of Pt is co-extracted with the Pd, and the ion exchange raffinate also contains <1 mg/L Pd (Lillkung & Aromaa, 2012). The implementation of CE practices at the Impala Platinum refinery has increased PGM recovery rates, reduced energy consumption, and minimized environmental impact, showcasing the economic and environmental benefits of CE models in the PGM industry.

7.4.5 Collaborative initiatives

There is a huge call for collaborative initiatives involving multiple stakeholders, including industry, government, and academia, to raise public awareness and implement local PGM circular value chains to push the overall CE agenda (Yakoumis *et al.*, 2021). Specifically, primary and secondary suppliers of PGM resources, the recycling companies, integrators of exhaust systems, manufacturers of automotive catalytic converters, and automotive companies are encouraged to collaborate in this endeavor, following green chemistry principles and cost-effective solutions within the PGM industry. Instead of the conventional pyrometallurgical processes, innovative hydrometallurgical alternatives ought to be championed. There are already various partnerships and recycling initiatives globally aiming to recover PGMs from scraps through partnerships between PGM producers, manufacturers, and recycling facilities. Successful implementation of these recycling initiatives will significantly contribute to the recovery of weighty amounts of PGMs from recycled wastes, helping to improve their circularity and reduce the overreliance on primary mining sources.

7.5 CONCLUSION

This chapter discussed in detail the critical strategies and methodologies aimed at enhancing the sustainability and circularity of PGMs. The imperative nature of these elements in modern industries, coupled with their scarcity and environmental impact of primary mining, necessitates a concerted effort to boost recovery and refining efforts in order to keep their continued supply to industry while increasing revenue and limiting their adverse environmental impacts. Throughout this chapter, the need to inculcate circularity into the PGM recovery and refinement processes has been highlighted. The pursuit of circularity not only involves efficient recovery but also the refinement of these metals to meet stringent purity standards, ensuring their successful reintroduction into the production cycle. In addition, conventional and emerging innovative technologies, ranging from mechanical, pyrometallurgical, hydrometallurgical, and biohydrometallurgical processes have been covered. These advanced technologies showcase the potential to recover PGMs from various sources, including electronic wastes, industrial residues and spent autocatalysts. Instances of recycling applications in real case PGM refineries have also been presented. The discussion has underscored the importance of interdisciplinary approaches, collaboration between academia and industry, and the incorporation of sustainable practices in refinement processes. As the world navigates the challenges of resource depletion and environmental degradation, the integration of advanced technologies and the adoption of CE and green chemistry principles become paramount for a more sustainable and responsible use of these precious resources. In the broader context of sustainable development, the insights presented in this chapter contribute to the ongoing dialogue on resource efficiency and the transition toward a more circular and environmentally conscious approach in the domain of PGMs. It is hoped that the knowledge shared herein will inspire further research, drive innovation, and urge industry practices that foster resilient CE practices, ensuring the longevity and responsible stewardship of PGMs for generations to come.

REFERENCES

Abdalla A. M., Abdullah M. F., Dawood M. K., Wei B., Subramanian Y., Azad A. T., Nourin S., Afroze S., Taweekun J. and Azad A. K.. (2023). Innovative lithium-ion battery recycling: sustainable process for recovery of critical materials from lithium-ion batteries. *Journal of Energy Storage*, **67**, 107551, https://doi.org/10.1016/j.est.2023.107551

Akinbile B. J., Makhubela B. C. and Ambushe A. A. (2023). Phytomining of valuable metals: status and prospective–a review. *International Journal of Environmental Analytical Chemistry*, **103**(16), 3913–3933, https://doi.org/10.1080/03067319.2021.1917557

Aktas S. (2011). Rhodium recovery from rhodium-containing waste rinsing water via cementation using zinc powder. *Hydrometallurgy*, **106**(1–2), 71–75, https://doi.org/10.1016/j.hydromet.2010.12.005

Baghalha M., Gh H. K. and Mortaheb H. R. (2009). Kinetics of platinum extraction from spent reforming catalysts in aqua-regia solutions. *Hydrometallurgy*, **95**(3–4), 247–253, https://doi.org/10.1016/j.hydromet.2008.06.003

Barnwal A. and Dhawan N. (2020). Recycling of discarded mobile printed circuit boards for extraction of gold and copper. *Sustainable Materials and Technologies*, **25**, e00164, https://doi.org/10.1016/j.susmat.2020.e00164

Bediako J. K., Kim S., Wei W. and Yun Y. S. (2016). Adsorptive separation of Pb (II) and Cu (II) from aqueous solutions using as-prepared carboxymethylated waste Lyocell fiber. *International Journal of Environmental Science and Technology*, **13**, 875–886, https://doi.org/10.1007/s13762-015-0926-7

Bediako J. K., Park S. W., Choi J. W., Song M. H. and Yun Y. S. (2019). High-performance and acid-tolerant polyethylenimine-aminated polyvinyl chloride fibers: fabrication and application for recovery of platinum from acidic wastewaters. *Journal of Environmental Chemical Engineering*, **7**(1), 102839, https://doi.org/10.1016/j.jece.2018.102839

Bediako J. K., Lim C. R., Repo E., Choi S. H. and Yun Y. S. (2023). Polyelectrolyte complex-derived adsorbent capsules capable of selective recovery of precious metal from multiple mixtures. *Chemical Engineering Science*, **274**, 118688, https://doi.org/10.1016/j.ces.2023.118688

Benson M., Bennett C. R., Harry J. E., Patel M. K. and Cross M. (2000). The recovery mechanism of platinum group metals from catalytic converters in spent automotive exhaust systems. *Resources, Conservation and Recycling*, **31**(1), 1–7, https://doi.org/10.1016/S0921-3449(00)00062-8

Bernardis F. L., Grant R. A. and Sherrington D. C. (2005). A review of methods of separation of the platinum-group metals through their chloro-complexes. *Reactive and Functional Polymers*, **65**(3), 205–217, https://doi.org/10.1016/j.reactfunctpolym.2005.05.011

Bilesan M. R., Makarova I., Wickman B. and Repo E. (2021). Efficient separation of precious metals from computer waste printed circuit boards by hydrocyclone and dilution-gravity methods. *Journal of Cleaner Production*, **286**, 125505, https://doi.org/10.1016/j.jclepro.2020.125505

Binnemans K. and Jones P. T. (2023). The twelve principles of circular hydrometallurgy. *Journal of Sustainable Metallurgy*, **9**, 1–25, https://doi.org/10.1007/s40831-022-00636-3

Bizzo W. A., Figueiredo R. A. and De Andrade V. F. (2014). Characterization of printed circuit boards for metal and energy recovery after milling and mechanical separation. *Materials*, **7**(6), 4555–4566, https://doi.org/10.3390/ma7064555

Burlakovs J., Vincevica-Gaile Z., Krievans M., Jani Y., Horttanainen M., Pehme K. M., Dace E., Setyobudi R. H., Pilecka J., Denafas G., Grinfelde I., Bhatnagar A., Rud V., Rudovica V., Mersky R. L., Anne O., Kriipsalu M., Ozola-Davidane R., Tamm T. and Klavins M. (2020). Platinum group elements in geosphere and anthroposphere: interplay among the global reserves, urban ores, markets and circular economy. *Minerals*, **10**(6), 558, https://doi.org/10.3390/min10060558

Changming D., Chao S., Gon X., Ting W. and Xiange W. (2018). Plasma methods for metals recovery from metal-containing waste. *Waste Management*, **77**, 373–387, https://doi.org/10.1016/j.wasman.2018.04.026

Chen Y., Qiao Q., Cao J., Li H. and Bian Z. (2021). Precious metal recovery. *Joule*, **5**(12), 3097–3115, https://doi.org/10.1016/j.joule.2021.11.002

Creamer M. (2022). Platinum group metals gaining ground in Australia where PGMs have critical status. https://www.miningweekly.com/article/platinum-group-metals-gaining-ground-in-australia-where-pgms-have-critical-mineral-status-2022-10-31 (accessed 19 January 2024)

Cui J. and Forssberg E. (2003). Mechanical recycling of waste electric and electronic equipment: a review. *Journal of Hazardous Materials*, **99**(3), 243–263, https://doi.org/10.1016/S0304-3894(03)00061-X

da Silva-Rêgo L. L., de Almeida L. A. and Gasparotto J. (2022). Toxicological effects of mining hazard elements. *Energy Geoscience*, **3**(3), 255–262, https://doi.org/10.1016/j.engeos.2022.03.003

Duan C., Wen X., Shi C., Zhao Y., Wen B. and He Y. (2009). Recovery of metals from waste printed circuit boards by a mechanical method using a water medium. *Journal of Hazardous Materials*, **166**(1), 478–482, https://doi.org/10.1016/j.jhazmat.2008.11.060

Estrada-Ruiz R. H., Flores-Campos R., Gámez-Altamirano H. A. and Velarde-Sánchez E. J. (2016). Separation of the metallic and non-metallic fraction from printed circuit boards employing green technology. *Journal of Hazardous Materials*, **311**, 91–99, https://doi.org/10.1016/j.jhazmat.2016.02.061

European Union. (2024). CORDIS–EU research results. https://cordis.europa.eu/ (accessed 9 January 2025)

Finnveden G. and Potting J. (2014). Life cycle assessment. Encyclopedia of Toxicology, **3**, 74–77, https://doi.org/10.1016/B978-0-12-386454-3.00627-8

Gałaś A., Kot-Niewiadomska A., Czerw H., Simić V., Tost M., Wårell L. and Gałaś S. (2021). Impact of COVID-19 on the mining sector and raw materials security in selected European countries. *Resources*, **10**(5), 39, https://doi.org/10.3390/resources10050039

Glaister B. J. and Mudd G. M. (2010). The environmental costs of platinum–PGM mining and sustainability: is the glass half-full or half-empty? *Minerals Engineering*, **23**(5), 438–450, https://doi.org/10.1016/j.mineng.2009.12.007

Granados-Fernández R., Montiel M. A., Díaz-Abad S., Rodrigo M. A. and Lobato J. (2021). Platinum recovery techniques for a circular economy. *Catalysts*, **11**(8), 937, https://doi.org/10.3390/catal11080937

Grilli M. L., Slobozeanu A. E., Larosa C., Paneva D., Yakoumis I. and Cherkezova-Zheleva Z. (2023). Platinum group metals: green recovery from spent auto-catalysts and reuse in new catalysts–a review. *Crystals*, **13**(4), 550, https://doi.org/10.3390/cryst13040550

Hagelüken C. (2006). Recycling of electronic scrap at Umicore precious metals refining. *Acta Metallurgica Slovaca*, **12**, 111–120.

Hagelüken C. and Goldmann D. (2022). Recycling and circular economy–towards a closed loop for metals in emerging clean technologies. *Mineral Economics*, **35**(3–4), 539–562, https://doi.org/10.1007/s13563-022-00319-1

Hu W., Feng S., Tong Y., Zhang H. and Yang H. (2020). Adaptive defensive mechanism of bioleaching microorganisms under extremely environmental acid stress: advances and perspectives. *Biotechnology Advances*, **42**, 107580, https://doi.org/10.1016/j.biotechadv.2020.107580

Hurskainen A., Bediako J. K., El Ouardi Y., Lamsayah M., Frimodig J. and Repo E. (2025). Understanding the mechanisms of gold (III) adsorption onto additively manufactured polyamide adsorbent, AM-N1. Chemical Engineering Science, **305**, 121130, https://doi.org/10.1016/j.ces.2024.121130

Ilyas S., Kim H. and Srivastava R. R. (2021). Separation of platinum group metals from model chloride solution using phosphonium-based ionic liquid. *Separation and Purification Technology*, **278**, 119577, https://doi.org/10.1016/j.seppur.2021.119577

Işıldar A., van Hullebusch E. D., Lenz M., Du Laing G., Marra A., Cesaro A., Pandah S., Akcilh A., Kucukeri M. A. and Kuchtag K. (2019). Biotechnological strategies for the recovery of valuable and critical raw materials from waste electrical and electronic equipment (WEEE)–a review. *Journal of Hazardous Materials*, **362**, 467–481, https://doi.org/10.1016/j.jhazmat.2018.08.050

Izatt S. R., McKenzie J. S., Izatt N. E., Bruening R. L., Izatt R. M. and Krakowiak K. E. (2016). Selective recovery of platinum group metals and rare earth metals from complex matrices using a green chemistry/molecular recognition technology approach. In: Metal Sustainability: Global Challenges, Consequences, and Prospects, R. M. Izatt (ed.), John Wiley & Sons, Hoboken, New Jersey, pp. 317–332.

Jenkin G. R., Al-Bassam A. Z., Harris R. C., Abbott A. P., Smith D. J., Holwell D. A., Chapman R. J. and Stanley C. J. (2016). The application of deep eutectic solvent ionic liquids for environmentally-friendly dissolution and recovery of precious metals. *Minerals Engineering*, **87**, 18–24, https://doi.org/10.1016/j.mineng.2015.09.026

Jowitt S. M. (2020). COVID-19 and the global mining industry. SEG Newsletter, **122**, 33–41.

Kamenetsky V. S., Park J. W., Mungall J. E., Pushkarev E. V., Ivanov A. V., Kamenetsky M. B. and Yaxley G. M. (2015). Crystallization of platinum-group minerals from silicate melts: evidence from Cr-spinel-hosted inclusions in volcanic rocks. *Geology*, **43**(10), 903–906, https://doi.org/10.1130/G37052.1

Kamisono M., Hanada T. and Goto M. (2023). Green leaching of metallic platinum using an ionic liquid with synergistic organic acid-chlorinating agent additives. *Journal of Chemical Engineering of Japan*, **56**(1), 2228363, https://doi.org/10.1080/00219592.2023.2228363

Kramer H. J. M. and van Rosmalen G. M. (2000). Crystallization. In: Encyclopedia of Separation Science, I. D. Wilson (ed.), Academic Press, Oxford, pp. 64–84.

Lillkung K. and Aromaa J. (2012). Hydrometallurgical Recovery of Platinum Group Metals. Aalto University Publication Series Science+Technology. Aalto University, Espoo, Helsinki, pp. 1–80.

Lin S., Wei W., Wu X., Zhou T., Mao J. and Yun Y. S. (2015). Selective recovery of Pd (II) from extremely acidic solution using ion-imprinted chitosan fiber: adsorption performance and mechanisms. *Journal of Hazardous Materials*, **299**, 10–17, https://doi.org/10.1016/j.jhazmat.2015.05.050

Liu M., Zhao Y., Tian M., Liu K., Hu G., Zhang J., Xue T., Qi T. and Zhang H. (2024). Extraction of PGMs from insoluble Fe-Si-PGMs alloy by combined thermochemical and mechanochemical leaching: sustainable, efficient, and scalable. *Separation and Purification Technology*, **333**, 125925, https://doi.org/10.1016/j.seppur.2023.125925

Løvik A. N., Hagelüken C. and Wäger P. (2018). Improving supply security of critical metals: current developments and research in the EU. *Sustainable Materials and Technologies*, **15**, 9–18, https://doi.org/10.1016/j.susmat.2018.01.003

Manesiotis P., Fitzhenry L., Theodoridis G. and Jandera P. (2012). Applications of SPE-MIP in the field of food analysis. In: Comprehensive Sampling and Sample Preparation, J. Pawliszyn (ed.), Academic Press, Oxford, pp. 457–471.

Mokhodoeva O., Maksimova V., Shishov A. and Shkinev V. (2023). Separation of platinum group metals using deep eutectic solvents based on quaternary ammonium salts. *Separation and Purification Technology*, **305**, 122427, https://doi.org/10.1016/j.seppur.2022.122427

Mudd G. M., Jowitt S. M. and Werner T. T. (2018). Global platinum group element resources, reserves and mining–a critical assessment. *Science of the Total Environment*, **622**, 614–625, https://doi.org/10.1016/j.scitotenv.2017.11.350

Mugavero S. J. III, Gemmill W. R., Roof I. P. and Zur Loye H. C. (2009). Materials discovery by crystal growth: lanthanide metal containing oxides of the platinum group metals (Ru, Os, Ir, Rh, Pd, Pt) from molten alkali metal hydroxides. *Journal of Solid State Chemistry*, **182**(7), 1950–1963, https://doi.org/10.1016/j.jssc.2009.05.006

Mulvaney D., Richards R. M., Bazilian M. D., Hensley E., Clough G. and Sridhar S. (2021). Progress towards a circular economy in materials to decarbonize electricity and mobility. *Renewable and Sustainable Energy Reviews*, **137**, 110604, https://doi.org/10.1016/j.rser.2020.110604

Muralikrishna I. V. and Manickam V. (2017). Life cycle assessment. *Environmental Management*, **11**(1), 57–75.

Nikoloski A. N. and Ang K. L. (2014). Review of the application of ion exchange resins for the recovery of platinum-group metals from hydrochloric acid solutions. *Mineral Processing and Extractive Metallurgy Review*, **35**(6), 369–389, https://doi.org/10.1080/08827508.2013.764875

O'Driscoll B. and González-Jiménez J. M. (2016). Petrogenesis of the platinum-group minerals. *Reviews in Mineralogy and Geochemistry*, **81**(1), 489–578, https://doi.org/10.2138/rmg.2016.81.09

Paiva A. P., Piedras F. V., Rodrigues P. G. and Nogueira C. A. (2022). Hydrometallurgical recovery of platinum-group metals from spent auto-catalysts–focus on leaching and solvent extraction. *Separation and Purification Technology*, **286**, 120474, https://doi.org/10.1016/j.seppur.2022.120474

Pourhossein F., Mousavi S. M. and Beolchini F. (2022). Innovative bio-acid leaching method for high recovery of critical metals from end-of-life light emitting diodes. *Resources, Conservation and Recycling*, **182**, 106306, https://doi.org/10.1016/j.resconrec.2022.106306

Power-technology.com. (2024). Umicore NV/SA patent: Recovery of platinum group metals from catalysts. https://www.ipmi.org/news/umicore-gets-grant-recovery-platinum-group-metals-catalysts-using-smelting (accessed 9 January 2025)

Reuter M. and Van Schaik A. (2016). Gold–a key enabler of a circular economy: recycling of waste electric and electronic equipment. In: Gold Ore Processing, Mike D. Adams (ed.), Elsevier, Fugue Pte Ltd, Singapore, pp. 937–958.

Saidani M., Kendall A., Yannou B., Leroy Y. and Cluzel F. (2019). Closing the loop on platinum from catalytic converters: contributions from material flow analysis and circularity indicators. *Journal of Industrial Ecology*, **23**(5), 1143–1158, https://doi.org/10.1111/jiec.12852

Sharma K. L. S. (2011). Automation strategies. In: Overview of Industrial Process Automation, Sonnini R. Yura (ed.), Elsevier, Oxford, United Kingdom, pp. 53–62.

Sinisalo P. and Lundström M. (2018). Refining approaches in the platinum group metal processing value chain–a review. *Metals*, **8**(4), 203, https://doi.org/10.3390/met8040203

Statista.com, Statista Research Department. (2023). Platinum-group metal reserves worldwide 2022, by country. https://www.statista.com/statistics/273624/platinum-metal-reserves-by-country/#:~:text=Total%20worldwide%20reserves%20of%20platinum,the%20United%20States%2C%20and%20Canada (accessed 18 January 2024)

Sun S., Jin C., He W., Li G., Zhu H. and Huang J. (2022). A review on management of waste three-way catalysts and strategies for recovery of platinum group metals from them. *Journal of Environmental Management*, **305**, 114383, https://doi.org/10.1016/j.jenvman.2021.114383

Suoranta T., Zugazua O., Niemelä M. and Perämäki P. (2015). Recovery of palladium, platinum, rhodium and ruthenium from catalyst materials using microwave-assisted leaching and cloud point extraction. *Hydrometallurgy*, **154**, 56–62, https://doi.org/10.1016/j.hydromet.2015.03.014

Tanaka S., Harada A., Nishihama S. and Yoshizuka K. (2012). Selective recovery of platinum group metals from spent automobile catalyst by integrated ion exchange methods. *Separation Science and Technology*, **47**(9), 1369–1373, https://doi.org/10.1080/01496395.2012.672526

Tang H., Peng Z., Tian R., Ye L., Zhang J., Rao M. and Li G. (2023). Platinum-group metals: demand, supply, applications and their recycling from spent automotive catalysts. Journal of Environmental Chemical Engineering, **11**, 110237, https://doi.org/10.1016/j.jece.2023.110237

Taylor P. R. and Pirzada S. A. (1993). Plasma technology in extractive and process metallurgy. *Mineral Processing and Extractive Metallurgy Review*, **12**(2–4), 257–269, https://doi.org/10.1080/08827509508935260

Trinh H. B., Lee J. C., Suh Y. J. and Lee J. (2020). A review on the recycling processes of spent auto-catalysts: towards the development of sustainable metallurgy. *Waste Management*, **114**, 148–165, https://doi.org/10.1016/j.wasman.2020.06.030

Van den Bossche A., Rodriguez N. R., Riaño S., Dehaen W. and Binnemans K. (2021). Dissolution behavior of precious metals and selective palladium leaching from spent automotive catalysts by trihalide ionic liquids. *RSC Advances*, **11**(17), 10110–10120, https://doi.org/10.1039/D1RA00695A

Wei W., Cho C. W., Kim S., Song M. H., Bediako J. K. and Yun Y. S. (2016a). Selective recovery of Au (III), Pt (IV), and Pd (II) from aqueous solutions by liquid–liquid extraction using ionic liquid Aliquat-336. *Journal of Molecular Liquids*, **216**, 18–24, https://doi.org/10.1016/j.molliq.2016.01.016

Wei W., Lin S., Reddy D. H. K., Bediako J. K. and Yun Y. S. (2016b). Poly (styrenesulfonic acid)-impregnated alginate capsule for the selective sorption of Pd (II) from a Pt (IV)-Pd (II) binary solution. *Journal of Hazardous Materials*, **318**, 79–89, https://doi.org/10.1016/j.jhazmat.2016.06.050

Wei W., Reddy D. H. K., Bediako J. K. and Yun Y. S. (2016c). Aliquat-336-impregnated alginate capsule as a green sorbent for selective recovery of gold from metal mixtures. *Chemical Engineering Journal*, **289**, 413–422, https://doi.org/10.1016/j.cej.2015.12.104

Wen Q., Ding Y., Zheng H., Liu B. and Zhang S. (2020). Process and mechanism of electrolytic enrichment of PGMs from Fe-PGMs alloy. *Journal of Cleaner Production*, **271**, 122829, https://doi.org/10.1016/j.jclepro.2020.122829

Wen J., Wu M., Peng J. and Zheng H. (2023). Purification of high-purity tin via vertical zone refining. *Separations*, **10**(7), 380, https://doi.org/10.3390/separations10070380

Xia J. and Ghahreman A. (2023). Platinum group metals recycling from spent automotive catalysts: metallurgical extraction and recovery technologies. Separation and Purification Technology, **311**, 123357, https://doi.org/10.1016/j.seppur.2023.123357

Xun D., Sun X., Liu Z., Zhao F. and Hao H. (2022). Comparing supply chains of platinum group metal catalysts in internal combustion engine and fuel cell vehicles: a supply risk perspective. *Cleaner Logistics and Supply Chain*, **4**, 100043, https://doi.org/10.1016/j.clscn.2022.100043

Yakoumis I., Panou M., Moschovi A. M. and Panias D. (2021). Recovery of platinum group metals from spent automotive catalysts: a review. *Cleaner Engineering and Technology*, **3**, 100112, https://doi.org/10.1016/j.clet.2021.100112

You G. B., Fang W. B., Li Q. B., Ma Y. B., Yang X. T. and Yang H. B. (2016). Study on enrichment method of platinum, palladium and rhodium in spent auto-catalysts. *Yejin Fenxi/Metallurgical Analysis*, **36**(5), 7Ą11.

Yu Z., Han H., Feng P., Zhao S., Zhou T., Kakade A., Kulshrestha S., Majeed S. and Li X. (2020). Recent advances in the recovery of metals from waste through biological processes. *Bioresource Technology*, **297**, 122416, https://doi.org/10.1016/j.biortech.2019.122416

Yuan Z., Liu H., Yong W. F., She Q. and Esteban J. (2022). Status and advances of deep eutectic solvents for metal separation and recovery. *Green Chemistry*, **24**(5), 1895–1929, https://doi.org/10.1039/D1GC03851F

Zhang X., Friedrich S., Liu B., Huang T. and Friedrich B. (2020a). Computation-assisted analyzing and forecasting on impurities removal behavior during zone refining of antimony. *Journal of Materials Research and Technology*, **9**(2), 1221–1230, https://doi.org/10.1016/j.jmrt.2019.11.049

Zhang H., Zhao J., Xu J., Li Y., Pu Z., Xu B. and Yang B. (2020b). Preparation of high-purity tin by zone melting. *Russian Journal of Non-Ferrous Metals*, **61**, 9–20, https://doi.org/10.3103/S1067821220010186

Zhang C., Chen F., Huang S. and Shen S. (2023). Resource nationalism and platinum group metals: Chinese strategies in a hydrogen-powered future. *The Extractive Industries and Society*, **15**, 101317, https://doi.org/10.1016/j.exis.2023.101317

Zheng H., Ding Y., Wen Q., Liu B. and Zhang S. (2021). Separation and purification of platinum group metals from aqueous solution: recent developments and industrial applications. *Resources, Conservation and Recycling*, **167**, 105417, https://doi.org/10.1016/j.resconrec.2021.105417

doi: 10.2166/9781789065084_0187

Chapter 8

Sustainable recovery of platinum group metals from spent automotive catalysts

M. Compagnone, J. J. González-Cortés and M. Ramírez*

Department of Chemical Engineering and Food Technologies, University of Cadiz, Wine and Agrifood Research Institute (IVAGRO), Cádiz, Spain
**Corresponding author: joaquin.gonzalez@uca.es*

ABSTRACT

Platinum group metals (PGMs) are essential for reducing vehicle emissions via automotive catalysts (ACs), yet their mining is environmentally and economically challenging. This chapter examines PGM recovery from spent ACs, with a particular focus on traditional pyrometallurgical and hydrometallurgical methods, as well as emerging biotechnological approaches such as bioleaching. Despite their potential for sustainability, biotechnological methods face efficiency and growth challenges. Despite the promising advances in recovery technologies, the high energy consumption and the management of waste remain significant challenges. The development of energy-efficient and waste-reducing processes, in addition to the implementation of green chemistry, is crucial for the establishment of a sustainable PGM recycling industry that can balance the economic and environmental needs.

Keywords: Automotive catalysts, biotechnological approaches, platinum group metals, recycling, sustainability

8.1 INTRODUCTION

Platinum group metals (PGMs) consist of six precious metals including platinum (Pt), palladium (Pd), rhodium (Rh), ruthenium (Ru), iridium (Ir), and osmium (Os). They are known for their durability, resistance to corrosion, high melting points, and catalytic properties (Hughes *et al.*, 2021). Primary supply involves extracting and selling PGMs directly from mining operations, while secondary supply refers to recovering PGMs through recycling processes, particularly open-loop recycling. In this context, open-loop recycling means that the original purchaser of the product containing the PGMs does not retain ownership of the metals after the product is discarded so the metal is available to the market once recovered. Outside the major markets of automotive, jewelry, and electronics, the amount of PGMs recovered through open-loop recycling is minimal. Despite the availability of secondary sources, the majority of PGMs still come from primary sources due to the complexity and cost involved in recycling these metals (Brown *et al.*, 2024). PGMs are extracted from sulfide ores through complex processes involving crushing, milling, flotation, smelting, and refining. In 2024, approximately 83% of PGMs were supplied by South Africa and Russia and 15% was produced from ore deposits situated in North America and Zimbabwe (Figure 8.1).

As a result, PGMs' low availability is a substantial issue for the European Union (EU), which in 2024 was one of the regions with the highest demand for Pd and Pt particularly for their utilization in the automotive catalyst (AC) industry. To address this supply challenge, in 2011 the European Commission included PGMs on the list of critical raw materials (CRMs) for the EU (Directorate-General for Enterprise & Industry (European Commission), 2013; Grohol & Veeh, 2023).

PGMs are used in jewelry, electronics, chemical, petroleum and automotive industries, and for medical devices (Rao & Reddi, 2000). Their unique properties, like corrosion resistance and biocompatibility, make them valuable across these fields. Since 1979, the automotive industry has been the principal consumer of PGMs. ACs stand out as the richest secondary source of PGMs. It has been estimated that the demand for PGMs for the automotive industry in 2024 represented >70% of the total demand (Brown *et al.*, 2024). Johnson Matthey's free-market review on the supply and demand for PGMs estimated that in 2024 the global gross demand for Pt, Pd, and Rh was 236.8, 302.7, and 33.1 t respectively, with less than 27% coming from secondary sources (Table 8.1).

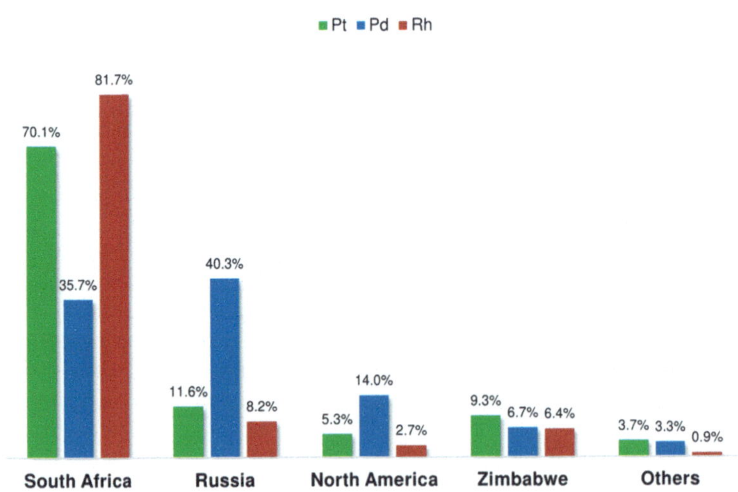

Figure 8.1 Pt, Pd, and Rh primary supply by Region in 2024 (Brown *et al.*, 2024).

Table 8.1 Pd, Pt, and Rh global demand and supply from primary and secondary sources from 2019 to 2024 (Brown *et al.*, 2024).

	2019	2020	2021	2022	2023	2024
Pt						
Total demand (t)	261.5	224.1	209.0	200.1	236.6	236.8
Total demand for ACs (t)	80.5	63.0	74.9	85.9	104.0	102.7
% for ACs	30.8%	28.1%	35.8%	42.9%	44.0%	43.4%
Total primary supply (t)	187.7	154.4	193.3	167.5	180.1	176.7
Total secondary supply (t)	65.1	52.8	51.2	47.2	40.4	41.5
% secondary supply	25.8%	25.5%	20.9%	22.0%	18.3%	19.0%
Pd						
Total demand (t)	357.2	311.1	319.4	310.2	322.6	302.7
Total demand for ACs (t)	301.0	266.6	264.4	263.1	272.0	253.4
% for ACs	84.3%	85.7%	82.8%	84.8%	84.3%	83.7%
Total primary supply (t)	221.4	192.7	212.9	187.4	203.6	200.7
Total secondary supply (t)	105.9	97.2	103.8	99.3	87.3	90.9
% secondary supply	32.4%	33.5%	32.8%	34.6%	30.0%	31.2%
Rh						
Total demand (t)	36.3	32.3	32	31.4	34.7	33.1
Total demand for ACs (t)	32.1	29.9	29.6	29.5	30.8	28.9
% for ACs	88.4%	92.6%	92.5%	93.9%	88.8%	87.3%
Total primary supply (t)	23.2	19	23.8	21.5	21.8	21.9
Total secondary supply (t)	11.1	10.5	11.2	10.4	8.9	9.2
% secondary supply	32.4%	35.6%	32.0%	32.6%	29.0%	29.6%

AC: automotive catalyst.

Supply constraints, extraction, and geopolitical risks impact the PGM market leading to price variability. The automotive sector's demand, geopolitical instability, and environmental regulations, especially for Pt, Pd, and Rh, drive market trends by impacting prices (Bloxham *et al.*, 2022).

ACs, found in catalytic converters, are designed to reduce harmful emissions from vehicle exhaust systems. They are indispensable for environmental protection, mitigating emissions that lead to issues like smog, acid rain, and ozone depletion. Consequently, they play a vital role in combating climate change and enhancing air quality (Nguyen *et al.*, 2021). Moreover, these catalysts are essential for public health as they reduce pollutants associated with respiratory and cardiovascular issues, thereby contributing to better health outcomes and lower healthcare costs (Pope *et al.*, 2009). The presence of PGMs in the composition of these catalysts is essential to transform toxic gases into less harmful substances before they are released into the atmosphere (Nguyen *et al.*, 2021). Specifically, ACs convert carbon monoxide (CO) to carbon dioxide (CO_2) through catalytic oxidation, nitrogen oxides (NO_x) are broken down to nitrogen (N_2) and oxygen (O_2) through catalytic reduction and unburned

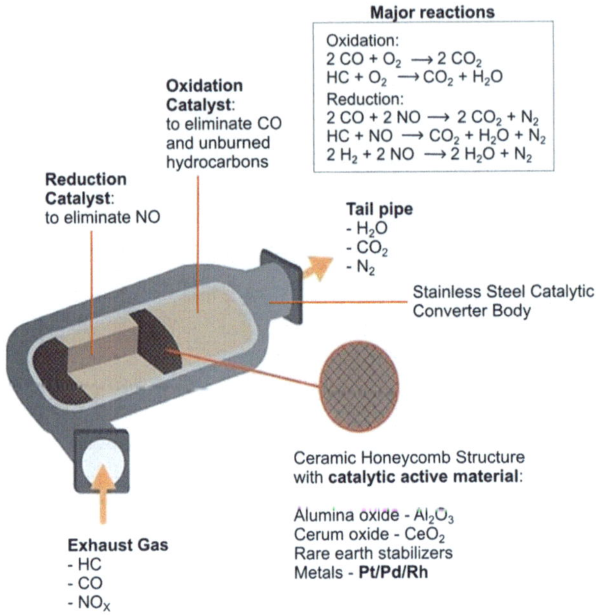

Figure 8.2 Scheme of automotive catalytic converter. Adapted from Moschovi *et al.* (2021).

hydrocarbons (HCs) are oxidized to CO_2 and water (H_2O) (Figure 8.2). Pt, Pd, and Rh are the most used PGMs in ACs. Pt is a key PGM in ACs, primarily used in diesel engines and some gasoline applications. It acts as an oxidizing catalyst, converting CO into CO_2 and unburned hydrocarbons into CO_2 and H_2O. Pd is especially used in gasoline engines. It performs a similar role to platinum but is often chosen for its lower cost and high reactivity. Rh is a critical component in ACs to reduce NO_x, breaking them down into nitrogen N_2 and O_2 (Nguyen *et al.*, 2021). This reduction process is essential for meeting strict environmental regulations and reducing air pollution.

PGMs are applied as a thin coating on a substrate within the catalytic converter, providing a large surface area for chemical reactions. The catalytic properties of PGMs allow these reactions to occur at lower temperatures and with greater efficiency, making them indispensable in the design and function of ACs (He *et al.*, 2020). The substrate serves as the foundational structure supporting catalytic materials, featuring a honeycomb-like design to maximize surface area and facilitate exhaust gas flow. Typically made from ceramic materials like cordierite or stainless steel, substrates offer thermal stability, heat resistance, and durability (Mehdipour *et al.*, 2024). A porous layer, known as a washcoat, is applied to the substrate to further increase the surface area for catalytic reactions. Composed of metal oxides such as alumina, ceria, or silica, the washcoat provides a stable foundation for catalytic materials. Commonly, honeycomb-type cordierite monoliths with a highly porous γ-alumina wash-coat on the inner surface are used in car catalytic converters. After use, this γ-alumina wash-coat transforms into α-Al_2O_3 containing up to 0.1% w/w of PGMs (De Jong *et al.*, 1997; Sun *et al.*, 2022). Furthermore, supplementary materials are incorporated into certain ACs to serve specific functions. Cerium, for instance, is frequently introduced to bolster the oxygen storage capacity, thereby optimizing catalytic efficiency. Conversely, zirconium finds application in enhancing the thermal stability and overall durability of the catalysts (Chasserio *et al.*, 2009).

Over the past few decades, the demand for PGMs has experienced a steady increase owing to their significant industrial significance, prompting concerns regarding economic and sustainability matters (Fajar *et al.*, 2020). To address environmental concerns and ensure sustainable PGM supply, recycling and sustainable recovery methods are gaining importance (Chidunchi *et al.*, 2024). From an environmental perspective, recycling PGMs plays a crucial role in alleviating the necessity for primary mining, a process frequently linked to considerable ecological harm such as deforestation, habitat destruction, and water pollution (Mpinga *et al.*, 2015). PGMs are scarcely present on the Earth's upper continental crust. Ore deposits typically contain 3–4 g of PGMs per ton of ore. ACs contain PGMs in considerably higher concentrations than natural ores (Aaltonen *et al.*, 2017; Thormann *et al.*, 2017). It has been estimated that by recycling 2 mg of spent ACs, the mining of 150 kg of PGM ores could be spared (Fornalczyk & Saternus, 2009). The reuse of PGMs extracted from end-of-life products such as ACs through recycling not only conserves natural resources, but also diminishes the environmental impact of mining activities. This decrease in primary mining activities results in reduced energy consumption and fewer greenhouse gas emissions, thereby aiding efforts to mitigate climate change (Norgate & Haque, 2010; Yang *et al.*, 2022). Furthermore, by extracting valuable metals from discarded products, the volume of waste sent to landfills decreases, minimizing soil and water contamination. This waste reduction is essential for promoting a circular economy, where resources are continually reused, enhancing sustainability.

From an economic perspective, PGM recycling has become increasingly important as a cost-effective alternative to primary extraction. The costs of mining, refining, and processing raw ores can be substantially higher than recycling existing material. This is especially true given the recent spikes in PGM prices, driven by factors like the COVID-19 pandemic and geopolitical tensions between Russia and Europe. New international conflicts or new pandemics can further increase prices (Skyle, 2022). These price increases have resulted from supply chain disruptions, mining interruptions, and trade uncertainties, emphasizing the economic advantages of recycling (Brown *et al.*, 2024). By utilizing recycled PGMs, industries can lower production costs, leading to more competitive pricing and supporting the development of greener technologies.

In short, it is clear that the primary production of PGMs must be integrated with the recycling of PGMs to balance economic necessities and the environmental footprint of PGM production (Sun *et al.*, 2022). Nowadays, the technologies to recover PGMs can be classified into pyrometallurgical, hydrometallurgical, and biotechnological processes (Chidunchi *et al.*, 2024). The existing industry for the recovery of PGMs mostly relies on pyrometallurgical operations due to the high recovery rates and the simplicity of the operations. However, pyrometallurgy involves pre-treatment steps combined with high-temperature processes that emit greenhouse gases and pollutants, making waste management challenging (Karim & Ting, 2021). On the other hand, hydrometallurgy has lower energy consumption compared to pyrometallurgy but involves the use of corrosive chemicals, posing environmental and health risks, and requires resource-intensive chemical production. Therefore, although effective, proper waste treatment and disposal are crucial in both methods to minimize their environmental footprint (Xia & Ghahreman, 2023). Hence, ongoing efforts focus on the development of more eco-friendly and efficient methods for PGM recovery, such as the use of biotechnological processes. Although the biometallurgical route acceptance in the industry is rather slow, it has raised great interest for its innovative and promising potential (Karim *et al.*, 2024).

In the following sections, the most important recovery techniques of PGMs will be described in more detail.

8.2 RECOVERY TECHNIQUES

This section explores the various techniques used to recover PGMs from ACs, focusing on chemical, physical, and biotechnological methods. Chemical methods such as leaching, solvent extraction, and precipitation/crystallization are detailed, explaining how they extract and purify PGMs. Physical methods like pyrometallurgical, hydrometallurgical, and electrometallurgical processes are discussed, highlighting their role in PGM recovery and associated environmental challenges. The chapter also examines biotechnological approaches, including bioleaching, bioaccumulation, and biomineralization, emphasizing their sustainability and lower environmental impact. By summarizing these recovery techniques, the section aims to comprehensively understand the current methods used in PGM recovery and their contributions to sustainable recycling practices. The general advantages and disadvantages of the various recovery techniques described are presented in Table 8.2.

8.2.1 Physico-chemical methods

The most widely used physico-chemical methods to recover PGMs from ACs encompass pyrometallurgical processes, hydrometallurgical processes, and electrometallurgical techniques (Figure 8.3). Each method has unique advantages and challenges, contributing to the overall efficiency and sustainability of PGM recovery. Proper environmental management and waste disposal are essential across all methods to ensure compliance with environmental regulations and promote sustainable practices in PGM recovery.

8.2.1.1 Pyrometallurgical processes

Pyrometallurgical processes are physical methods used to separate PGMs from their substrates and other materials by high-temperature treatment. Pyrometallurgical processes for recovering PGMs typically begin with

Table 8.2 Advantages and disadvantages of metallurgical methods to recover PGMs.

Method	Advantages	Disadvantages
Pyrometallurgical	High metal recovery rates. Simplicity of operation. Well-established technology. Suitable for high-grade materials. Can recover multiple metals simultaneously.	High energy consumption. Generation of air pollutants. High temperature operation may degrade catalyst materials
Hydrometallurgical	Lower energy consumption compared to pyrometallurgy. Can handle low-grade materials. Selective recovery of metals. Can be combined with other separation techniques. Generally lower operating temperatures	Requires extensive water usage. Generation of chemical waste. Complex process requiring careful control of parameters. May require additional steps for complete metal recovery.
Biometallurgical	Environmental friendly. Low energy consumption. Selective recovery of metals. Utilizes natural biological processes. Can operate at ambient temperatures. Potential for integration with other biotechnological processes for sustainable PGM recovery.	Slower process compared to traditional methods. Requires optimization for industrial scale implementation. May require additional steps for complete metal recovery.

pretreatment of the ACs. This step involves crushing and grinding the ACs into smaller particles to increase the specific surface area, allowing for more effective treatment. The pretreated material is then subjected to high temperatures in a furnace or reactor to initiate the pyrometallurgical recovery operation in a process called smelting (Karim & Ting, 2022). In this process, the crushed spent ACs material is heated to temperatures above 1200°C in a smelting furnace. The high heat causes the PGMs and other metals to melt and separate from the gangue (waste material). A flux, such as silica, lime, or borax, is often added to the smelting process to bind with impurities and form a slag. This slag floats on top of the molten metal and can be removed, leaving behind a molten matte containing PGMs (Chidunchi *et al.*, 2024; Sun *et al.*, 2022).

After smelting, the next step is refining. This involves additional heating and chemical treatments to further separate and purify the PGMs from other metals in the molten matte. Refining may include processes like fire assay and cupellation, which help isolate PGMs from base metals like lead or copper. The resulting PGM-rich material can be processed further to achieve the desired purity levels (Chidunchi *et al.*, 2024; Sun *et al.*, 2022). Converting is another pyrometallurgical technique used to recover PGMs from ACs. In this process, the molten matte from the smelting stage is subjected to high-temperature

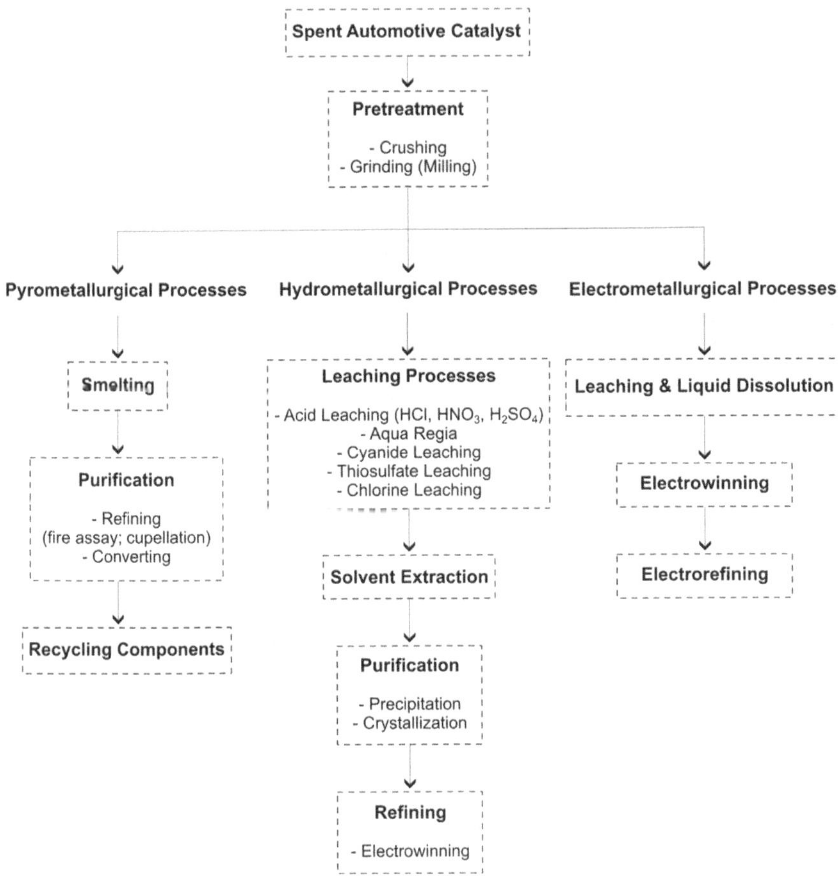

Figure 8.3 Physico-chemical methods to recover PGMs from ACs.

oxidation, which removes sulfur and other impurities. The converter's high temperatures and oxygen injection ensure a thorough separation of PGMs from base metals and residual impurities (Xu *et al.*, 2022).

Pyrometallurgical processes, while effective for PGM recovery from ACs, entail environmental and safety risks. The high temperatures involved result in notable energy consumption and greenhouse gas emissions. Furthermore, smelting and converting processes can emit hazardous gases, necessitating strict controls and pollution abatement systems to mitigate environmental impact. Nonetheless, efforts are underway to enhance sustainability and energy efficiency for these processes. Recycling components, such as off-gas recovery systems and waste heat recovery systems, are designed to capture and reuse energy and materials from high-temperature procedures, contributing to a more sustainable and energy-efficient recovery technique.

8.2.1.2 Hydrometallurgical processes

Hydrometallurgical processes rely on chemical reactions in aqueous solutions to extract, separate, and purify PGMs from solid materials. The first step in hydrometallurgical recovery is again pretreatment, where ACs are crushed and ground to increase the surface area, which facilitates the dissolution of PGMs in a liquid solution. The ground catalyst is then subjected to leaching processes. Common leaching agents include acids such as hydrochloric acid (HCl), sulfuric acid (H_2SO_4), or nitric acid (HNO_3) (Paiva *et al.*, 2022). Aqua regia, a mixture of HCl and HNO_3 in a 3:1 ratio, is particularly effective for dissolving PGMs such as Pt and Pd (Jha *et al.*, 2013; Nguyen *et al.*, 2021). Cyanide-based solutions are often used for their high selectivity and efficiency in dissolving PGMs. However, the use of cyanide carries significant environmental and safety risks, requiring stringent controls to prevent contamination and ensure safe handling (Mpinga *et al.*, 2015; Pourhossein & Mousavi, 2023). Alternatively, thiosulfate leaching offers a more environmental-friendly alternative to cyanide. This method uses sodium thiosulfate or ammonium thiosulfate to dissolve PGMs, providing a less toxic and more sustainable approach to leaching. Thiosulfate leaching is gaining traction due to its reduced environmental impact and lower risk of hazardous by-products (Compagnone *et al.*, 2023a, 2023b; Mpinga *et al.*, 2015). Chlorine leaching, based on the use of chlorine gas or chlorine-based solutions, is effective for recovering a range of PGMs, but it requires careful management to avoid the release of harmful chlorine gas during the process (Nguyen *et al.*, 2021).

Once the PGMs are in solution, the next step involves separating and purifying the metals from other impurities. This is achieved through various techniques, each contributing to the hydrometallurgical recovery process, such as solvent extraction, ion exchange, precipitation, and crystallization.

Solvent extraction, also known as liquid–liquid extraction, involves using an organic solvent that has a high affinity for specific PGMs (Jha *et al.*, 2013; Pietrelli & Fontana, 2013). This organic solvent is mixed with the aqueous leachate containing the dissolved PGMs, allowing the desired metals to transfer into the organic phase while leaving behind other impurities (Pianowska *et al.*, 2023; Wiecka *et al.*, 2022). One commonly used solvent extraction technique involves organic phosphorus-based extractants, like tri-n-butyl phosphate or tributyl phosphate, which can selectively extract PGMs such as Pt and Pd from the aqueous phase. These extractants form complexes with the PGMs, facilitating their transfer into the organic solvent (Nguyen *et al.*, 2021). After extraction, the organic solvent containing the PGMs is separated from the aqueous phase. The PGMs are then stripped from the organic solvent into a new aqueous solution using a specific stripping agent. This process allows for concentrated recovery of the PGMs in a smaller volume, facilitating further purification and refining (Paiva, 2017; Yamada *et al.*, 2018). Overall, solvent extraction techniques are effective in selectively recovering PGMs from ACs, offering a scalable and efficient approach to recycling these valuable metals.

Following solvent extraction, the isolated PGMs can be subjected to additional processes like precipitation or crystallization to achieve higher purity levels and convert the PGMs into solid forms for further processing or recycling. These techniques ultimately allow for the recovery of high-purity PGMs, which can be reused in various applications, including new ACs and industrial processes (Paiva, 2017; Yamada *et al.*, 2018).

Precipitation methods involve adding specific reagents to the leachate obtained in the solvent extraction process to form insoluble compounds with PGMs. This process allows the PGMs to separate from the solution as a solid precipitate, which can be filtered or centrifuged for collection. Common precipitation reagents include sodium chloride (NaCl), ammonium chloride (NH_4Cl), or sodium hydroxide (NaOH). Depending on the target PGM, different reagents are used to ensure selective precipitation. An example is the precipitation of Pt and Pd as ammonium chloroplatinate and ammonium chloropalladate, respectively (Jha *et al.*, 2013; Mpinga *et al.*, 2015; Pianowska *et al.*, 2023). By adding NH_4Cl to a solution containing dissolved Pt and Pd, the metals form insoluble ammonium salts, which can be filtered and recovered. This process allows for the separation of PGMs from other impurities in the leachate.

Crystallization methods involve reducing the solubility of dissolved PGMs in a solution, leading to the formation of crystals. This can be achieved by altering the temperature, concentration, or pH of the solution. As the conditions change, PGMs begin to crystallize, allowing for their separation and collection. One common approach involves evaporative crystallization, where the solution containing dissolved PGMs is heated to remove water, thereby concentrating the PGMs and leading to their crystallization. This method is effective for obtaining high-purity PGM crystals that can be further processed or refined. After precipitation or crystallization, the recovered PGMs are typically subjected to refining processes to achieve the desired level of purity.

After separation and purification, the recovered PGMs typically undergo refining processes to produce high-purity metals suitable for reuse. The most common refining process is electrowinning. In electrowinning, an electric current is used to deposit the PGMs onto a cathode, further refining and solidifying the metals (Cao *et al.*, 2006; Paiva *et al.*, 2022; Saguru *et al.*, 2018).

Hydrometallurgical processes are generally more environmental-friendly than pyrometallurgical methods, as they operate at lower temperatures and produce fewer emissions. However, they still require careful management to avoid hazardous waste and chemical contamination. Proper treatment and disposal of waste solutions are crucial to minimize environmental impact (Grilli *et al.*, 2023).

8.2.1.3 Electrometallurgical techniques

Electrometallurgical techniques involve the use of electrical energy to extract and purify PGMs from leachates or other solutions. After pretreatment, which involves again the crushing and grinding of the AC, the first step in electrometallurgical recovery is typically leaching, where PGMs are dissolved into a liquid solution, for example H_2SO_4, HCl, HNO_3 or $HClO_4$ (Ghalehkhondabi *et al.*, 2021). This step is crucial to ensure that the metals are in a form that can be subjected to electrometallurgical processes.

Once the PGMs are in solution, electrowinning is a commonly used electrometallurgical technique to recover PGMs. Electrowinning involves passing an electric current through the solution to deposit the dissolved PGMs onto a cathode, typically made of a conductive metal like copper or stainless steel. As the current flows, PGMs are attracted to the cathode and begin to plate out, forming a solid layer of recovered metal. This process is effective in concentrating and recovering PGMs with high purity (Chidunchi *et al.*, 2024; Rai *et al.*, 2021).

Electrorefining is used to further purify PGMs after initial recovery via electrowinning. The process involves using an electrolytic cell where an impure anode containing the recovered PGMs is dissolved in an electrolyte solution, while the purified PGMs are deposited onto a cathode. This process can help separate PGMs from base metals and other impurities, resulting in higher purity metals suitable for industrial use (Liu *et al.*, 2021; Rai *et al.*, 2021).

Electrometallurgical processes can be highly efficient and produce high-purity PGMs, but they require careful control of electrical parameters, such as current density and voltage, to ensure optimal results. These processes also generate waste, such as sludge and spent electrolytes, which must be managed to minimize environmental impact. Proper treatment and disposal of these by-products are crucial to ensure compliance with environmental regulations.

8.2.2 Biotechnological approaches

Biotechnological approaches are emerging as innovative methods for recovering PGMs from ACs. These approaches utilize biological processes and microorganisms to extract, concentrate, or induce the formation of PGM precipitates, offering environmental-friendly and sustainable alternatives to traditional methods. Below is a detailed exploration of bioleaching, bioaccumulation, and biomineralization, explaining how they can contribute to PGM recovery from ACs (Figure 8.4).

8.2.2.1 Bioleaching

Bioleaching represents the most promising technique for recovering PGMs from secondary sources such as ACs. Bioleaching involves the use of bacteria or other microorganisms that can facilitate the dissolution and extraction of PGMs from solid substrates. These microorganisms, such as certain strains of bacteria and fungi, can produce organic acids or other compounds that help break down the structure of the ACs (Figure 8.5), releasing the PGMs into a solution (Gavrilescu, 2022; Mwase *et al.*, 2012).

Bacteria such as *Acidithiobacillus ferrooxidans* and *Acidithiobacillus thiooxidans* produce sulfuric acid and ferric iron, which help in metal dissolution (Ferreira *et al.*, 2017; Hedrich *et al.*, 2020). Fungi like *Aspergillus niger* produce citric and oxalic acids, which are effective in breaking down metal structures (Adetunji *et al.*, 2023; Ferreira *et al.*, 2017; Pathak *et al.*, 2022). Additionally, bacteria like *Gluconobacter oxydans* produce gluconic acid, aiding in the leaching process (Machado *et al.*, 2023). *Escherichia coli* can also be involved in bioleaching by producing biogenic cyanide, which forms metal-cyanide complexes to facilitate metal extraction (Kumar *et al.*, 2023; Shin

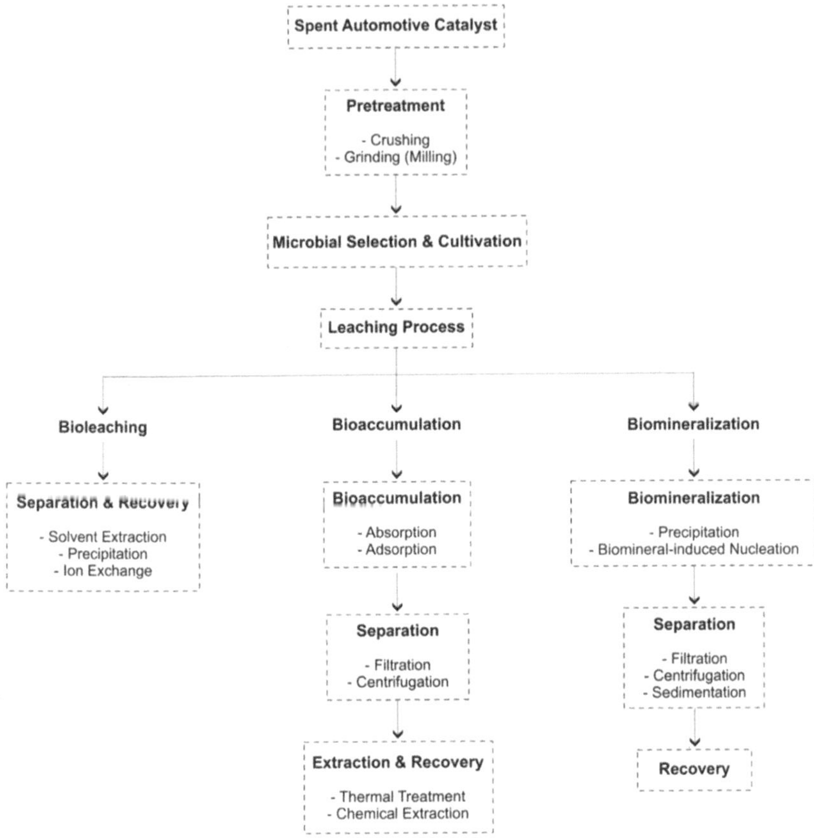

Figure 8.4 Biotechnological methods to recover PGMs from ACs.

et al., 2015). These acids and compounds break down the metal-containing substrates, releasing the metals into solution, making bioleaching an efficient and sustainable method for metal recovery. The bioleaching process typically includes several steps. The first step is to cultivate the specific microorganisms to be used for bioleaching. This involves providing a suitable growth environment, including nutrients and appropriate conditions (such as temperature and pH) for optimal microbial activity (Ferreira *et al.*, 2017; Malekian *et al.*, 2019). Once the microorganisms are cultured, they are added to the crushed and ground ACs. The bacteria or microbes produce the acids or chelates that react with the solid substrate, dissolving the PGMs into an aqueous solution. This process is slower than chemical leaching, but it operates at lower (ambient) temperatures and reduces the need for harsh chemicals, making it more environmentally friendly (Grilli *et al.*, 2023; Rendón-Castrillón *et al.*, 2023). After bioleaching, the solution contains dissolved PGMs that need to be separated and purified. Techniques such as solvent extraction, precipitation, or ion exchange can be

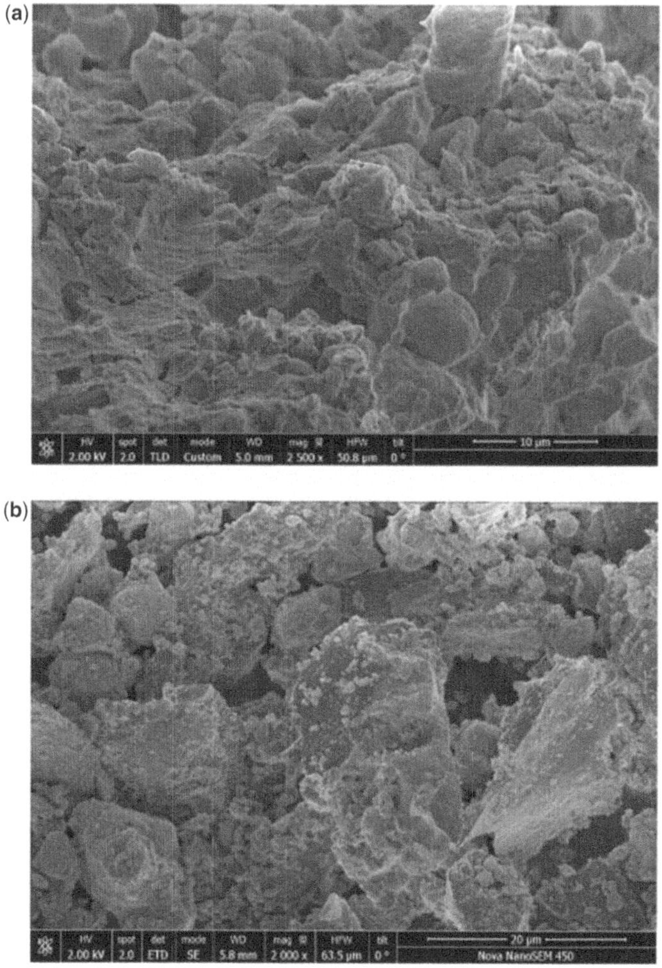

Figure 8.5 SEM images of a spent AC before (a) and after (b) bioleaching treatment through *Acidithiobacillus thiooxidans* with a magnification of ×2500 and ×2000, respectively (Compagnone *et al.*, 2023a, 2023b).

used to isolate the PGMs from the leachate. Once isolated, the PGMs can be refined and converted into a usable form (Zupanc *et al.*, 2023).

Bioleaching offers several advantages over traditional chemical leaching processes. It operates at lower temperatures, reducing energy consumption and greenhouse gas emissions. Additionally, it minimizes the use of hazardous chemicals, decreasing the risk of environmental contamination (Xolo *et al.*, 2021; Yakoumis *et al.*, 2021). Bioleaching can also be more cost-effective, as it relies on natural biological processes rather than complex chemical reactions. However, the process is generally slower, requiring extended time for the microorganisms to break down the substrate and release the PGMs. Additionally, maintaining optimal conditions for microbial activity can be complex, requiring careful control of factors like temperature, pH, and nutrient levels (Compagnone *et al.*, 2023a, 2023b).

Overall, it is a promising biotechnological approach for recovering PGMs from ACs. It provides a more sustainable and environmental-friendly alternative to traditional chemical methods, reducing the environmental impact of PGM recovery while contributing to a circular economy. With further research and development, bioleaching could become a widely adopted approach for PGM recovery, supporting sustainable practices in the automotive and materials industries.

8.2.2.2 Bioaccumulation

Bioaccumulation relies on the ability of certain microorganisms, like bacteria, fungi, or algae, to accumulate PGMs from leachates or other aqueous sources. These organisms can bind or concentrate PGMs within their cells or on their surfaces, allowing for subsequent recovery and purification. The process starts with selecting the appropriate microorganisms that exhibit high bioaccumulation capabilities for PGMs. It is the case for *Desulfovibrio desulfuricans* NCIMB 8307 which can biosorb and bioreduce Pd from AC (Yong *et al.*, 2002). The biosorption capacities of Pd and Pt were studied by De Vargas *et al.* (2004) in three species of *Desulfovibrio*: *Desulfovibrio desulfuricans*, *Desulfovibrio fructosivorans*, and *Desulfovibrio vulgaris*, with *Desulfovibrio desulfuricans* showing the highest efficiency. Differences in biosorption parameters among the species suggest variations in metal speciation-dependent sorption mechanisms. Godlewska-Żyłkiewicz *et al.* (2019) explored the ability of *Aspergillus* sp. (fungi) and *Saccharomyces* sp. (yeast) to remove Pt(IV) and Pd(II) from environmental samples. The process was best described by the Langmuir isotherm and pseudo-second-order kinetics. *Aspergillus* sp. showed higher adsorption capacities compared to *Saccharomyces* sp., demonstrating efficient removal of these metals from wastewater and potential for Pd recovery from industrial waste. Pinto *et al.* (2021) assessed the ability of six macroalgae species (*Ulva lactuca*, *Ulva intestinalis*, *Fucus spiralis*, *Fucus vesiculosus*, *Osmundea pinnatifida* and *Gracilaria* sp.) to remove PGMs from contaminated seawater. Among the tested algae, *Ulva lactuca* showed the highest removal rates, especially for Pd followed by Rh, Ru, Ir, and Pt. The removal efficiency varied with time, initial concentration, and species, with brown macroalgae being the least effective.

The microorganisms are cultivated in controlled conditions to ensure optimal growth and functionality. Conditions such as temperature, pH, and nutrient availability are carefully monitored. To facilitate bioaccumulation, PGMs must first be dissolved into a solution which is typically achieved through a previous leaching process. Once the PGMs are in solution, they can interact with the microorganisms (Roy *et al.*, 2024; Zhuang *et al.*, 2015). The cultivated microorganisms are introduced to the leachate containing dissolved PGMs to start the bioaccumulation process which involves the absorption or adsorption of PGMs by the microorganisms (Figure 8.6). This can occur through various mechanisms, including complexation with cellular components, binding to cell walls, or internalization within the cells (De Vargas *et al.*, 2004; Godlewska-Żyłkiewicz *et al.*, 2019; Gómez-Ramírez *et al.*, 2020; Jeyakumar *et al.*, 2022; Pinto *et al.*, 2021; Roy *et al.*, 2024).

After bioaccumulation, the microorganisms containing accumulated PGMs are separated from the solution, often through filtration or centrifugation. The accumulated PGMs can then be extracted from the microorganisms through further processing, such as by thermal treatment or chemical extraction, to

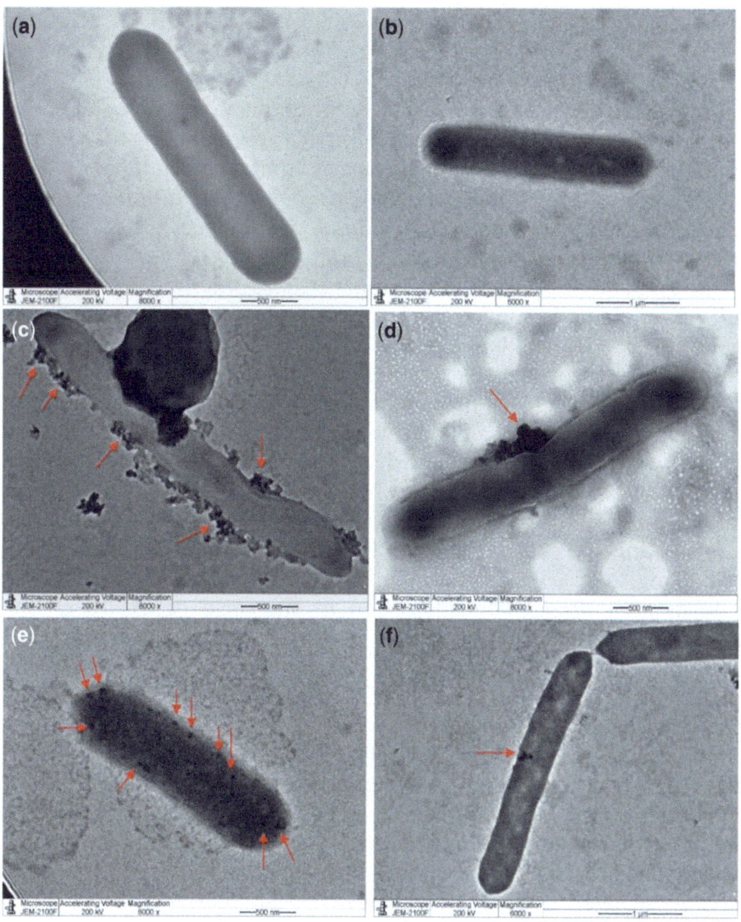

Figure 8.6 Transmission electron microscopy images of (a) *P. fluorescens* and (b) *B. megaterium*; biosorption of PGM by (c) *P. fluorescens* and (d) *B. megaterium* on day 2 of bioleaching, bioaccumulation of PGM by (e) *P. fluorescens* and (f) *B. megaterium* on day 5 of bioleaching from a spent AC. Reproduced from Karim and Ting (2022).

recover the metals (Assunção *et al.*, 2016; Golnaraghi Ghomi *et al.*, 2020). Bioaccumulation has several advantages over traditional recovery methods, including being environmentally friendly by operating at low temperatures and reducing the need for harsh chemicals. It also lowers the energy consumption and greenhouse gas emissions, making the process more sustainable and potentially more cost-effective due to its reliance on natural biological processes to accumulate PGMs (Mpinga *et al.*, 2015; Sedlakova-Kadukova *et al.*, 2020). However, it also has some limitations, such as being slower than traditional methods, thus requiring longer timeframes for microorganisms to accumulate sufficient quantities of PGMs. Additionally, maintaining optimal conditions for microbial growth and activity can be complex and requires careful environmental control.

8.2.2.3 Biomineralization

Biomineralization engages the use of microorganisms or biological processes that facilitate the formation of mineralized structures, enabling the recovery and separation of PGMs from aqueous solutions or other media. It relies on the ability of bacteria such as *Shewanella*, *Escherichia coli*, *Clostridium pasteurianum*, *Micrococcus luteus*, *Arthrobacter oxyidans*, *Staphylococcus sciuri*, *Bacillus sphaericus*, *Rhodobacter capsulatus*, *Cupriavidus metallidurans*, *Cupriavidus necator* and *Pseudomonas putid*a, fungi such as micromycete of the genus *Penicillium*, or algae, to induce the formation of mineralized compounds that incorporate PGMs (Karim & Ting, 2021). The first step in biomineralization is cultivating microorganisms that can induce mineral formation. This requires providing an appropriate environment with nutrients, temperature, and pH conditions conducive to microbial growth and activity. Then, PGMs must first be dissolved from the ACs through leaching. Once the microorganisms are cultivated and the PGMs are in solution, the bioactivity of the microorganisms can initiate the formation of mineralized structures. This can occur through various mechanisms, such as precipitation or biomineral-induced nucleation (Karim & Ting, 2021). The microorganisms create a microenvironment where PGMs form mineralized deposits, often in the form of oxides or sulfides, depending on the specific conditions. The resulting mineralized structures containing PGMs are separated from the solution. This can be achieved through filtration, centrifugation, or sedimentation. The separated mineralized materials are then processed to extract and purify the PGMs.

Biomineralization offers several benefits for PGM recovery. It operates at lower temperatures, reducing energy consumption compared to pyrometallurgical processes. Additionally, it minimizes the need for harsh chemicals, contributing to a more environmental-friendly recovery process. Biomineralization also aligns with the principles of a circular economy by using natural biological processes to recover and recycle valuable metals (Karim & Ting, 2021; Lou *et al.*, 2021; Rawlings *et al.*, 2003). However, like other biological processes, biomineralization can be slower than conventional methods, as it relies on microbial activity and natural mineral formation. Also, maintaining optimal conditions for biomineralization can be complex, requiring careful control of environmental factors like pH and temperature.

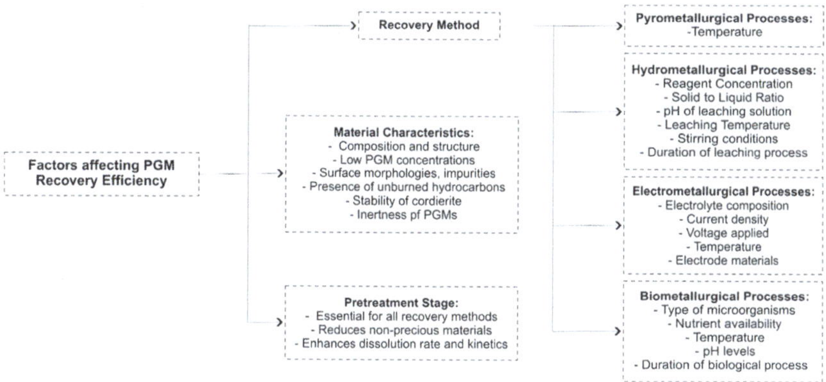

Figure 8.7 Factors affecting PGM recovery efficiency from spent ACs in physico-chemical and biotechnological approaches.

8.3 CHALLENGES IN PGM RECOVERY

The recovery of PGMs from ACs is influenced by several factors that impact efficiency, environmental safety, and economic viability (Figure 8.7). By understanding these factors, pathways to improve the sustainability of PGM recovery can be identified while addressing critical environmental and economic challenges.

8.3.1 Factors affecting PGM recovery efficiency

Firstly, the selection of a recovery method has a significant impact on the leaching of PGMs. Pyrometallurgical methods are commonly employed in the AC recycling industry to recover PGMs, offering promising recovery yields through the consistent application of elevated temperatures to facilitate efficient metal separation. Nevertheless, this technology typically necessitates additional preliminary steps to separate metals from slag and other by-products that could negatively impact recovery rates (Grilli *et al.*, 2023; Karim *et al.*, 2024). Hydrometallurgical methods rely on chemical reactions to leach PGMs from materials. The effectiveness of these processes is contingent upon the concentration of the leaching reagents. At low concentrations, the reagents may not be sufficiently potent, while concentrations that are too high could result in the formation of unwanted by-products. The solid-to-liquid ratio between the solid material being leached and the liquid solution also needs to be carefully controlled. Other factors that play a role include the pH value of the leaching solution, the leaching temperature, the stirring conditions, and the duration of the leaching process. The right balance of these variables determines how efficiently PGMs are dissolved and eventually recovered (Chidunchi *et al.*, 2024; Paiva *et al.*, 2022).

Electrometallurgical methods utilize electrical processes to recover metals. These methods can be highly efficient but depend on various factors such as electrolyte composition, current density, voltage applied, temperature, and the materials of the electrodes. The precise control of these variables is crucial to optimize the recovery rates of PGMs (Rai *et al.*, 2021).

Biometallurgical methods involve the use of biological agents, such as bacteria or fungi, to leach PGMs from the source material. The effectiveness of these methods is influenced by the type of microorganisms used, nutrient availability, temperature, pH levels, and the duration of the biological process. While this method is environmentally friendly, the recovery rates can vary significantly depending on the conditions and the microorganisms employed (Chidunchi *et al.*, 2024; Grilli *et al.*, 2023).

The characteristics of the material also play a pivotal role in the efficiency of the recovery process. The composition and structure of PGMs extracted from ACs differ from those found in electronic waste or industrial scrap. The type and age of the source material exert a significant influence on the concentration of PGMs and the complexity of the recovery process. The principal disadvantage of the recovery of PGMs from spent ACs wastes arises from the nature of car catalysts (Mohallem *et al.*, 2011). Surface morphologies, impurities, and pollutants vary among spent ACs, which contain PGMs in very low concentrations (up to 0.1% w/w). Consequently, pretreatment procedures are essential to reduce the quantity of non-precious materials that can lead to PGM enrichment and to enhance the dissolution rate and kinetics (Sun *et al.*, 2022). The deposition of unburned hydrocarbons renders the catalyst surface incapable of undergoing chemical treatment (Xia & Ghahreman, 2023). The remarkable stability of cordierite, coupled with the high inertness of PGMs, presents a significant challenge to the complete recycling of PGMs from ACs (Mohallem *et al.*, 2011). Therefore, regardless of the method employed to recover PGMs, a pretreatment stage is always necessary (Grilli *et al.*, 2023; Karim & Ting, 2020, 2021; Wei *et al.*, 2019; Wiecka *et al.*, 2022; Yakoumis *et al.*, 2021).

8.3.2 Environmental concerns associated with conventional extraction methods

The recycling of PGMs from secondary sources represents a pivotal aspect of the metal supply chain. Nevertheless, the extraction of these metals is challenging, and the traditional methods employed to do so have significant environmental impacts, which are listed below.

One significant issue is air pollution. The high temperatures used in traditional pyrometallurgical extraction processes result in the release of harmful substances, including sulfur dioxide and particulate matter, into the air. This not only affects air quality but also contributes to climate change, as these processes also result in the release of carbon dioxide (Van Leeuwen *et al.*, 2019).

In certain extraction methods, powerful oxidizing acidic solutions, including aqua regia and hydrochloric acid with chlorine gas, are employed to dissolve metals, resulting in the generation of a toxic wastewater by-product that contributes to the deterioration of water quality. In the absence of adequate treatment, the wastewater may percolate into rivers and lakes, thereby endangering fish and other aquatic life. Furthermore, the contamination of local water sources may pose a risk to the health of those who reside in the vicinity (Al-Enezi *et al.*, 2011; Alsukaibi, 2022).

The management of solid waste is similarly complex. The metal extraction process generates a considerable quantity of waste, particularly slag, which is a by-product of smelting. Slag can contain residual heavy metals and other hazardous materials (Samanta *et al.*, 2023). It is of the utmost importance to ensure the proper disposal of this waste in order to prevent contamination of soil and water. If not handled with the requisite care, these hazardous substances have the potential to cause long-term environmental damage.

Furthermore, traditional extraction methods consume a considerable amount of energy. The energy required to operate high-temperature furnaces and other equipment contributes to a larger carbon footprint. Furthermore, the use of hazardous chemicals in these processes presents a risk to workers and the environment. In the event that safety precautions are not rigorously adhered to, the aforementioned chemicals have the potential to inflict harm (Chidunchi *et al.*, 2024; Grilli *et al.*, 2023).

In addition, it is important to consider the regulatory requirements and compliance to ensure environmental protection, proper waste management, safe chemical handling, energy efficiency, and adherence to legal standards. In light of the increasingly stringent environmental regulations, conventional extraction methods frequently encounter difficulties in meeting the new standards. Companies must navigate a complex landscape of permits and environmental reports, which can be particularly challenging for smaller businesses due to their limited resources.

Finally, there are social and health risks to consider. The pollution generated by these extraction processes can have a detrimental impact on the health of nearby communities, leading to a range of respiratory problems and other health issues. Furthermore, workers in these environments are at risk if proper safety measures are not in place (Lyatuu *et al.*, 2021; Schrecker *et al.*, 2018).

8.3.3 Economic barriers to sustainable recovery

Economic barriers to the recovery of PGMs and other critical resources are significant and can impede the adoption of more environmental-friendly practices. These barriers span from high initial costs and market volatility to infrastructure limitations and regulatory uncertainty, posing challenges for industries and policymakers.

The high initial investment required for more sustainable recovery methods is a key economic barrier. Recovery processes often require advanced technology, specialized equipment, and skilled labor, all of which demand substantial upfront capital. This financial commitment can be a deterrent, particularly for smaller companies or those with tight budgets.

Market volatility is another significant barrier. Prices for PGMs and other resources can fluctuate widely due to various factors, such as supply chain disruptions, geopolitical events, pandemics, and changes in industrial demand. This volatility creates uncertainty, making it difficult for companies to justify large investments in sustainable recovery, especially when the return on investment is unpredictable (Brown *et al.*, 2023).

Infrastructure limitations can hinder the scalability and effectiveness of sustainable recovery. Existing facilities may not be equipped to handle advanced recycling technologies or processes, and can require significant upgrades. In regions with inadequate infrastructure, the cost of building new facilities or modifying existing ones can be prohibitive, slowing the transition to sustainable practices (Anshassi & Townsend, 2023).

Regulatory uncertainty also plays a role. Environmental regulations vary across regions and can change rapidly, impacting the feasibility of sustainable recovery. Companies may be reluctant to invest in sustainable methods if they are uncertain about future regulatory requirements. Additionally, inconsistent regulations can create an uneven playing field, making it challenging for companies to compete on a global scale.

Supply chain complexities further complicate sustainable recovery. Recycling and sustainable recovery often require collecting, sorting, and processing a wide variety of waste materials, which can be logistically challenging and costly. Effective supply chain management and coordination are essential, and any disruption can increase costs and reduce recovery efficiency.

Lack of incentives or financial support from governments and institutions is another barrier. Sustainable recovery initiatives often require external support to be economically viable, especially in their early stages. Without incentives like tax breaks, subsidies, or grants, companies may find it difficult to justify the costs associated with sustainable recovery.

In summary, these barriers must be addressed to promote the wider adoption of sustainable practices and ensure a more environmentally conscious approach to resource recovery.

8.4 TECHNOLOGICAL INNOVATIONS AND ENVIRONMENT

8.4.1 Recent advancements in PGM recovery technology

Recent advancements in the recovery of PGMs from ACs have focused on enhancing the efficiency, sustainability, and cost-effectiveness of traditional methods while simultaneously developing new methods. These advancements are driven by the increasing demand for PGMs in various industries and the growing need for more environmental-friendly and cost-efficient recovery processes.

Pyrometallurgical processes have seen significant advancements that move toward reducing the environmental footprint. New smelting techniques utilize optimized fluxes and off-gas recovery systems to minimize waste and emissions. Waste heat recovery systems contribute to energy efficiency, making these high-temperature processes more sustainable.

The development of more efficient leaching techniques has been significant in recent years. Traditional leaching processes relied on harsh chemicals, which pose environmental risks. New methods, such as thiosulfate leaching, offer a more sustainable approach, dissolving PGMs without using dangerous substances like cyanide (Compagnone *et al.*, 2023a, 2023b; McNeice *et al.*, 2022). These eco-friendly leaching agents are effective in extracting PGMs from ACs, contributing to reduced environmental impact. Although cyanide is a hazardous compound, it can be generated biologically and used for PMG leaching (Kumar *et al.*, 2023; Shin *et al.*, 2015).

Improvements in solvent extraction have also been instrumental in advancing PGM recovery technologies. Modern solvents, such as the thiophosphate-based extractant (Yamada *et al.*, 2020) and the dithiophenol-based extractant with three sulfur atoms (Yamada *et al.*, 2023), which achieve selective extraction of 90–100% of Pd, along with alkoxyimine-1-propylpyridinium derivatives with over 90% extraction efficiency for Pt(IV) and Pd(II) (Wiecka *et al.*, 2023) are more selective and efficient, allowing for better separation of PGMs from other metals. This selectivity enhances the purity of the recovered PGMs and reduces contamination risks. Additionally, new solvent extraction processes operate with lower energy consumption, supporting a more sustainable approach (Paiva, 2023; Zheng *et al.*, 2021).

The use of electrometallurgical techniques has evolved as well, with advancements in electrowinning and electrorefining. Electrowinning, which involves using electrical currents to recover PGMs from leachates, has become more precise with improved electrode materials and designs. Electrorefining allows for further purification of PGMs by separating them from base metals, leading to high-purity outputs (Dong *et al.*, 2017; Sethurajan *et al.*, 2019).

In recent years, there has been a notable increase in the use of biotechnological approaches. These technologies are emerging as innovative alternatives to conventional methods. These approaches employ the power of microorganisms to extract and concentrate PGMs from ACs through bioleaching, bioaccumulation, and biomineralization. These biotechnological methods operate at lower temperatures and with fewer hazardous chemicals, offering significant environmental benefits (Asghari *et al.*, 2013; Karim & Ting, 2021). These advancements in PGM recovery technology reflect a growing commitment to sustainability and environmental responsibility. They collectively address the challenges associated with conventional methods and pave the way for more efficient, sustainable, and cost-effective PGM recovery practices. As these technologies continue to evolve, the automotive and recycling industries can anticipate a more sustainable future.

8.4.2 Novel catalysts and materials

The use of novel materials has significantly improved the efficiency of catalysts. These developments aim to offer better performance, reduce costs, and minimize environmental impact. Nanotechnology has transformed catalyst design by allowing for the creation of catalysts with specific structural characteristics. Nanocatalysts, with their high surface area-to-volume ratios, are more reactive, thus enhancing catalytic efficiency. They can be engineered to have specific shapes, sizes, and compositions, which improves selectivity and reactivity in chemical processes. In PGM recovery, nanocatalysts contribute to more efficient recovery methods by maximizing contact with target metals and reducing raw material usage (Wang *et al.*, 2022). Biomimetic catalysts, inspired by natural enzymatic processes, offer high specificity and efficiency in chemical reactions. These catalysts mimic the behavior of enzymes, providing precise control over chemical transformations and reducing waste production.

Their application in PGM recovery and other processes helps achieve greater efficiency with fewer by-products.

Hybrid catalysts combine different types of catalytic materials to create more versatile and efficient systems. A typical hybrid catalyst might consist of a metal-based component combined with an organic or inorganic material. This approach can lead to improved catalytic performance and a broader range of applications. The flexibility in designing hybrid catalysts makes them suitable for various industrial processes (Qin *et al.*, 2022).

Graphene and graphene oxide have emerged as promising materials for novel catalysts. Their high electrical conductivity, large surface areas, and flexibility in chemical modifications make them valuable for energy storage, fuel cell applications, and chemical synthesis. Graphene-based catalysts offer improved efficiency and reduced environmental impact, adding to the growing arsenal of advanced catalytic materials (Alshammari *et al.*, 2021; Li *et al.*, 2012).

Researchers are also exploring alternatives to platinum-based catalysts due to the high cost and limited availability of PGMs. Catalysts made from abundant materials such as iron, cobalt, or nickel can deliver similar performance while reducing costs and reliance on scarce resources (Qin *et al.*, 2022). This shift is especially relevant in applications like fuel cells, where non-platinum catalysts are driving innovation and cost reduction (Qin *et al.*, 2022).

Overall, these advancements in novel catalysts and materials are reshaping chemical processes, including PGM recovery from ACs. The potential applications are extensive, covering energy production, environmental remediation, and beyond.

8.5 CONCLUSIONS AND FUTURE PERSPECTIVES

The automotive industry heavily relies on PGMs for ACs, which are critical for reducing harmful vehicle emissions. As regulations tighten around emissions and air quality, the demand for PGMs is expected to increase. Recycling PGMs from ACs offers a sustainable way to meet this demand while reducing the environmental impact associated with primary mining. By recycling, the automotive industry can help minimize resource depletion, energy consumption, and emissions, aligning with broader environmental sustainability goals. Furthermore, as the automotive industry transitions toward electric vehicles, the demand for certain PGMs used in ACs may shift. This transition could influence the types of PGMs recovered and the recycling strategies employed. The industry's ability to adapt to these changes, while maintaining a commitment to sustainability, will be crucial in the coming years.

By focusing on sustainable recovery methods, the automotive industry and other PGM-dependent sectors can reduce their environmental footprint, promote a circular economy, and contribute to a cleaner, greener future. The chapter's exploration of recovery techniques, from chemical and physical methods to biotechnological approaches, demonstrates that sustainable PGM recovery is both feasible and beneficial. By embracing these methods, industries can minimize waste, reduce energy consumption, and decrease reliance on primary mining, all while meeting the growing demand for PGMs.

To further advance sustainable PGM recovery, future research should focus on several key areas. Improving the efficiency of recovery processes is crucial, with a focus on reducing energy consumption and waste generation. Exploring biotechnological approaches and novel catalysts can offer more environmental-friendly solutions. Research should also consider the impact of the transition to electric vehicles on PGM recovery. Understanding how this shift will affect the types of PGMs used and the recycling strategies employed is essential. Finally, collaboration across industries, academia, and policymakers will be critical to drive innovation and ensure the adoption of sustainable practices in PGM recovery.

ACKNOWLEDGEMENTS

This work has been co-financed by the 2014–2020 ERDF Operational Programme and by the Department of Economy, Knowledge, Business and University of the Regional Government of Andalusia, Spain (Project references: FEDER-UCA18-106138 and P18-FR-4128).

REFERENCES

Aaltonen M., Peng C., Wilson B. P. and Lundström M. (2017). Leaching of metals from spent lithium-ion batteries. *Recycling*, **2**(4), 20, https://doi.org/10.3390/recycling2040020

Adetunji A. I., Oberholster P. J. and Erasmus M. (2023). Bioleaching of metals from e-waste using microorganisms: a review. *Minerals*, **13**, 828, https://doi.org/10.3390/min13060828

Al-Enezi G., Hamoda M. F. and Fawzi N. (2011). Ion exchange extraction of heavy metals from wastewater sludges. *Journal of Environmental Science and Health, Part A*, **39**(2), 455–464, https://doi.org/10.1081/ese-120027536

Alshammari A. S., Alfatah A. A. and Alabdi M. M. (2024). Boosting the catalytic performance of Au-Pd/graphene oxide nanocomposites via the controlled oxidation of graphene sheets. *Journal of Composites Science*, **8**(3), 82, https://doi.org/10.3390/jcs8030082

Alsukaibi A. K. D. (2022). Various approaches for the detoxification of toxic dyes in wastewater. *Processes*, **10**(10), 1968, https://doi.org/10.3390/pr10101968

Anshassi M. and Townsend T. G. (2023). The hidden economic and environmental costs of eliminating kerb-side recycling. *Nature Sustainability*, **6**, 919–928, https://doi.org/10.1038/s41893-023-01122-8

Asghari I., Mousavi S. M., Amiri F. and Tavassoli S. (2013). Bioleaching of spent refinery catalysts: a review. *Journal of Industrial and Engineering Chemistry*, **19**(4), 1069–1081, https://doi.org/10.1016/j.jiec.2012.12.005

Assunção A., Matos A., Da Costa A. M. R., Candeias A. and Costa M. C. (2016). A bridge between liquid-liquid extraction and the use of bacterial communities for palladium and platinum recovery as nanosized metal sulphides. *Hydrometallurgy*, **163**, 40–48, https://doi.org/10.1016/j.hydromet.2016.03.012

Bloxham L., Brown S., Cole L., Cowley A., Fujita M., Girardot N., Jiang J., Raithatha R., Ryan M., Tang B. and Wang A. (2022). PGM market report, retrieved from internet: https://matthey.com/pgm-market-report-2022

Brown S., Cole L., Cowley A., Fujita M., Girardot N., Grant S., Jiang J., Raithatha R., Ryan M. and Tang B. (2023). PGM market report, 361–363, retrieved from internet:

https://matthey.com/documents/161599/404086/PGM+Market+Report+May23.pdf/2f048a72-74a8-8b23-f18e-c875000ed76b?t=1684144507321

Brown S., Cole L., Cowley A., Dickson J., Fujita M., Girardot N., Grant S., Jiang J., Raithatha R., Ryan M., Underwood M. and Wang A. (2024). *PGM market report*, 1–48, retrieved from internet: https://matthey.com/media/2024/johnson-matthey-publishes-latest-pgm-market-report-2024

Cao Y., Harjanto S., Shibayama A., Naitoh I., Nanami T., Kasahara K., Okumura Y. and Fujita T. (2006). Kinetic study on the leaching of Pt, Pd and Rh from automotive catalyst residue by using chloride solutions. *Materials Transactions*, **47**(8), 2015–2024, https://doi.org/10.2320/matertrans.47.2015

Chasserio N., Guillemet-Fritsch S., Lebey T. and Dagdag S. (2009). Ceramic substrates for high-temperature electronic integration. *Journal of Electronic Materials*, **38**(June), 164–174, https://doi.org/10.1007/s11664-008-0571-8

Chidunchi I., Kulikov M., Safarov R. and Kopishev E. (2024). Extraction of platinum group metals from catalytic converters. *Heliyon*, **10**(3), e25283, https://doi.org/10.1016/j.heliyon.2024.e25283

Compagnone M., González-Cortés J. J., Pilar Yeste M., Cantero D. and Ramírez M. (2023a). Sustainable recovery of platinum group metals from spent automotive three-way catalysts through a biogenic thiosulfate-copper-ammonia system. *Molecules*, **28**(24), 8078, https://doi.org/10.3390/molecules28248078

Compagnone M., González-Cortés J. J., del Yeste M. P., Cantero D. and Ramírez M. (2023b). Bioleaching of the α-alumina layer of spent three-way catalysts as a pretreatment for the recovery of platinum group metals. *Journal of Environmental Management*, **345**, 118825, https://doi.org/10.1016/J.JENVMAN.2023.118825

De Jong G. A. H., Hazeu W. and Kuenen J. G. (1997). Isolation of the tetrathionate hydrolase from *Thiobacillus acidophilus*. *European Journal of Biochemistry*, **243**(3), 678–683, https://doi.org/10.1111/j.1432-1033.1997.00678.x

De Vargas I., Macaskie L. E. and Guibal E. (2004). Biosorption of palladium and platinum by sulfate-reducing bacteria. *Journal of Chemical Technology and Biotechnology*, **79**(1), 49–56, https://doi.org/10.1002/jctb.928

Directorate-General for Enterprise and Industry (European Commission) (2013). Tackling the Challenges in Commodity Markets and on Raw Materials–COM(2011) 25 Final, Official Journal of the European Union. Publication office of the European Union, Luxembourg. https://doi.org/10.2769/74549

Dong Z., Jiang T., Xu B., Yang Y. and Li Q. (2017). Recovery of gold from pregnant thiosulfate solutions by the resin adsorption technique. *Metals*, **7**(12), 1–, https://doi.org/10.3390/met7120555

Fajar A. T. N., Hanada T., Firmansyah M. L., Kubota F. and Goto M. (2020). Selective separation of platinum group metals via sequential transport through polymer inclusion membranes containing an ionic liquid carrier. *ACS Sustainable Chemistry and Engineering*, **8**(30), 11283–11291, https://doi.org/10.1021/acssuschemeng.0c03205

Ferreira P. F., Sérvulo E. F. C., Da Costa A. C. A., Ferreira D. M., Godoy M. L. D. P. and Oliveira F. J. S. (2017). Bioleaching of metals from a spent diesel hydrodesulfurization catalyst employing *Acidithiobacillus thiooxidans* FG-01. *Brazilian Journal of Chemical Engineering*, **34**(1), 119–129, https://doi.org/10.1590/0104-6632.20170341s20150208

Fornalczyk A. and Saternus M. (2009). Removal of platinum group metals from the used auto catalytic converter. *Metalurgija*, **48**(2), 133–136. https://hrcak.srce.hr/32004

Gavrilescu M. (2022). Microbial recovery of critical metals from secondary sources. *Bioresource Technology*, **344**(PA), 126208, https://doi.org/10.1016/j.biortech.2021.126208

Ghalehkhondabi V., Fazlali A. and Daneshpour F. (2021). Electrochemical extraction of palladium from spent heterogeneous catalysts of a petrochemical unit using the leaching and flat plate graphite electrodes. *Separation and Purification Technology*, **258**, 117527, https://doi.org/10.1016/j.seppur.2020.117527

Godlewska-Żyłkiewicz B., Sawicka S. and Karpińska J. (2019). Removal of platinum and palladium from wastewater by means of biosorption on fungi *Aspergillus* sp. and yeast *Saccharomyces* sp. *Water*, **11**(7), 1522, https://doi.org/10.3390/w11071522

Golnaraghi Ghomi A., Asasian-Kolur N., Sharifian S. and Golnaraghi A. (2020). Biosorpion for sustainable recovery of precious metals from wastewater. *Journal of Environmental Chemical Engineering*, **8**(4), 103996, https://doi.org/10.1016/j.jece.2020.103996

Gómez-Ramírez M., Moreno-Villanueva F. and Rojas-Avelizapa N. G. (2020). *Acidithiobacillus thiooxidans* DSM 26636: an alternative for the bioleaching of metallic burrs. *Catalysts*, **10**(11), 1–11, https://doi.org/10.3390/catal10111230

Grilli M. L., Slobozeanu A. E., Larosa C., Paneva D., Yakoumis I. and Cherkezova-Zheleva Z. (2023). Platinum group metals: green recovery from spent auto-catalysts and reuse in new catalysts–a review. *Crystals*, **13**(4), 550, https://doi.org/10.3390/cryst13040550

Grohol M. and Veeh C. (2023). Study on the critical raw materials for the EU 2023–Final report, retrieved from internet: https://data.europa.eu/doi/10.2873/725585

He L., Fan Y., Luo L., Bellettre J. and Yue J. (2020). Preparation of $Pt/\gamma\text{-}Al_2O_3$ catalyst coating in microreactors for catalytic methane combustion. *Chemical Engineering Journal*, **380**, 122424 Contents, https://doi.org/10.1016/j.cej.2019.122424

Hedrich S., Kraemer D., Junge M., Marbler H., Bau M. and Schippers A. (2020). Bioprocessing of oxidized platinum group element (PGE) ores as pre-treatment for efficient chemical extraction of PGE. *Hydrometallurgy*, **196**(June), 105419, https://doi.org/10.1016/j.hydromet.2020.105419

Hughes A. E., Haque N., Northey S. A. and Giddey S. (2021). Platinum group metals: a review of resources, production and usage with a focus on catalysts. *Resources*, **10**(9), 1–40, https://doi.org/10.3390/resources10090093

Jeyakumar P., Debnath C., Vijayaraghavan R. and Muthuraj M. (2022). Trends in bioremediation of heavy metal contaminations. *Environmental Engineering Research*, **28**(4), 220631–0, https://doi.org/10.4491/eer.2021.631

Jha M. K., Lee J. C., Kim M. S., Jeong J., Kim B. S. and Kumar V. (2013). Hydrometallurgical recovery/recycling of platinum by the leaching of spent catalysts: a review. *Hydrometallurgy*, **133**(2013), 23–32, https://doi.org/10.1016/j.hydromet.2012.11.012

Karim S. and Ting Y. P. (2020). Ultrasound-assisted nitric acid pretreatment for enhanced biorecovery of platinum group metals from spent automotive catalyst. *Journal of Cleaner Production*, **255**, 120199, https://doi.org/10.1016/j.jclepro.2020.120199

Karim S. and Ting Y. P. (2021). Recycling pathways for platinum group metals from spent automotive catalyst: a review on conventional approaches and bio-processes. *Resources, Conservation and Recycling*, **170**(2021), 105588, https://doi.org/10.1016/j.resconrec.2021.105588

Karim S. and Ting Y. P. (2022). Bioleaching of platinum, palladium, and rhodium from spent automotive catalyst using bacterial cyanogenesis. *Bioresource Technology Reports*, **18**(April), 101069, https://doi.org/10.1016/j.biteb.2022.101069

Karim S., Saw H. M. and Ting Y. P. (2024). Evaluation of green chemistry metrics for sustainable recycling of platinum group metals from spent automotive catalysts via bioleaching. *Green Chemistry*, **26**(7), 4112–4126, https://doi.org/10.1039/D3GC03918H

Kumar A., Shemi A., Chipise L., Moodley S., Yah C. S. and Ndlovu S. (2023). Can microbial Bio-CN be a sustainable alternative to the chemical cyanidation of precious metals? An update and way forward. *Renewable and Sustainable Energy Reviews*, **188**(March), 113892, https://doi.org/10.1016/j.rser.2023.113892

Li Y., Sun C., Yu C., Wang C., Liu Y. and Song Y. (2012). Graphene oxide and its applications in catalysis. *Advanced Materials Research*, **476–478**(February), 1488–1495, https://doi.org/10.4028/www.scientific.net/AMR.476-478.1488

Liu C., Sun S., Zhu X. and Tu G. (2021). Metals smelting-collection method for recycling of platinum group metals from waste catalysts: a mini review. *Waste Management & Research*, **39**(1), 43–52, https://doi.org/10.1177/0734242X20969795

Lou Y., Chang W., Cui T., Qian H., Huang L., Ma L., Hao X. and Zhang D. (2021). Microbiologically influenced corrosion inhibition of carbon steel via biomineralization induced by *Shewanella putrefaciens*. *NPJ Materials Degradation*, **5**(1), 1–11, https://doi.org/10.1038/s41529-021-00206-0

Lyatuu I., Loss G., Farnham A., Lyatuu G. W., Fink G. and Winkler M. S. (2021). Associations between natural resource extraction and incidence of acute and chronic health conditions: evidence from Tanzania. *International Journal of Environmental Research and Public Health*, **18**, 6052, https://doi.org/10.3390/ijerph18116052

Machado J. V., da Silva M. L. A., Silva C. L. S., Correia M. C. G., da Silva Ruy A. D. and Pontes L. A. M. (2023). Catalysts and processes for gluconic and glucaric acids production: A comprehensive review of market and chemical routes. *Catalysis Communications*, **182**(July), 106740, https://doi.org/10.1016/j.catcom.2023.106740

Malekian H., Salehi M. and Biria D. (2019). Investigation of platinum recovery from a spent refinery catalyst with a hybrid of oxalic acid produced by *Aspergillus niger* and mineral acids. *Waste Management*, **85**, 264–271, https://doi.org/10.1016/j.wasman.2018.12.045

McNeice J., Mahandra H. and Ghahreman A. (2022). Application of biogenic thiosulfate produced by *Methylophaga sulfidovorans* for sustainable gold extraction. *ACS Sustainable Chemistry and Engineering*, **10**(30), 10034–10046, https://doi.org/10.1021/acssuschemeng.2c02872

Mehdipour F., Delrieux T., Maurer F., Grunwaldt J. D., Klahn C. and Dittmeyer R. (2024). Opportunities and limitations of metal additive manufacturing of structured catalytic converters. *Catalysis Communications*, **187**(October), 106873, https://doi.org/10.1016/j.catcom.2024.106873

Mohallem N. D. S., Viana M. M. and Silva R. A. (2011). Automotive catalysts: performance, characterization and development. In: New Trends and Developments in Automotive Industry, InTechOpen, London, United Kingdom. pp. 347–364, https://doi.org/10.5772/13303

Moschovi A. M., Zagoraiou E., Polyzou E. and Yakoumis I. (2021). Recycling of critical raw materials from hydrogen chemical storage stacks (PEMWE), membrane electrode assemblies (MEA) and electrocatalysts. *IOP Conference Series: Materials Science and Engineering*, **1024**(1), 12008, https://doi.org/10.1088/1757-899X/1024/1/012008

Mpinga C. N., Eksteen J. J., Aldrich C. and Dyer L. (2015). Direct leach approaches to platinum group metal (PGM) ores and concentrates: a review. *Minerals Engineering*, **78**, 93–113, https://doi.org/10.1016/j.mineng.2015.04.015

Mwase J. M., Petersen J. and Eksteen J. J. (2012). A conceptual flowsheet for heap leaching of platinum group metals (PGMs) from a low-grade ore concentrate. *Hydrometallurgy*, **111–112**(1), 129–135, https://doi.org/10.1016/j.hydromet.2011.11.012

Nguyen V. T., Riaño S., Aktan E., Deferm C., Fransaer J. and Binnemans K. (2021). Solvometallurgical recovery of platinum group metals from spent automotive catalysts. *ACS Sustainable Chemistry and Engineering*, **9**(1), 337–350, https://doi.org/10.1021/acssuschemeng.0c07355

Norgate T. and Haque N. (2010). Energy and greenhouse gas impacts of mining and mineral processing operations. *Journal of Cleaner Production*, **18**(3), 266–274, https://doi.org/10.1016/j.jclepro.2009.09.020

Paiva A. P. (2017). Recycling of palladium from spent catalysts using solvent extraction–some critical points. *Metals*, **7**(11), 505, https://doi.org/10.3390/met7110505

Paiva A. P. (2023). Recent advances in platinum and palladium solvent extraction from real leaching solutions of spent catalysts. *Catalysts*, **13**(7), 1146, https://doi.org/10.3390/catal13071146

Paiva A. P., Piedras F. V., Rodrigues P. G. and Nogueira C. A. (2022). Hydrometallurgical recovery of platinum-group metals from spent auto-catalysts–focus on leaching and solvent extraction. *Separation and Purification Technology*, **286**(January), 120474, https://doi.org/10.1016/j.seppur.2022.120474

Pathak A., Al-Sheeha H., Navvamani R., Kothari R., Marafi M. and Rana M. S. (2022). Recycling of platinum group metals from exhausted petroleum and automobile catalysts using bioleaching approach: a critical review on potential, challenges, and outlook. *Reviews in Environmental Science and Biotechnology*, **21**(4), 1035–1059, https://doi.org/10.1007/s11157-022-09636-x

Pianowska K., Kluczka J., Benke G., Goc K., Malarz J., Ochmański M. and Leszczyńska-Sejda K. (2023). Solvent extraction as a method of recovery and separation of platinum group metals. *Materials*, **16**(13), 1–16, https://doi.org/10.3390/ma16134681

Pietrelli L. and Fontana D. (2013). Automotive spent catalysts treatment and platinum recovery. *International Journal of Environment and Waste Management*, **11**(2), 222–232, https://doi.org/10.1504/IJEWM.2013.051845

Pinto J., Lopes C. B., Henriques B., Couto A. F., Ferreira N., Carvalho L., Costa M., Torres J. M. P., Vale C. and Pereira E. (2021). Platinum-group elements sorption by living macroalgae under different contamination scenarios. *Journal of Environmental Chemical Engineering*, **9**(2), 105100, https://doi.org/10.1016/j.jece.2021.105100

Pope C. A., Ezzati M. and Dockery D. W. (2009). Fine-particulate air pollution and life expectancy in the United States. *New England Journal of Medicine*, **360**(4), 376–386, https://doi.org/10.1056/nejmsa0805646

Pourhossein F. and Mousavi S. M. (2023). Improvement of gold bioleaching extraction from waste telecommunication printed circuit boards using biogenic thiosulfate by *Acidithiobacillus thiooxidans*. *Journal of Hazardous Materials*, **450**(February), 131073, https://doi.org/10.1016/j.jhazmat.2023.131073

Qin C., Tian S., Wang W. and Jiang Z. (2022). Advances in platinum-based and platinum-free oxygen reduction reaction catalysts for cathodes in direct methanol fuel cells. *Frontiers in Chemistry*, **10**(November), 1–19, https://doi.org/10.3389/fchem.2022.1073566

Rai V., Liu D., Xia D., Jayaraman Y. and Gabriel J. C. P. (2021). Electrochemical approaches for the recovery of metals from electronic waste: a critical review. *Recycling*, **6**(3), 1–28, https://doi.org/10.3390/recycling6030053

Rao C. R. M. and Reddi G. S. (2000). Platinum group metals (PGM); occurrence, use and recent trends in their determination. *Trends in Analytical Chemistry*, **19**(9), 565–586, https://doi.org/10.1016/S0165-9936(00)00031-5

Rawlings D. E., Dew D. and Du Plessis C. (2003). Biomineralization of metal-containing ores and concentrates. *Trends in Biotechnology*, **21**(1), 38–44, https://doi.org/10.1016/S0167-7799(02)00004-5

Rendón-Castrillón L., Ramírez-Carmona M., Ocampo-López C. and Gómez-Arroyave L. (2023). Bioleaching techniques for sustainable recovery of metals from solid matrices. *Sustainability (Switzerland)*, **15**(13), 10222, https://doi.org/10.3390/su151310222

Roy R., Samanta S., Pandit S., Naaz T., Banerjee S., Rawat J. M., Chaubey K. K. and Saha R. P. (2024). An overview of bacteria-mediated heavy metal bioremediation strategies. *Applied Biochemistry and Biotechnology*, **196**(3), 1712–1751, https://doi.org/10.1007/s12010-023-04614-7

Saguru C., Ndlovu S. and Moropeng D. (2018). A review of recent studies into hydrometallurgical methods for recovering PGMs from used catalytic converters. *Hydrometallurgy*, **182**(October), 44–56, https://doi.org/10.1016/j.hydromet.2018.10.012

Samanta N. S., Das P. P., Dhara S. and Purkait M. K. (2023). An overview of precious metal recovery from steel industry slag: recovery strategy and utilization. *Industrial & Engineering Chemistry Research*, **62**, 9006–9031, https://doi.org/10.1021/acs.iecr.3c00604

Schrecker T., Birn A. and Aguilera M. (2018). How extractive industries affect health: political economy underpinnings and pathways. *Health & Place*, **52**, 135–147, https://doi.org/10.1016/j.healthplace.2018.05.005

Sedlakova-Kadukova J., Marcincakova R., Luptakova A., Vojtko M., Fujda M. and Pristas P. (2020). Comparison of three different bioleaching systems for Li recovery from lepidolite. *Scientific Reports*, **10**(1), 1–8, https://doi.org/10.1038/s41598-020-71596-5

Sethurajan M., Van Hullebusch E. D., Fontana D., Akcil A., Deveci H., Batinic B., Leal J. P., Gasche T. A., Kucuker A., Kuchta K., Neto I. F. F., Soares H. M. V. M., Chmielarz A., Ao J. and Leal P. (2019). Recent advances on hydrometallurgical recovery of critical and precious elements from end of life electronic wastes–a review. *Critical Reviews in Environmental Science and Technology*, **49**(3), 212–275, https://doi.org/10.1080/10643389.2018.1540760

Shin D., Park J., Jeong J. and Kim B. S. (2015). A biological cyanide production and accumulation system and the recovery of platinum-group metals from spent automotive catalysts by biogenic cyanide. *Hydrometallurgy*, **158**, 10–18, https://doi.org/10.1016/j.hydromet.2015.09.021

Skyle D. (2022). WHO Pathogen X conference. *The Lancet Infectious Diseases*, **22**(11), 1541, https://doi.org/10.1016/S1473-3099(22)00670-3

Sun S., Jin C., He W., Li G., Zhu H. and Huang J. (2022). A review on management of waste three-way catalysts and strategies for recovery of platinum group metals from them. *Journal of Environmental Management*, **305**(December), 114383, https://doi.org/10.1016/j.jenvman.2021.114383

Thormann L., Buchspies B., Mbohwa C. and Kaltschmitt M. (2017). PGE Production in Southern Africa, part I: production and market trends. *Minerals*, **7**(11), 224, https://doi.org/10.3390/min7110224

Van Leeuwen C., Destrac-irvine A., Dubernet M., Duch E., Gowdy M., Marguerit E., Pieri P., Parker A., De Ress L. and Ollat N. (2019). An update on the impact of climate change in viticulture and potential adaptations. *Agronomy*, **10**, 1–20, https://doi.org/10.3390/agronomy9090514

Wang W., Nadagouda M. N. and Mukhopadhyay S. M. (2022). Advances in matrix-supported palladium nanocatalysts for water treatment. *Nanomaterials*, **12**(20), 3593, https://doi.org/10.3390/nano12203593

Wei X., Liu C., Cao H., Ning P., Jin W., Yang Z., Wang H. and Sun Z. (2019). Understanding the features of PGMs in spent ternary automobile catalysts for development of cleaner recovery technology. *Journal of Cleaner Production*, **239**, 118031, https://doi.org/10.1016/j.jclepro.2019.118031

Wiecka Z., Rzelewska-Piekut M. and Regel-Rosocka M. (2022). Recovery of platinum group metals from spent automotive converters by leaching with organic and inorganic acids and extraction with quaternary phosphonium salts. *Separation and Purification Technology*, **280**, 119933, https://doi.org/10.1016/j.seppur.2021.119933

Wiecka Z., Rzelewska-Piekut M., Reis M. T. A., Ismael M. R. C., Wieszczycka K. and Regel-Rosocka M. (2023). Pd(II) and Pt(IV) dispersive or non-dispersive extraction from model and real leach solutions with alkoxyimine-1-propylpyridinium derivatives. *Separation and Purification Technology*, **317**(February), 123800, https://doi.org/10.1016/j.seppur.2023.123800

Xia J. and Ghahreman A. (2023). Platinum group metals recycling from spent automotive catalysts: metallurgical extraction and recovery technologies. *Separation and Purification Technology*, **311**(February), 123357, https://doi.org/10.1016/j.seppur.2023.123357

Xolo L., Moleko-Boyce P., Makelane H., Faleni N. and Tshentu Z. R. (2021). Status of recovery of strategic metals from spent secondary products. *Minerals*, **11**(7), 1–28, https://doi.org/10.3390/min11070673

Xu B., Chen Y., Zhou Y., Zhang B., Liu G., Li Q., Yang Y. and Jiang T. (2022). A review of recovery of palladium from the spent automobile catalysts. *Metals*, **12**, 533, https://doi.org/10.3390/ met12040533

Yakoumis I., Panou M., Moschovi A. M. and Panias D. (2021). Recovery of platinum group metals from spent automotive catalysts: a review. *Cleaner Engineering and Technology*, **3**, 100112, https://doi.org/10.1016/j.clet.2021.100112

Yamada M., Gandhi M. R. and Shibayama A. (2018). Rapid and selective recovery of palladium from platinum group metals and base metals using a extractant in environmentally friendly hydrocarbon fluids. *Scientific Reports*, **8**(July), 1–11, https://doi.org/10.1038/s41598-018-35026-x

Yamada M., Rajiv Gandhi M., Kunda U. M. R., Mori T., Haga K. and Shibayama A. (2020). Recovery of Pd(II) from leach solutions of automotive catalysts by solvent extraction with new thiophosphate extractants. *Hydrometallurgy*, **191**(May), 105221, https://doi.org/10.1016/j.hydromet.2019.105221

Yamada M., Ohira T., Watanabe N., Katagiri H., Shibayama A. and Hamada F. (2023). Recovery of Pd(II) by solvent extraction with a dithiophenol-based extractant from the undiluted leachate of spent automotive catalysts followed by water scrubbing and thiourea stripping. *Hydrometallurgy*, **215**(April), 105986, https://doi.org/10.1016/j.hydromet.2022.105986

Yang X., Hu M., Zhang C. and Steubing B. (2022). Urban mining potential to reduce primary material use and carbon emissions in the Dutch residential building sector. *Resources, Conservation and Recycling*, **180**, 106215, https://doi.org/10.1016/J.RESCONREC.2022.106215

Yong P., Rowson N. A., Farr J. P. G., Harris I. R. and Macaskie L. E. (2002). Bioaccumulation of palladium by *Desulfovibrio desulfuricans*. *Journal of Chemical Technology & Biotechnology*, **77**(5), 593–601, https://doi.org/10.1002/jctb.606

Zheng H., Ding Y., Wen Q., Liu B. and Zhang S. (2021). Separation and purification of platinum group metals from aqueous solution: recent developments and industrial applications. *Resources, Conservation and Recycling*, **167**(December), 105417, https://doi.org/10.1016/j.resconrec.2021.105417

Zhuang W. Q., Fitts J. P., Ajo-Franklin C. M., Maes S., Alvarez-Cohen L. and Hennebel T. (2015). Recovery of critical metals using biometallurgy. *Current Opinion in Biotechnology*, **33**, 327–335, https://doi.org/10.1016/j.copbio.2015.03.019

Zupanc A., Install J., Jereb M. and Repo T. (2023). Sustainable and selective modern methods of noble metal recycling. *Angewandte Chemie–International Edition*, **62**(5), 26, https://doi.org/10.1002/anie.202214453

doi: 10.2166/9781789065084_0217

Chapter 9

Platinum group elements recycling processes

*Ankur Srivastava[1], Harish Kumar Trivedi[2] and Arunabh Meshram[1]**

[1]Department of Materials Science and Engineering, Indian Institute of Technology Kanpur, Kanpur, Uttar Pradesh, India
[2]Materials Science Programme, Indian Institute of Technology Kanpur, Kanpur, Uttar Pradesh, India
**Corresponding author: arunabhm@iitk.ac.in*

ABSTRACT

The recycling of platinum group elements (PGEs) has attracted much attention in recent years. This chapter describes the importance of recycling PGE secondary resources and reutilization of the recovered metallic values. Although PGEs are used in numerous applications, the dominance in automobile catalyst production is significant. This results in a large share of PGEs in spent catalysts. The methods of recycling PGEs are elaborated with alternative and newer processes that enhance material recovery. Pretreatment processes such as thermal and hydrogen treatments assist in achieving a higher overall metal recovery. Some important investigations have been included for recovery of platinum, palladium, and rhodium to describe recycling processes. Industrial scaling up of processes and the challenges associated with it are also presented. Holistic views of material recycling and recovery of PGEs have been outlined.

Keywords: Biometallurgy, hydrometallurgy, platinum group elements, pyrometallurgy, recycling.

9.1 INTRODUCTION

9.1.1 Applications of platinum group elements

Platinum group elements (PGEs) consist of six elements: platinum (Pt), palladium (Pd), iridium (Ir), osmium (Os), rhodium (Rh), and ruthenium (Ru). PGEs are divided as light PGEs (Ru, Rh, and Pd) which belong to the fifth period, and the last electron is filled in the 4d orbital; and as heavy PGEs (Os, Ir, and Pt) which belong to the sixth period, and the last electron is filled in the 5d orbital (Hughes *et al.*, 2021). PGEs are members of the transition metals and are typically found together in nature due to their similar atomic structures. PGEs have relatively lower natural occurrence compared to other elements. Russia and South Africa contain ~96% of the reserve of PGEs (Tang *et al.*, 2022). America, Canada, and Zimbabwe also have significant PGE reserves (Karim & Ting, 2021; Liu *et al.*, 2021). PGEs have similar physical and chemical properties such as high melting points, high corrosion resistance, high electrical conductivity, stable chemical properties, and good selectivity as catalysts for chemical reactions (Tang *et al.*, 2022; Yakoumis *et al.*, 2021).

Figure 9.1 shows all PGE along with their various applications. Initially, PGEs were used in making coins and jewellery. However, the unique properties of these elements made them suitable for other important applications as well (Xia & Ghahreman, 2023). Their economic importance relates to their use in various industrial applications, particularly in catalysis and electronics. PGEs are critical in numerous industries, with catalytic converters in automobiles being one of the major uses and contributing >60% of the total PGE demand (Liu *et al.*, 2021; Pathak *et al.*, 2022; Tang *et al.*, 2022).

Platinum, palladium, and rhodium are metals that are used for the treatment of hazardous gases generated in automobiles. Modern industry considers Pt, Pd, and Rh metals as crucial for future development, and their shares in various applications are shown in Table 9.1 (Sun *et al.*, 2022). However, ruthenium, osmium, and iridium are not suitable for automobile catalytic converter purposes due to the volatility of their oxides (Xia & Ghahreman, 2023).

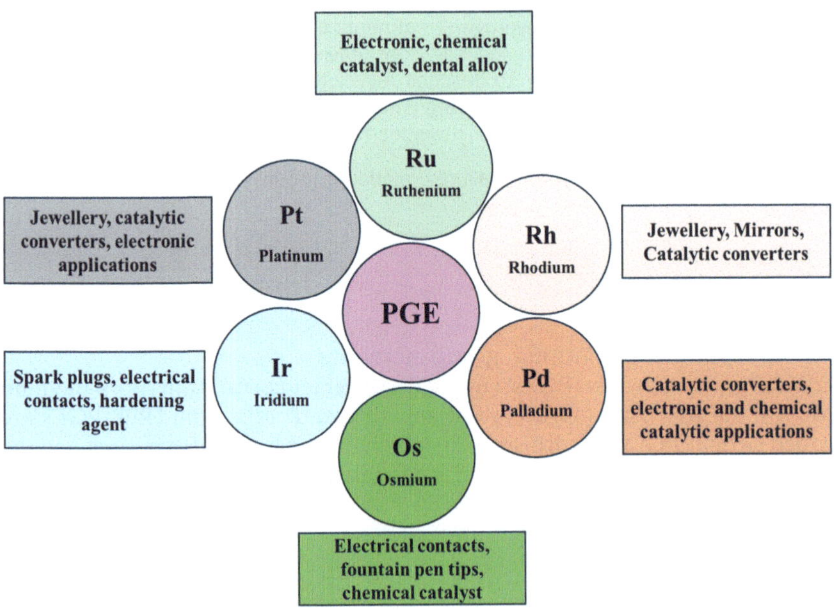

Figure 9.1 PGE and their applications.

Table 9.1 Uses of platinum (Pt), palladium (Pd), and rhodium (Rh) for various applications (in %).

	Auto Catalyst	Industrial	Investment	Jewellery
Platinum (Pt)	38.87	31.32	0.86	28.95
Palladium (Pd)	76.73	16.86	5.03	1.39
Rhodium (Rh)	84.91	15.09	0	0

Source: Hughes *et al.* (2021).

For glassmaking, dyes for producing fibreglass and crucibles use Pt–Rh. Hard disks, filters, capacitors, mobile phone electronics, connectors, fuel cells, and electrolysers use a variety of PGEs. Similarly, medical inserts, pacemakers, and dental fillings use Pt and Ir. High-temperature thermocouples use PGEs for the determination of temperature in various furnaces (Hughes *et al.*, 2021). Therefore, a wide range of applications of PGE demonstrate the broad spectrum of wastes that contain discarded PGEs.

9.1.2 Mining versus recycling PGEs

PGEs are economically valuable, and their scarcity contributes to their high market prices. PGEs are expensive to mine, making recycling essential as an alternate source of these elements. Recycling ensures a steady supply, stabilizes prices, and supports sustainable economic growth. Dependence on primary sources can lead to supply chain disruptions. Recycling of PGEs is an alternate source to supply demand for these metals in the future (Liu *et al.*, 2021). Spent petroleum and automobile catalysts are chief secondary sources of PGE, containing significant concentrations (2000–15,000 g/t) in contrast to primary sources (2–10 g/t) (Tang *et al.*, 2022; Taninouchi & Okabe, 2023). Recycling PGE wastes aligns with the global PGE supply policies (Liu *et al.*, 2021; Pathak *et al.*, 2022). Mining and refining PGEs are environmentally damaging processes. Due to the shortage of high-grade ores, mining is becoming more costly and energy-intensive compared to the recovery of PGEs from recycling of automobile catalyst waste (Xia & Ghahreman, 2023). PGEs can also be found in low concentrations in other ores and their separation is not easy (Karim & Ting, 2021).

Compared to the extraction of PGEs from the ores which involves a complex combination of pyrometallurgical and hydrometallurgical processes, the recycling of various secondary resources of PGEs can be viewed as more economical. Recycling of PGEs also mitigates the release of harmful pollutants associated with mining and refining processes. Recycling of automobile catalysts is not only significant for PGE recovery, but also compulsory from an environmental perspective. Recycling PGEs from automobile catalysts reduces the need for new mining, saving costs and reducing price volatility (Liu *et al.*, 2021). PGEs are used as automobile catalytic converters to convert toxic emissions into other products that are less hazardous (Karim & Ting, 2021). To control emissions and to follow upcoming standards for exhaust emissions from automobiles, a larger quantity of PGE usage is projected in the near future

(Tang *et al.*, 2022). In addition, PGE demand will escalate in the near future due to their uses in hydrogen production (Hughes *et al.*, 2021). Discarded automobile catalysts contain heavy metals lead (Pb) and chromium (Cr) as well as other hazardous organic substances which are not allowed into landfills without proper treatment (Liu *et al.*, 2021).

9.2 SOURCING PGEs FROM SECONDARY SOURCES

9.2.1 Secondary PGE sources

Recovery of economically valuable and hazardous metals from industrial waste, e-waste, and automobile scrap is essential for circular economy (Srivastava & Meshram, 2023a; Trivedi *et al.*, 2023a, 2023b). Recycling of PGEs has several unique challenges due to their properties and applications. PGE-containing materials often exist in complex forms such as catalysts, electronic components, and jewellery (Hagelüken 2012; Mooiman *et al.*, 2016). Separating and recovering these metals from such diverse materials can be technically challenging and costly. PGEs are typically present in relatively low concentrations in recycled materials, making their recovery economically unfeasible without efficient extraction technologies. Recycled PGE-containing materials may contain various contaminants, including other metals and organic compounds, which can interfere with the recovery process and reduce the purity of the recovered PGEs. Many PGE recycling processes require high temperatures and energy inputs, increasing the environmental footprint and operational costs. PGEs often form complex chemical compounds with other elements, making their separation and purification challenging and requiring specialized chemical processes (Hagelüken, 2012; Mooiman *et al.*, 2016; Xia & Ghahreman, 2023).

Addressing these challenges requires innovative technologies, efficient processes, and collaboration among stakeholders, including industry, academia, and government agencies, to promote sustainable and economically viable PGE recycling practices. Improper recycling leads to increased demand for new mining, contributing to environmental degradation. Additionally, inefficient recovery processes release pollutants and waste. Figure 9.2 illustrates the essential steps involved in the recycling/recovery process of PGE.

Recycling of PGE is crucial for resource conservation, environmental protection, and economic sustainability. Implementing efficient recycling methods and technologies is essential to reap the benefits of PGE recycling and avoid the negative consequences of improper recycling practices. The supply–management–recycling of the secondary resources can ensure the continuation of the availability of the metals in the market (Tang *et al.*, 2023). The recycling of end-of-life automobile catalyst converters includes collection,

Figure 9.2 Steps involved in the recycling of secondary resources of PGE.

sorting, dismantling to recover catalyst monolith, and finally crushing and grinding. Following this step, a homogeneous concentrated powder of PGE materials is recovered which is further treated as mentioned below (Yakoumis *et al.*, 2021).

9.2.2 Methods and technologies for PGE recycling

Pyrometallurgy is the most popular method for recycling spent automotive catalysts (SAC). It involves high-temperature processes such as smelting and refining. Spent catalysts are heated to extract PGEs, forming alloys with other metals. This method is, however, energy-intensive and requires large investment for plant set up. It may also result in some PGE losses (Liu *et al.*, 2021; Tang *et al.*, 2022).

Hydrometallurgy uses aqueous solutions of acid or base, cyanide, aqua regia, hydrogen peroxide, and cupric chloride to dissolve PGE from spent catalysts (Tang *et al.*, 2022). It offers better purity but requires careful management of toxic chemicals (Liu *et al.*, 2021).

Bioleaching uses microorganisms (such as cyanogenic bacteria and fungi) to mobilize PGEs from spent catalysts. Bioleaching produces lixiviants (e.g. cyanide or organic acids) that extract PGEs. This process is also in demand due to the high Rh recovery with low energy demands (Tang *et al.*, 2022).

Although pyrometallurgy has been a cornerstone of metal extraction and refining for millennia, it has some challenges. High temperatures inherent to the process elevate safety risks. Additionally, pyrometallurgy struggles with low-grade ores and waste materials containing minimal recoverable metals. To surmount these challenges, advancements in hydrometallurgy and biometallurgy offer promising alternatives. These methods excel at extracting valuable materials even from low-grade sources.

9.3 RECYCLING OF PGEs

Recycling PGEs presents numerous and multifaceted benefits such as reduced energy consumption, diminished environmental burden, and lower production costs (Pokhrel *et al.*, 2020). SACs exhibit significantly higher PGE grades than primary ore sources, resulting in decreased energy requirements, operational expenses, and waste generation (Ilyas *et al.*, 2021; Trinh *et al.*, 2020).

To illustrate, 1 kg of platinum production necessitates 150 tonnes of primary ore. The same quantity can be reclaimed by recycling only 2 tonnes of SACs. Moreover, the energy consumption involved in the recycling of PGEs is $\sim$70–100 times lower when juxtaposed with the energy requirements of primary ore mining (Xia & Ghahreman, 2023; Yakoumis *et al.*, 2021).

In a typical passenger car, a catalytic converter uses 2–6 g PGEs, whereas a sport vehicle or light truck may require 6–30 g PGEs (Wilburn & Bleiwas, 2004). In SACs, platinum can be found at levels of up to 900 ppm, a stark contrast to the primary ore's median grade, which typically attains only $\sim$5 ppm. Similarly, Pd and Rh concentrations in SACs can attain 1600 and 120 ppm, respectively, surpassing the much lower levels of 5 and 1 ppm found in primary ores (Ilyas *et al.*, 2021; Trinh *et al.*, 2020; Xia & Ghahreman, 2023).

Therefore, recycling emerges as a suitable and effective method for recovering PGEs from SACs. By adopting recycling practices, the extraction of PGEs can be maximized from various sources, contributing to resource sustainability, and minimizing environmental impacts associated with primary mining operations.

9.3.1 Pretreatment of SACs

Pretreatment of the raw feed is crucial for material recovery. Some of these important pretreatment processes are described in Figure 9.3. SACs comprise many elements such as ferrous metals, non-ferrous metals, rare-earth metals, and precious metals. Hence, the diverse composition of SACs poses challenges in selectively recovering PGEs. To optimize energy efficiency, minimize capital investments, and enhance recovery selectivity, SAC samples are sorted and subjected to comminution, crushing, and grinding (Barakat & Mahmoud, 2004). This is because smaller particles offer a larger effective surface area, facilitating a more efficient interaction with chemicals for extraction (Meshram & Singh, 2018; Srivastava & Meshram, 2023a, 2023b). Therefore, this preprocessing stage is indispensable for effective pretreatment prior to initiating the PGE recovery

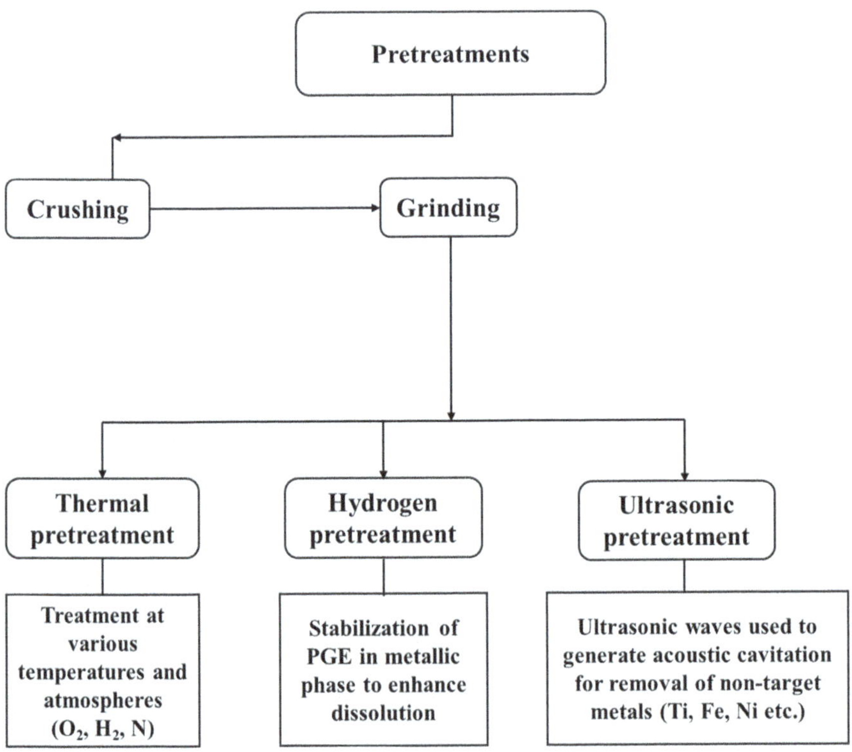

Figure 9.3 Key pretreatment operations essential for PGE recovery from secondary resources.

process. Various technologies, including hydrogen pretreatment, thermal pretreatment, and ultrasonic pretreatment are employed to pretreat SACs.

9.3.1.1 Hydrogen pretreatment

Hydrogen plays a crucial role in stabilizing PGEs by maintaining the phase, thereby enhancing their tendency to react in the subsequent processes that are carried out to recover PGEs. Its effectiveness is pronounced when dealing with PGEs existing as oxides. For instance, efficient recovery of rhodium oxides from SAC waste streams can be achieved by exposure of SACs to H_2 or CO at different temperatures and pressures. Investigations aiming at determining the effect of H_2 in PGE pretreatment show that rhodium oxide can be recovered (<5 s) with H_2 partial pressures (1–760 torr) and temperatures (50–400°C). In contrast, lower H_2 partial pressures (~7.6 torr) will require higher temperatures (200°C) for similar results, whereas higher pressures (>76 torr) significantly improve the recovery of PGE oxides (Williams *et al.*, 1998). This observation was supported by a study examining the influence of hydrogen pretreatment on PGE extraction from catalytic converters. The study showed that ~7% H_2 at varying temperatures (25–250°C) resulted in increased metal recovery (Jimenez De Aberasturi *et al.*, 2011).

9.3.1.2 Thermal pretreatment

The utilization of roasting pretreatment has been shown to significantly boost the leaching efficiency, especially in the case of rhodium, which exhibits slower leaching kinetics. Thermal pretreatment of catalytic materials occurs in environments containing oxygen, nitrogen, or hydrogen to eliminate non-target elements. The removal of iron as Fe_2O_3 and carbon as CO_2 gas from SACs is carried out without affecting other elements.

9.3.1.3 Microwave pretreatment

Microwave-assisted pretreatment has emerged as a promising technique to enhance the extraction of PGEs from waste materials. This method leverages the unique heating properties of microwaves to induce rapid and uniform heating, leading to structural changes in the material that improve leaching efficiency (Abo Atia & Spooren, 2021). Studies have demonstrated that microwave-assisted leaching in hydrochloric acid can achieve PGE recoveries comparable to traditional methods, with the added benefits of reduced processing time and lower chemical consumption. For instance, it has been shown that using microwave heating at 150°C in 6 M HCl can extract over 90% of PGEs from SACs without the need for additional oxidizing agents (Spooren *et al.*, 2021). Similarly, PGEs leaching of ~80% can be achieved using a microwave-assisted leaching process in 15 min (Suoranta *et al.*, 2015). The leach liquor used for the reduction of PGE was produced using a bioreactor. The recovery efficiency for metallic Pt and Pd can attain 90%, whereas it can surpass 70% for Rh (Yong *et al.*, 2003).

9.3.2 Recycling routes

Figure 9.4 shows the classification of recycling routes for PGE recovery. Two primary methods, namely pyrometallurgy and hydrometallurgy, are utilized

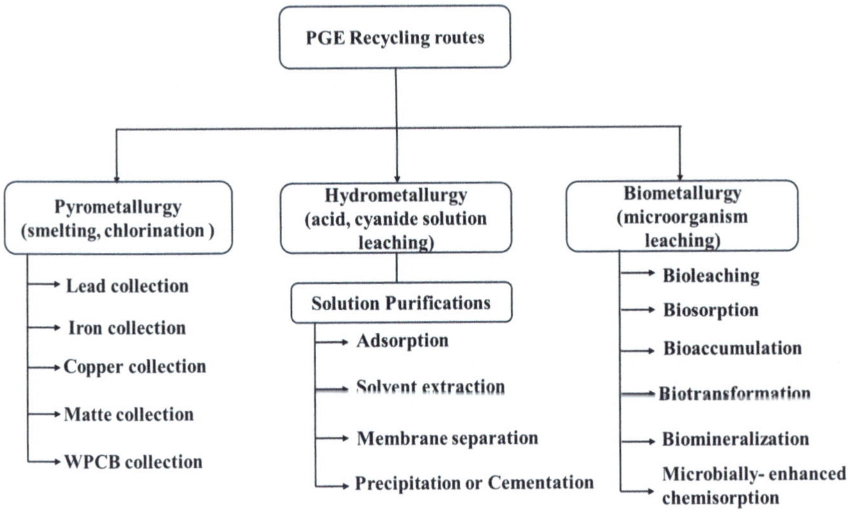

Figure 9.4 Major recycling routes for PGE recovery (WPCB: waste printed circuit board).

for the recovery of valuable metals. Solution purification technologies are instrumental in extracting metals from the PGE leachate. Biometallurgy is another route of PGE recycling that is being explored at various scales for improving the recovery from secondary resources. This section delves into recycling advancements and recovery technologies. It also emphasizes exploring future prospects by developing innovative recycling methods.

9.3.2.1 Pyrometallurgical route of extracting PGEs

The conventional route for PGE recovery from spent catalysts in the pyrometallurgical process includes crushing, batching, granulation, smelting, and subsequent separation (Dong *et al.*, 2015; Liu *et al.*, 2021; Peng *et al.*, 2017). Within the domain of high-temperature transformations, pyrometallurgical processes involve the physical or chemical alteration of PGEs. In this route, a range of processes is utilized, including smelting, chlorination, treatment of metal vapours, and heating–quenching. These processes can be executed with or without the addition of substances such as collectors (e.g. Cu or Fe) and fluxes (e.g. SiO_2 or CaO) (Trinh *et al.*, 2020). Two commercialized pyrometallurgical processing routes are elucidated in Figure 9.5 with other essential steps involved in the PGE recovery.

9.3.2.1.1 Smelting processes

After physically separating the components, the remaining stream is blended with fluxes, reductants, and metal collectors to facilitate the smelting process. Following this step, PGEs undergo dissolution and are accumulated within the matte. Carriers of PGEs, such as ceramic or alumina, are then directed into the slag, ensuring that the collector content is maintained below 0.2%

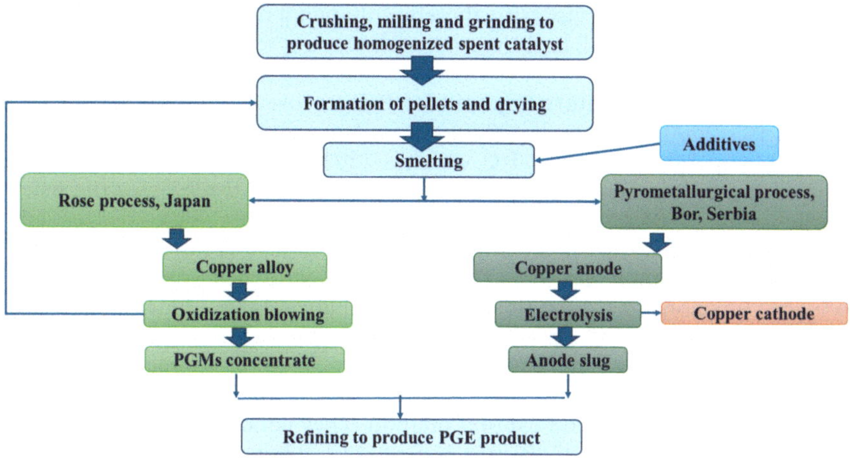

Figure 9.5 Pyrometallurgical processes for the PGE recovery.

(Ghodrat *et al.*, 2018; Karim & Ting, 2021; Panda *et al.*, 2018). The resulting PGE-bearing matte is cast as an anode, marking the commencement of subsequent electrorefining processes, achieving a recovery rate of up to 99% (Peng *et al.*, 2017).

Smelting is a state-of-the-art technique to extract metals not only from primary ores but also from the recycling perspective. The circular three-electrode electric furnaces or rectangular six-in-line submerged arc electric furnaces for smelting PGE sulfidic ores have been used for gangue separation (Jones, 1999; Thethwayo, 2018).

During smelting, end-of-life products are melted in furnaces along with fluxes and collectors. Spent SAC components, together with collectors and fluxes, undergo high-temperature treatment. This process results in the extraction of PGEs in alloy form, which can then undergo a dissolution process to separate valuable elements. The remainder phases are discarded as slag, combined with fluxes (Dong *et al.*, 2015). It is crucial to select a collector as it should have mutual solubility, melting point, and chemical properties with PGEs. Typically, copper, nickel, lead, iron, and matte are preferred collectors (Fornalczyk & Saternus, 2009).

Kolliopoulos *et al.* (2014) elucidated the mechanism behind PGE extraction, wherein they detailed how liquid copper forms minute droplets, attracting dispersed PGE microparticles using 'wetting' and 'setting' stages. These microparticles amalgamate with the molten copper to form alloys, after which solid PGE particles settle due to their high density and gravitational forces, subsequently mingling with the molten copper. Platinum exhibits superior recovery owing to its higher density compared to palladium and rhodium. The viscosity of the slag phase plays a pivotal role in separation, with PGEs settling more readily in slag with lower viscosity. Higher temperatures (1450–1600°C) result in a decreased slag viscosity, thereby enhancing recovery rates. However, smelting durations of <2 h show almost linear increases in PGE recovery (Kolliopoulos *et al.*, 2014).

Typically, the temperature required for PGE recovery via the iron smelting process is ~2000°C. Morcali (2020) devised a method where the required smelting temperature is only 950°C, which is <50% of the classical iron smelting temperature requirements. They achieved this by using three times the usual amount of fluxes (0.72 wt./wt., B_2O_3/Na_2O) from spent catalytic converters and 10 g of FeS_2. The recovery percentage for platinum and palladium attained 99%, whereas for rhodium it was 97%. Despite the greater expense in materials due to higher flux consumption, the environmental advantages of lower operating temperatures surpass this drawback (Morcali, 2020).

Similarly, Zheng *et al.* (2022) developed a method to recycle leached residues generated from hydrometallurgical processes to recover low-grade Pt, Pd, and Rh through the iron capture method. They conducted optimization of smelting conditions (1400°C, 30 min, basicity of 1.1) to ensure that the slag had a low melting point, density, and viscosity, ultimately selecting a $CaO–Al_2O_3–SiO_2–Na_2O–B_2O_3–CaF_2$ slag system. Under optimal conditions, they achieved a recovery efficiency of over 92% for all PGEs, with only a small amount of Pt, Pd, and Rh collected in the slag at concentrations of 0.83, 4.99, and 1.47 g/t, respectively. Zheng *et al.* (2022) also assessed the cost of the process, which amounted to ~1% of the total revenue, affirming the economic viability of the process (Zheng *et al.*, 2022).

There are different types of metal collection processes in pyrometallurgy that large-scale companies use to extract PGE. The PGE–Pb collection method is one of the older methods (Benson *et al.*, 2000; Panda *et al.*, 2018) of collecting PGE from secondary resources. The essence of this process lies in its simplicity and its reliance on maintaining a low smelting temperature. However, a significant drawback is the use of Pb, which poses health hazards to humans, coupled with the challenge of achieving high recovery rates of Ir and Rh. Prior to the 1980s, major companies such as Impala, Johnson Matthey, and Inco employed this technique (Benson *et al.*, 2000; Juvonen *et al.*, 2002; Keyworth, 1983).

Similarly, the PGE–Cu collection method is highly effective in recovering PGEs. At 1200°C, Pt, Pd, and Rh exhibit good solubility (30–50 wt.%) in Cu. In the treatment process of secondary sources of PGEs, spent exhaust catalysts are subjected to an electric arc furnace alongside various reagents such as collectors ($CuCO_3$ and CuO), fluxes (e.g. CaO and SiO_2), and reductants such as petroleum coke. The PGEs are enriched within the temperature range of 1450–1600°C, and under a mildly reducing atmosphere. The copper collection mechanism, involving 'wetting' and 'setting' stages, has been elucidated by Kolliopoulos *et al.* (2014). A significant drawback of this method, however, is the costly process of separating PGEs from copper and recycling the copper itself. Nippon PGE Co., a prominent company in Japan, adopts the rose process, which is considered one of the premier methods for copper collection. This technique involves enriching PGEs by blowing off copper alloys leading to oxidation during the smelting process. This involves the blowing of metal gaseous vapours through the feed of catalytic converters. Approximately 30% of PGE concentrate is yielded, while the copper oxide produced in the process can be recycled for further use (Fornalczyk & Saternus, 2009, 2013). Umicore based in Belgium is one of the key industrial recycling companies of PGEs from industrial wastes, auto-catalysts, printed circuit boards, and electronic boards (Hagelüken, 2006).

The Institute of Mining and Metallurgy in Bor (Serbia) has devised a process that integrates pyrometallurgy and electrometallurgy to extract PGE using copper as a collector from spent automobile catalysts. Before pyrometallurgical operations, beneficiation processes such as crushing, milling, and grinding are performed to prepare the material. During pyrometallurgy, molten copper collects the PGE, whereas carriers such as alumina, cordierite, or mullite form a low-viscosity slag along with fluxes. The molten copper, enriched with PGEs, is cast into an anode plate and subjected to electrolysis. PGEs are recovered from the anode sludge, whereas the pure copper cathode obtained can be reintroduced into the smelting of similar applications. The purity of the recovered copper cathode is 99.99%. The process has a yield of up to 99% recovery of PGEs (Ivanović *et al.*, 2011).

The PGE–Fe method offers an attractive way to capture PGEs using iron as a collector. Iron's strong affinity to PGEs and the low cost of Fe make it cost-effective (He *et al.*, 2016). Similar to other collection processes, this method involves pretreating the waste with fluxes and reducing agents. However, it stands out by utilizing high-temperature plasma arc melting furnaces (1500–2000°C). This intense heat separates the PGEs, which gets trapped in the slag layer floating on top of the molten metal due to their different densities (Dong *et al.*, 2014; Shen *et al.*, 2021). The method achieves a PGE recovery efficiency of 95%. The process is hampered by the short lifespan of the plasma gases and refractory materials.

Similar to the above-discussed processes, there are relatively newer approaches used to recover PGEs. The matte collection method concentrates the waste material into a matte phase (Eksteen, 2011; Jones & Geldenhuys, 2011; Van Schalkwyk *et al.*, 2013). The melting temperature varies from 1000 to 1450°C with the addition of Ni or NiS along with fluxes used to recover the PGEs from the end-of-life wastes. In total, 90% Pt, 93% Pd, and 88% Rh have been reported to be recovered (You *et al.*, 2016). Another innovative method involves the crushing and pretreatment of waste printed circuit boards (WPCBs) and spent catalysts (Kim *et al.*, 2013; Willner *et al.*, 2014). The mixture undergoes subsequent incineration at 600°C for 30–90 min to eliminate organic substances, followed by smelting at 1250–1700°C. The composition of the slag is critical in the recovery of PGEs and other precious metals such as Au present in WPCBs. This method facilitates the recovery of Au (99%), Pt (98.5%), Pd (98%), Cu and Sn (~88%), and Fe or Pb (up to 50%) achieved with a certain slag ratio. In this collection method, the distribution ratios of elements between the formed alloy of Cu–Sn and slag are very high (Kim *et al.*, 2004; Nagamori & Mackey, 1978).

9.3.2.1.2 Chlorination process

In pyrometallurgy, metals undergo chlorination at high temperatures, transforming them into their corresponding chlorides. These metal chlorides are then separated based on their varying volatility or through an adsorption process onto a carbon bed (Allen *et al.*, 1992). Xu and Xu (1998) conducted a chlorination process on a spent catalyst, incorporating NaCl at temperatures ranging from 600 to 700°C. The resulting melt undergoes treatment with steam and hot water, yielding chloride salt solutions containing PGEs. The process achieved a recovery efficiency of ~85–90% for Pt, Pd, and Rh. However, its industrial application was hindered by significant drawbacks, including high corrosion rates, the necessity for elevated temperatures, and the requirement for sophisticated instrumentation (Xu & Xu, 1998).

9.3.2.2 Hydrometallurgical route of extracting PGEs

Hydrometallurgy presents a compelling approach for metal recovery from waste, boasting distinct advantages over pyrometallurgy. These include operating at lower temperatures, providing better process control, minimizing energy consumption, achieving higher recovery yields, scalability, and yielding higher purity in the extracted materials. Figure 9.6 shows the different steps involved in the hydrometallurgy route. In this route, metals are leached in acid or alkaline solutions either with or without pretreatment operations in the presence of O_2, I_2, Br_2, Cl_2, or H_2O_2 (Angelidis, 2001; Duyvesteyn *et al.*, 1994; Jha *et al.*, 2013; Yoo, 1998). There are also several solution purification techniques used to purify the leaching solutions such as adsorption, solvent extraction, precipitation, cementation, and membrane separation (Xia & Ghahreman, 2023).

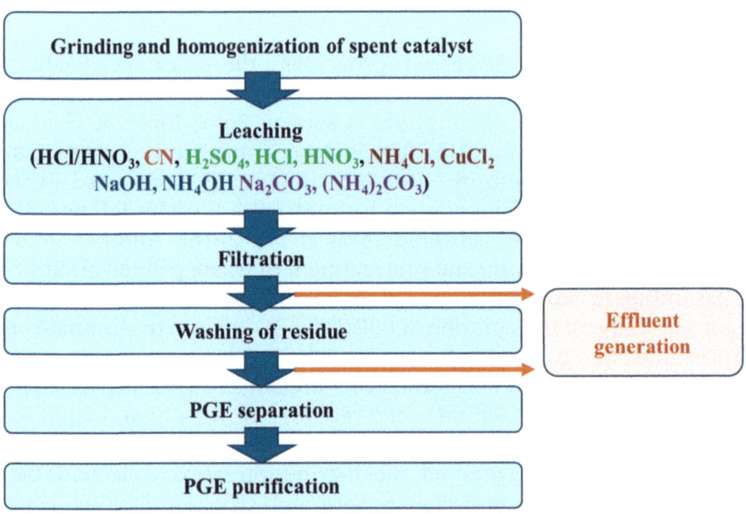

Figure 9.6 Stages involved in hydrometallurgical processes for PGE recovery.

The selection of a proper reagent for metal recovery is essential for PGE recycling. The recovery of PGEs also depends on leaching parameters such as agitation, concentration of the reagent, time, and temperature. A diverse array of leaching reagents has been explored in hydrometallurgical extraction. These include aqua regia (a mixture of hydrochloric acid and nitric acid), cyanides, strong mineral acids (H_2SO_4, HCl, and HNO_3), chlorides (ammonium chloride, copper chloride, and iron chloride), hydroxides (sodium hydroxide and ammonium hydroxide), and carbonates (sodium carbonate and ammonium carbonate) (Rzelewska & Regel-Rosocka, 2019).

Commonly employed leaching solutions for PGE recovery from spent car catalysts comprise a blend of high-concentration HCl and an oxidizing agent such as H_2O_2 (Ilyas *et al.*, 2020), nitric acid (Niemelä *et al.*, 2012), or sodium chlorate (Trinh *et al.*, 2019). This is attributed to the enhanced stability of PGE–chloride in acidic solutions. Aqua regia has been used for the leaching of spent Pt gauge dust, spent catalysts, and other sources, whereas combinations of strong acids such as HCl, HNO_3, and hot H_2SO_4 have been used for Pt–Re–Al_2O_3 catalyst, reforming and isomerization catalyst, Pt–aluminosilicate catalyst (Jha *et al.*, 2013).

9.3.2.2.1 Significance of pretreatment on leaching of PGEs

A roasting pretreatment prior to the leaching process can significantly enhance the leaching efficiency. Various roasting methods are commonly employed for this pretreatment. Reduction roasting, for instance, serves to eliminate the passive layer formed over PGE, whereas oxidation roasting facilitates the formation of metallic phases or alloys of PGE and the conversion of metal sulfides to metal oxides (De Sá Pinheiro *et al.*, 2004). Additionally, calcination at temperatures exceeding 800°C prior to leaching aids in enhancing leaching kinetics and enhances the specific surface area of PGE available for leaching reagents, thereby increasing their chemical reactivity and overall process efficiency (Ficeriová & Baláž, 2010).

Sasaki and Maeda (2014) conducted experiments with spent automotive catalytic converters, wherein they proposed pretreating them with zinc vapour. This operation showed this pretreatment helps in reducing time and chemical consumption. Subsequently, they subjected the pretreated catalysts to leaching with aqua regia. Their results revealed that the Zn-vapour treatment notably enhanced the dissolution of the PGE, with particularly effective outcomes for rhodium (Sasaki & Maeda, 2014).

9.3.2.2.2 Alkaline roasting and the addition of oxidants

Trinh *et al.* (2019) employed NaOH roasting followed by a cementation process, utilizing leaching with an HCl and $NaClO_3$ solution to recover PGE. The optimum conditions for leaching were 1.0 M H_2SO_4, 90°C, and 1 h, yielding >95% Al and Mg. In the cementation process, Al^0 powder was used and >99% PGE were precipitated within 15 min. With this method, a leaching efficiency of 97.5% for Pt, 98.8% for Pd, and 98.2% for Rh was achieved (Trinh *et al.*, 2019).

Similarly, Yakoumis *et al.* (2020) introduced a straightforward and effective leaching system for extracting PGE. Their method involved utilizing a solution mixture of 3 M HCl, 4.5 M NaCl, and 1% H_2O_2. They applied this mild leachate to 25 commercial SAC, which were subjected to comminution. The leaching was carried out for 2 h at 70°C. This approach yielded complete extraction of Pt, 92% of Pd, and 61% of Rh. Subsequent investigations revealed that the dissolution kinetics of Rh were slower compared to Pt and Pd due to passivation (Yakoumis *et al.*, 2020).

Leaching of PGEs leads to the formation of complexes in HCl and various oxidants such as hydrogen peroxide (H_2O_2), nitric acid (HNO_3), sodium hypochlorite (NaClO), and sodium chlorate ($NaClO_3$) (De Sá Pinheiro *et al.*, 2004; Marinho *et al.*, 2011; Sun & Lee, 2011). Additionally, the high concentration of chloride ions facilitates the formation of metal–chloro complexes. The relevant reactions are (Jha *et al.*, 2013; Karim & Ting, 2021):

$$3Pt + 18HCl + 4HNO_3 \rightarrow 3PtCl_6^{2-} + 6H^+ + 4NO + 8H_2O \tag{9.1}$$

$$3Pd + 12HCl + 2HNO_3 \rightarrow 3PdCl_4^{2-} + 6H^+ + 2NO + 4H_2O \tag{9.2}$$

$$Rh + 6HCl + HNO_3 \rightarrow RhCl_6^{3-} + 3H^+ + NO + 2H_2O \tag{9.3}$$

Nogueira *et al.* (2014) conducted hydrometallurgical extraction on waste spent catalysts utilizing aqua regia and cupric chloride as an oxidant. The study identified optimal leaching conditions at 6 M HCl, 80°C for 4 h, and employing 0.3 M Cu^{2+} ions, leading to 95% Pd and 86% Rh extraction (Nogueira *et al.*, 2014). Pietrelli and Fontana (2013) performed experiments to explore the impact of temperature (20–100°C), particle size (0.5–2 mm), and liquid-to-solid ratios (ranging from 1 to 10) on Pt recovery from SAC using aqua regia. Pt leaching was significantly enhanced at elevated temperatures, smaller particle sizes, and a liquid-to-solid ratio of 2 (Pietrelli & Fontana, 2013).

9.3.2.2.3 Cyanide leaching

Various studies have demonstrated the effectiveness of cyanide solutions in leaching PGE, facilitating the dissolution of a significant quantity of PGE. In this regard, NaCN has emerged as a potent lixiviant that stabilizes metal complexes, particularly at elevated temperatures and pressures. However, it has been noted that at lower temperatures and pressures, cyanide leaching yields less favourable outcomes. This is due to the formation of a passivation layer of metal oxides and the higher metallic bonding strength of PGEs (Chen & Huang, 2006; Shin *et al.*, 2015; Trinh *et al.*, 2020). The reactions of PGEs and sodium cyanide are (Chen & Huang, 2006; Karim & Ting, 2021):

$$2Pt + 8NaCN + 2H_2O + O_2 \leftrightarrow 2Na_2\left[Pt(CN)_4\right] + 4NaOH \tag{9.4}$$

$$2Pd + 8NaCN + 2H_2O + O_2 \leftrightarrow 2Na_2\left[Pd(CN)_4\right] + 4NaOH \tag{9.5}$$

$$4Rh + 24NaCN + 6H_2O + 3O_2 \leftrightarrow 4Na_3\left[Rh(CN)_6\right] + 12NaOH \tag{9.6}$$

Shams *et al.* (2004) demonstrated effective platinum recovery through cyanide leaching, achieving an 85% recovery rate under optimized conditions. These conditions included a NaCN to catalyst weight ratio of 2:1, operating at temperatures between 140 and 180°C, and maintaining a pH of 8–9 for 1 h. The process involved utilizing an anion exchange resin to extract the dissolved platinum complex. Subsequently, the resin loaded with the platinum complex was subjected to drying and incineration under an oxidizing atmosphere at temperatures ranging from 800 to 850°C. This resulted in the production of predominantly platinum metallic powder. The presence of coke did not hinder platinum recovery in cyanide leaching, suggesting that decoking of spent catalysts was unnecessary when employing this method (Shams *et al.*, 2004).

Although cyanide leaching is frequently employed for efficient recovery of PGEs from spent catalysts, it is important to note that cyano compounds pose potential toxicity risks (Cui & Zhang, 2008). To address this concern, various treatment methods are utilized to decrease the cyanide concentration. These techniques include biologically mediated cyanide oxidation, as well as chemical processes such as oxidation using air or alternative oxidants such as hydrogen peroxide or chlorine. Additionally, electrochemical methods are also employed to reduce the cyanide concentration. These approaches help mitigate environmental and health hazards associated with cyanide-containing solutions (Donato *et al.*, 2007).

Similar to cyanide leaching, leaching with halides is also very effective. Cl_2, Br_2, and I_2 have been used as substitutes for cyanide. A milder method for extracting Pt and Rh from SAC, utilizing a two-step process, has been proposed. Initially, leaching takes place in the presence of H_2SO_4 and $N_2H_6SO_4$, followed by leaching with HCl and $AlCl_3$. In this process, HCl and $AlCl_3$ provide the required chloride ions for dissolving PGE (Angelidis, 2001). Recovery efficiencies ranging from 90 to 99% for both Pt and Rh have been achieved. Another investigation utilized $AlCl_3$ along with HNO_3 as the chloride ion source and oxidant, respectively. This study highlighted the advantage of aluminium chloride over HCl, as it enabled the reduction in reagent usage while maintaining the effectiveness of platinum recovery (Angelidis & Skouraki, 1996).

The dissolution of PGEs in different halides has been explored by Dawson and Kelsall (2006). They showed that Pt and Pd iodide leaching presents less acidic and aggressive conditions, whereas the bromide solutions form bromo complexes having a more alkaline pH. Due to this, the leaching efficiencies in the case of bromide leaching decrease, therefore PGE leaching in bromide solutions are favourable at more acidic pH (Dawson & Kelsall, 2006). The use of ultrasound assistance in the leaching leads to the generation of fragments in the raw feed due to microjets produced after the bursting of bubbles near particles. This leads to improved exposure of the raw feed particles to the leach liquor used, in turn improving the leaching efficiency (Karim & Ting, 2021).

From this section, it can be deduced that the leaching of PGE-bearing raw feed is dependent on leaching reagents and process parameters. There is a great scope of research for alkali treatments to recover PGEs. There remain numerous unexplored avenues in this domain, particularly in the promising realm of recovering PGEs from SACs, especially when dealing with minimal metal quantities.

9.3.2.3 Biometallurgical route of extracting PGEs

9.3.2.3.1 Bioprocesses in metallurgy

Biometallurgy or biohydrometallurgy emerges as one of the most sustainable and promising extraction methods. Figure 9.7 shows a schematic outline of the important steps involved in biometallurgy. Here, microorganisms such as fungi and bacteria assist in recovering metals. The microorganisms leverage their metabolism to sequester, remove, or alter metals in the waste into non-hazardous or less hazardous forms (Mulligan *et al.*, 2001). Therefore, this approach is more environmental-friendly compared to conventional methods. Biometallurgy includes six major remediation techniques including bioleaching, biosorption, bioaccumulation, biotransformation, biomineralization, and microbially enhanced chemisorption (Patel & Kasture, 2014). These technologies have already been shown to be effective in the recovery of Au and are considered feasible routes for PGE recycling from SACs and e-waste (Frangoul, 2020; Narayanasamy *et al.*, 2022; Rizki *et al.*, 2019).

9.3.2.3.2 Biosorption

Similar to bioleaching, biosorption is the process by which soluble pollutants, including ionized hazardous metals are concentrated and bound to the surface of biological structures. In contrast to active metabolism, biosorption

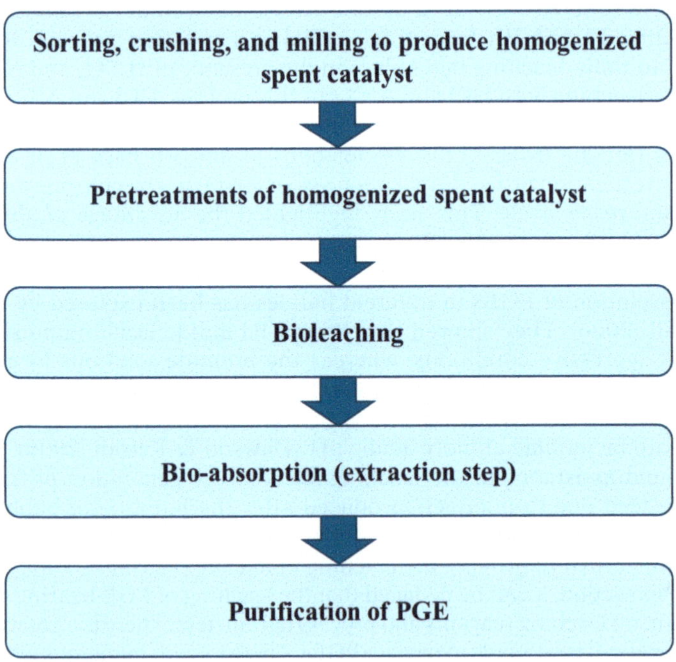

Figure 9.7 Biometallurgy operations for PGE recovery.

proceeds without the need for metabolic activity (Volesky & Holan, 1995). Biotransformation denotes the process through which a substance undergoes conversion from one chemical form to another through chemical reactions. Specifically concerning toxic metals, biotransformation involves altering their oxidation state by either adding or removing electrons, consequently modifying their chemical properties (Patel & Kasture, 2014; Satyanarayana *et al.*, 2012).

9.3.2.3.3 Bioleaching using acid-producing microorganisms

The capacity of a wide range of heterotrophic and chemolithotrophic bacteria to extract base metals such as Ni, Cu, Zn, and Fe, as well as Pt, Pd, Rh, Ag, and Au from solid sources has been studied (Ilyas & Lee, 2014). Because they are good at oxidizing iron and/or sulfur, mesophilic bacteria from genera such as *Acidithiobacillus* (which includes *Acidithiobacillus ferrooxidans, Acidithiobacillus thiooxidans*, and *Acidithiobacillus caldus*) and *Leptospirillum* are frequently used in bioleaching procedures. These bacteria are found in acidic environments and are active in a pH range of 1.5–2.5 and a temperature range of 20–55°C. They are autotrophic, meaning they use inorganic carbon (CO_2) as their carbon source, and chemoautotrophic, meaning they oxidize ferrous iron, metal sulfides or reduced sulfur compounds to provide energy for growth and vital metabolic processes (Ilyas & Lee, 2014).

9.3.2.3.4 Bioleaching using biogenic cyanide

Karim and Ting (2020) used a pretreatment method to enhance the bioleaching of PGEs from SACs. They studied the effect of acid pretreatment on increasing the recovery of PGEs by cyanogenic bacteria. Ultrasonic-assisted 6 M HNO_3 leaching was used in the pretreatment of SACs for 50 min, which led to the removal of base metals including 60% copper and 70% zinc. A bioleaching approach employing naturally generated cyanide from *Pseudomonas fluorescens* and *Bacillus megaterium* was employed to treat PGEs at 30°C. Leaching rates of Pt, Pd, and Rh all increased by pretreating the SACs with ultrasonic-assisted HNO_3 pretreatment. Although the extraction efficiencies of Pt and Pd were low (<40%), 91% Rh was recovered (Karim & Ting, 2020).

Chromobacterium violaceum was used to generate biogenic cyanide for the pressure leaching of SACs. Around 95% recovery of Pt and Pd was recorded for this pressure leaching technique in which 14 bar pressure was used for 2 h at a temperature of 150°C. The advantage of this process was to reuse the spent bacteria from the Pt/Pd-depleted bleed solution, illustrating zero discharge of harmful substances by the process (Ilyas *et al.*, 2022). Numerous factors are linked to bioleaching, all of which can be adjusted to enhance the process. These factors include pulp density, solution pH, oxygen levels, and the composition of the growth medium. Among these, oxygen stands out as the most critical control factor in bioleaching, as the dissolution of precious metals in the solution occurs at a specific dissolved oxygen level (0–0.25 mmol/L). Oxygen serves two primary functions in the solution: it sustains the life of the microorganisms and acts as an electron acceptor in the process (Kita *et al.*, 2006).

9.4 METHODS OF PGE RECOVERY

To ascertain the composition of the PGE raw feed plays a pivotal role in developing the recycling process. With a wide variety of catalysts using different PGEs, the determination of the chemical composition of the raw feed is necessary. For example, catalysts used in different chemical industries such as oil refining, hydrogenation, and oxidation can employ PGEs (Dong *et al.*, 2015). The chemical composition of the raw feed should be determined by analysing the source of the catalyst.

9.4.1 Precipitation

Most of the hydrometallurgical processes involve precipitation of the PGEs by suitable reagents. Precipitation and separation using solvent extraction of PGEs using chemical reagents have to be carried out at various operational conditions that include variations in pH, temperature, and stripping reagents. An example of recovering platinum from the leach liquor containing palladium (1292 mg/L) and platinum (486 mg/L) involves the use of an NH_4Cl solution. The precipitated product has been characterized as $(NH_4)_2[PtCl_6]$, an ammonium–chloro–platinum complex (Ilyas *et al.*, 2020). Similarly, precipitation of other important metals can be carried out using different reagents.

9.4.2 Complexation

The formation of complexes using various chemical reagents can lead to the separation of PGEs from the leach liquor containing multiple metallic ions. Another method of recovering Pt, Pd and Rh is by using sodium cyanide, leading to the formation of $[Pt(CN)_4]^{2-}$, $[Pd(CN)_4]^{2-}$, and $[Rh(CN)_6]^{3-}$. The key disadvantage of using cyanide reagents for recovering PGEs lies in the decomposition of the recovered cyanide complexes containing PGEs. High temperature (250–270°C) and pressure (2758–6206 kPa) are required for successful stripping of PGEs (Liu *et al.*, 2021).

9.4.3 Supported ionic liquid phases

In recent years, the use of alternative methods for metal recovery has been extensively investigated. One of the key areas of exploration is supported ionic liquid phases (SILP). Poly-SILP has been used as the sorbent medium for the extraction of PGEs. Using glass wool for immobilizing, the leach liquor was fed to poly-SILP and solutions of thiourea and HCl with varying concentrations were used for extraction of PGEs (Lanaridi *et al.*, 2021).

Different organic compounds are being developed and examined for the recovery of Pd and Pt from aqua regia leach liquor. One such investigation involves the development of bis((1*H*-benzimidazol-2-yl)methyl) sulfide which is a ligand capable of recovering PGEs. It was synthesized by treating a mixture of 2,2′-thiodipropionic and *o*-phenylenediamine in 4 M HCl solution at 200°C for 24 h. The resultant product was further treated with excess ammonia to produce the ligand. It has been reported that using this compound, 177 and 131 mmol/g of Pd were recovered in the first and second cycles, whereas 5.85 and 3.79 mmol/g of Pt were recovered in the two cycles (Moleko-Boyce *et al.*, 2022).

9.4.4 Solvent extraction

Conventional solvent extraction is one of the most commonly followed processes for the extraction of Pt and Pd. Some of the important organic solvents used for the process are Cyanex® 471X and trioctyldodecyl phosphonium chloride, and stripping reagents are formed by varying the combinations of $CS(NH_2)_2$, HCl, and HNO_3. Similar to the SILP, the development of polymer inclusion membranes (PIM) shows promising results in the recovery of PGEs. Isotropic-dense PIM achieves a higher metal recovery, whereas anisotropic porous PIM has rapid mass transport kinetics (Xia & Ghahreman, 2023).

Some other important materials have been used in the separation of PGEs from different sources. Oxygen-containing porous materials such as hydrazine monohydrochloride-functionalized zeolite and bentonite functionalized with 3-aminopropyl(diethoxy)methylsilane can be used for the recovery of Pt, Pd, Ir, and Rh. Similarly, organic polymer-based porous materials such as polyethylenimine/polyvinyl chloride can be used for Pt recovery and nano-Al_2O_3 can recover Os and Ir (Iftekhar *et al.*, 2022).

9.5 CHALLENGES IN SCALING-UP OF PGE RECYCLING PROCESSES

Numerous research ventures have been made in the direction of improving the existing technologies and processes of PGE recycling or developing newer and alternative processes that may get industrially accepted. Although most of the processes are explored at a laboratory scale, only a few of the recycling processes are scaled up to the industrial level. The key challenges that are hurdles to industrial scale/pilot scale establishment for a given process include numerous factors which can be economical, socio-economical, technological, and/or environmental. For a process to be adopted on an industrial scale, the technological foundations of the recycling process need to be on par with the production scale of the desired product.

Laboratory experiments usually involve smaller units of mass and volumes, whereas industrial production is at a much larger scale (in tonnes). Various operational parameters such as the size of the reaction vessel, temperature, agitation, time, volume of raw materials, choice of reagents, energy inputs, and source of energy are some important aspects that need to be thoroughly considered during scaling up. The technological feasibility of the process needs to be validated by simulation and plant scale trials.

The availability of various resources essential for an industrial scale establishment of a recycling facility is important. With a surplus of water resources and energy supply (dependable energy source in the vicinity), the greatest challenge is the steady supply of raw materials. As mentioned in section 9.3.1, the pretreatment of raw materials is an important step that ascertains the quality of raw materials. The pretreatment ensures a steady stream of concentrated feed to the industry.

Instances of lower economic value of the resultant product compared to higher investment in various segments of the process imply that the process becomes economically infeasible. The processes may have good technological performance, but if the energy and material inputs in the process do not lead to efficient material recovery that is sufficient for business, the process may not be industrially applicable. The techno-economic analysis, including the operation expenditure and capital expenditure, also needs to be carried out for successful industrial translation of the recycling process.

The most important aspect of any recycling process is to ensure efficient resource utilization from the raw materials and increase the material recovery. Resource spillage/wastage due to unorganized and improper recycling of materials is observed for many industrial wastes. The environmental burden due to the raw material should be reduced by recycling. Landfilling of raw materials can lead to adverse effects on the soil and water resources near the landfilling site. The recycling process should be developed such that further generation of waste is avoided. Else the development of other recycling processes for the newly generated wastes of the recycling process will be required.

9.6 CONCLUSION

This chapter describes the importance of PGEs and the necessity of recirculating the metallic values in secondary resources by efficient recovery. The sustainability of many industries depends on the contribution of PGEs, that is, PGEs are vital to numerous industrial sectors. Understanding the economic value of the secondary resources becomes pivotal to the development of the recycling process. SACs are an outstanding raw material for PGE recycling.

Pretreatment of the identified raw feed to increase the PGE concentrations is the first step to recycling PGEs. The selection of the pretreatment method depends on the chief route of recycling and recovery of the PGEs. Pyrometallurgy and hydrometallurgy are the two main routes that can achieve higher PGE recovery. Some recent investigations in biometallurgy have promising results. Due to the complex nature of recycling processes, the separation of valuable resources is a challenge and there is a wide scope of investigating alternative methods for PGE recycling. The scaling up of the existing and newer processes requires extensive technological and economic analysis to ensure their feasibility. The processes developed should ensure the reduction of the environmental burden by efficiently utilizing the raw feed and reducing waste generation.

REFERENCES

Abo Atia T. and Spooren J. (2021). Fast microwave leaching of platinum, rhodium and cerium from spent non-milled autocatalyst monolith. *Chemical Engineering and Processing – Process Intensification*, **164**, 108378, https://doi.org/10.1016/j.cep.2021.108378

Allen R. J., Foller P. C. and Giallombardo J. (1992). Two-step method for recovering dispersed noble metals. U.S. Patent 510632.

Angelidis T. N. (2001). Development of a laboratory scale hydrometallurgical procedure for the recovery of Pt and Rh from spent automotive catalysts. *Topics in Catalysis*, **16–17**, 419–423, https://doi.org/10.1023/a:1016641906103

Angelidis T. N. and Skouraki E. (1996). Preliminary studies of platinum dissolution from a spent industrial catalyst. *Applied Catalysis A: General*, **142**, 387–395, https://doi.org/10.1016/0926-860X(96)00088-9

Barakat M. A. and Mahmoud M. H. H. (2004). Recovery of platinum from spent catalyst. *Hydrometallurgy*, **72**, 179–184, https://doi.org/10.1016/S0304-386X(03)00141-5

Benson M., Bennett C. R., Harry J. E., Patel M. K. and Cross M. (2000). The recovery mechanism of platinum group metals from catalytic converters in spent automotive exhaust systems. *Resources, Conservation and Recycling*, **31**, 1–7, https://doi.org/10.1016/S0921-3449(00)00062-8

Chen J. and Huang K. (2006). A new technique for extraction of platinum group metals by pressure cyanidation. *Hydrometallurgy*, **82**, 164–171, https://doi.org/10.1016/j.hydromet.2006.03.041

Cui J. and Zhang L. (2008). Metallurgical recovery of metals from electronic waste: A review. *Journal of Hazardous Materials*, **158**, 228–256, https://doi.org/10.1016/j.jhazmat.2008.02.001

Dawson R. J. and Kelsall G. H. (2006). Recovery of platinum group metals from secondary materials. I. Palladium dissolution in iodide solutions. *Journal of Applied Electrochemistry*, **37**, 3–14, https://doi.org/10.1007/s10800-006-9218-8

De Sá Pinheiro A. A., Dc Lima T. S., Campos P. C. and Afonso J. C. (2004). Recovery of platinum from spent catalysts in a fluoride-containing medium. *Hydrometallurgy*, **74**, 77–84, https://doi.org/10.1016/j.hydromet.2004.01.001

Donato D. B., Nichols O., Possingham H., Moore M., Ricci P. F. and Noller B. N. (2007). A critical review of the effects of gold cyanide-bearing tailings solutions on wildlife. *Environment International*, **33**, 974–984, https://doi.org/10.1016/j.envint.2007.04.007

Dong H. G., Zhao J. C., Chen J. L., Fan X. X., Fu G. and Yang H. Q. (2014). Recovery of platinum group metal secondary resource by iron trapping method based on solid state. *The Chinese Journal of Nonferrous Metals*, **24**, 2692–2697.

Dong H., Zhao J., Chen J., Wu Y. and Li B. (2015). Recovery of platinum group metals from spent catalysts: a review. *International Journal of Mineral Processing*, **145**, 108–113, https://doi.org/10.1016/j.minpro.2015.06.009

Duyvesteyn S., Liu H. and Duyvesteyn W. P. C. (1994). Recovery of platinum group metals from oxide ores-TML process. In: Hydrometallurgy '94. Springer Netherlands, Dordrecht, pp. 887–912.

Eksteen J. J. (2011). A mechanistic model to predict matte temperatures during the smelting of UG2-rich blends of platinum group metal concentrates. *Minerals Engineering*, **24**, 676–687, https://doi.org/10.1016/j.mineng.2010.10.017

Ficeriová J. and Baláž P. (2010). Leaching of gold from a mechanically and mechanochemically activated waste. *Acta Montanistica Slovaca*, **15**, 183–187.

Fornalczyk A. and Saternus M. (2009). Removal of platinum group metals from the used auto catalytic converter. *Metalurgija*, **48**, 133–136.

Fornalczyk A. and Saternus M. (2013). Vapour treatment method against other pyro- and hydrometallurgical processes applied to recover platinum from used auto catalytic converters. *Acta Metallurgica Sinica (English Letters)*, **26**, 247–256, https://doi.org/10.1007/s40195-012-0125-1

Frangoul A. (2020). In New Zealand, microbes are extracting gold from electronic waste. https://www.cnbc.com/2020/10/01/in-new-zealand-microbes-are-extracting-gold-from-electronic-waste.html (accessed 9 April 2024)

Ghodrat M., Sharafi P. and Samali B. (2018). Recovery of platinum group metals out of automotive catalytic converters scrap: A review on Australian trends and challenges. *Minerals, Metals & Materials Series Part F*, **2**, 149–161, https://doi.org/10.1007/978-3-319-72131-6_13

Hagelüken C. (2006). Recycling of electronic scrap at Umicore precious metals refining. *Acta Metallurgica Slovaca*, **12**, 111–120.

Hagelüken B. C. (2012). Recycling the platinum group metals: a European perspective. *Platinum Metals Review*, **56**, 29–35, https://doi.org/10.1595/147106712X611733

He X., Li Y., Wu X., Zhao Y., Wang H. and Liu W. (2016). Study on the process of enrichment platinum group metals by plasma melting technology. *Precious Metals*, **37**, 1–5.

Hughes A. E., Haque N., Northey S. A. and Giddey S. (2021). Platinum group metals: A review of resources, production and usage with a focus on catalysts. *Resources*, **10**, 1–40, https://doi.org/10.3390/resources10090093

Iftekhar S., Heidari G., Amanat N., Zare E. N., Asif M. B., Hassanpour M., Lehto V. P. and Sillanpaa M. (2022). Porous materials for the recovery of rare earth elements, platinum group metals, and other valuable metals: a review. *Environmental Chemistry Letters*, **20**, 3697–3746, https://doi.org/10.1007/s10311-022-01486-x

Ilyas S. and Lee J. C. (2014). Biometallurgical recovery of metals from waste electrical and electronic equipment: A review. *ChemBioEng Reviews*, **1**, 148–169, https://doi.org/10.1002/cben.201400001

Ilyas S., Srivastava R. R., Kim H. and Cheema H. A. (2020). Hydrometallurgical recycling of palladium and platinum from exhausted diesel oxidation catalysts. *Separation and Purification Technology*, **248**, 117029, https://doi.org/10.1016/j.seppur.2020.117029

Ilyas S., Kim H. and Srivastava R. R. (2021). Sustainable Urban Mining of Precious Metals, 1st edn. CRC Press, Boca Raton, https://doi.org/10.1201/9781003055068

Ilyas S., Srivastava R. R. and Kim H. (2022). Mobilization of platinum and palladium from exhausted catalytic converters using bio-cyanide and an ionic-liquid as mass transport carriers. *Green Chemistry*, **24**, 5204–5218, https://doi.org/10.1039/d2gc00874b

Ivanović S. Z., Trujić V. K., Gorgievski M. D., Mišić L. D. and Božić D. S. (2011). Removal of Platinum Group Metals (PGMs) from the Spent Automobile Catalyst by the Pyrometallurgical Process. 15th International Research/Expert Conference "Trends in the Development of Machinery and Associated Technology", pp. 701–704.

Jha M. K., Lee J. C., Kim M. S., Jeong J., Kim B. S. and Kumar V. (2013). Hydrometallurgical recovery/recycling of platinum by the leaching of spent catalysts: a review. *Hydrometallurgy*, **133**, 23–32, https://doi.org/10.1016/j.hydromet.2012.11.012

Jimenez De Aberasturi D., Pinedo R., Ruiz De Larramendi I., Ruiz De Larramendi J. I. and Rojo T. (2011). Recovery by hydrometallurgical extraction of the platinum-group metals from car catalytic converters. *Minerals Engineering*, **24**, 505–513, https://doi.org/10.1016/j.mineng.2010.12.009

Jones R. T. (1999). Platinum smelting in South Africa. *South African Journal of Science*, **95**, 525–534.

Jones R. T. and Geldenhuys I. J. (2011). The pros and cons of reductive matte smelting for PGMs. *Minerals Engineering*, **24**, 495–498, https://doi.org/10.1016/j.mineng.2011.03.007

Juvonen R., Lakomaa T. and Soikkeli L. (2002). Determination of gold and the platinum group elements in geological samples by ICP-MS after nickel sulphide fire assay: Difficulties encountered with different types of geological samples. *Talanta*, **58**, 595–603, https://doi.org/10.1016/S0039-9140(02)00330-2

Karim S. and Ting Y. (2020). Ultrasound-assisted nitric acid pretreatment for enhanced biorecovery of platinum group metals from spent automotive catalyst. *Journal of Cleaner Production*, **255**, 120199, https://doi.org/10.1016/j.jclepro.2020.120199

Karim S. and Ting Y.-P. (2021). Recycling pathways for platinum group metals from spent automotive catalyst: A review on conventional approaches and bio-processes.

Resources, Conservation and Recycling, **170**, 105588, https://doi.org/10.1016/j.resconrec.2021.105588

Keyworth B. (1983). Role of Pyrometallurgy in the Recovery of Precious Metals from Secondary Materials. In: Precious Metals 1982. Pergamon Press Canada Ltd, Toronto, Ontario, Canada, pp 509–537.

Kim B. S., Lee J. C., Seo S. P., Park Y. K. and Sohn H. Y. (2004). A process for extracting precious metals from spent printed circuit boards and automobile catalysts. *JOM*, **56**, 55–58, https://doi.org/10.1007/s11837-004-0237-9

Kim B. S., Lee J. C., Jeong J., Yang D. H., Shin D. and Lee K. I. (2013). A novel process for extracting precious metals from spent mobile phone PCBs and automobile catalysts. *Materials Transactions*, **54**, 1045–1048, https://doi.org/10.2320/matertrans.M2013051

Kita Y., Nishikawa H. and Takemoto T. (2006). Effects of cyanide and dissolved oxygen concentration on biological Au recovery. *Journal of Biotechnology*, **124**, 545–551, https://doi.org/10.1016/j.jbiotec.2006.01.038

Kolliopoulos G., Balomenos E., Giannopoulou I., Yakoumis I. and Panias D. (2014). Behavior of platinum group during their pyrometallurgical recovery from spent automotive catalysts. *Open Access Library Journal*, **01**, 1–9, https://doi.org/10.4236/oalib.1100736

Lanaridi O., Sahoo A. R., Limbeck A., Naghdi S., Eder D., Eitenberger E., Csendes Z., Schnürch M. and Bica-Schröder K. (2021). Toward the recovery of platinum group metals from a spent automotive catalyst with supported ionic liquid phases. *ACS Sustainable Chemistry & Engineering*, **9**, 375–386, https://doi.org/10.1021/acssuschemeng.0c07384

Liu C., Sun S., Zhu X. and Tu G. (2021). Metals smelting-collection method for recycling of platinum group metals from waste catalysts: A mini review. *Waste Management & Research: The Journal for a Sustainable Circular Economy*, **39**, 43–52, https://doi.org/10.1177/0734242X20969795

Marinho R. S., da Silva C. N., Afonso J. C. and da Cunha J. W. S. D. (2011). Recovery of platinum, tin and indium from spent catalysts in chloride medium using strong basic anion exchange resins. *Journal of Hazardous Materials*, **192**, 1155–1160, https://doi.org/10.1016/j.jhazmat.2011.06.021

Meshram A. and Singh K. K. (2018). Recovery of valuable products from hazardous aluminum dross: A review. *Resources, Conservation and Recycling*, **130**, 95–108, https://doi.org/10.1016/j.resconrec.2017.11.026

Moleko-Boyce P., Makelane H. and Tshentu Z. R. (2022). Recovery of platinum group metals from leach solutions of spent catalytic converters using custom-made resins. *Minerals*, **12**, 361, https://doi.org/10.3390/min12030361

Mooiman M. B., Sole K. C. and Dinham N. (2016). The precious metals industry: Global Challenges, Responses, and Prospects. In: R. M. Izatt (ed.), Metal Sustainability: Global Challenges, Consequences, and Prospects. John Wiley & Sons, Inc., Hoboken, New Jersey, USA, pp. 361–396.

Morcali M. H. (2020). A new approach to recover platinum-group metals from spent catalytic converters via iron matte. *Resources, Conservation and Recycling*, **159**, 104891, https://doi.org/10.1016/j.resconrec.2020.104891

Mulligan C. N., Yong R. N. and Gibbs B. F. (2001). Remediation technologies for metal-contaminated soils and groundwater: An evaluation. 60.

Nagamori M. and Mackey P. J. (1978). Thermodynamics of copper matte converting: part II. Distribution of Au, Ag, Pb, Zn, Ni, Se, Te, Bi, Sb and As between copper, matte and slag in the Noranda process. *Metallurgical Transactions B*, **9**, 567–579, https://doi.org/10.1007/BF03257205

Narayanasamy M., Dhanasekaran D. and Thajuddin N. (2022). *Frankia* consortium extracts high-value metals from e-waste. *Environmental Technology & Innovation*, **28**, 102564, https://doi.org/10.1016/j.eti.2022.102564

Niemelä M., Pitkäaho S., Ojala S., Keiski R. L. and Perämäki P. (2012). Microwave-assisted aqua regia digestion for determining platinum, palladium, rhodium and lead in catalyst materials. *Microchemical Journal*, **101**, 75–79, https://doi.org/10.1016/j.microc.2011.11.001

Nogueira C. A., Paiva A. P., Oliveira P. C., Costa M. C. and da Costa A. M. R. (2014). Oxidative leaching process with cupric ion in hydrochloric acid media for recovery of Pd and Rh from spent catalytic converters. *Journal of Hazardous Materials*, **278**, 82–90, https://doi.org/10.1016/j.jhazmat.2014.05.099

Panda R., Jha M. K. and Pathak D. D. (2018). Commercial processes for the extraction of platinum group metals (PGMs). *Minerals, Metals & Materials Series Part F*, **5**, 119–130, https://doi.org/10.1007/978-3-319-72350-1_11

Patel S. and Kasture A. (2014). Review Article: E (electronic) waste management using biological systems – Overview. *International Journal of Current Microbiology and Applied Sciences*, **3**, 495–504.

Pathak A., Al-Sheeha H., Navvamani R., Kothari R., Marafi M. and Rana M. S. (2022). Recycling of platinum group metals from exhausted petroleum and automobile catalysts using bioleaching approach: A critical review on potential, challenges, and outlook. *Reviews in Environmental Science and Bio/Technology*, **21**, 1035–1059, https://doi.org/10.1007/s11157-022-09636-x

Peng Z., Li Z., Lin X., Tang H., Ye L., Ma Y., Rao M., Zhang Y., Li G. and Jiang T. (2017). Pyrometallurgical recovery of platinum group metals from spent catalysts. *JOM*, **69**, 1553–1562, https://doi.org/10.1007/s11837-017-2450-3

Pietrelli L. and Fontana D. (2013). Automotive spent catalysts treatment and platinum recovery. *International Journal of Environment and Waste Management*, **11**, 222–232, https://doi.org/10.1504/IJEWM.2013.051845

Pokhrel P., Lin S. L. and Tsai C. T. (2020). Environmental and economic performance analysis of recycling waste printed circuit boards using life cycle assessment. *Journal of Environmental Management*, **276**, 111276, https://doi.org/10.1016/j.jenvman.2020.111276

Rizki I. N., Tanaka Y. and Okibe N. (2019). Thiourea bioleaching for gold recycling from e-waste. *Waste Management*, **84**, 158–165, https://doi.org/10.1016/j.wasman.2018.11.021

Rzelewska M. and Regel-Rosocka M. (2019). Wastes generated by automotive industry-spent automotive catalysts. *Physical Sciences Reviews*, **3**, 1–27, https://doi.org/10.1515/psr-2018-0021

Sasaki H. and Maeda M. (2014). Zn-vapor pretreatment for acid leaching of platinum group metals from automotive catalytic converters. *Hydrometallurgy*, **147–148**, 59–67, https://doi.org/10.1016/j.hydromet.2014.04.019

Satyanarayana T., Johri B. N. and Prakash A. (2012). Microorganisms in Environmental Management, 1st edn. Springer Dordrecht, New York.

Shams K., Beiggy M. R. and Shirazi A. G. (2004). Platinum recovery from a spent industrial dehydrogenation catalyst using cyanide leaching followed by ion exchange. *Applied Catalysis A: General*, **258**, 227–234, https://doi.org/10.1016/j.apcata.2003.09.003

Shen H., Liu B., Ekberg C. and Zhang S. (2021). Harmless disposal and resource utilization for secondary aluminum dross: A review. *Science of The Total Environment*, **760**, 143968, https://doi.org/10.1016/j.scitotenv.2020.143968

Shin D., Park J., Jeong J. and Kim B. S. (2015). A biological cyanide production and accumulation system and the recovery of platinum-group metals from spent automotive catalysts by biogenic cyanide. *Hydrometallurgy*, **158**, 10–18, https://doi.org/10.1016/j.hydromet.2015.09.021

Spooren J., Atia T. A. and Kukurugya F. (2021). Microwave assisted leaching approaches to recover platinum group metals from waste and intermediate streams. 18th International Conference on Microwave and High Frequency Applications AMPERE, Gothenburg, Sweden, pp. 1–8.

Srivastava A. and Meshram A. (2023a). On trending technologies of aluminium dross recycling: A review. *Process Safety and Environmental Protection*, **171**, 38–54, https://doi.org/10.1016/j.psep.2023.01.010

Srivastava A. and Meshram A. (2023b). A hydrometallurgical perspective of aluminium dross recycling. *Mater Today Proc*, **112**, 13–17, https://doi.org/10.1016/j.matpr.2023.07.225

Sun P. P. and Lee M. S. (2011). Separation of Pt from hydrochloric acid leaching solution of spent catalysts by solvent extraction and ion exchange. *Hydrometallurgy*, **110**, 91–98, https://doi.org/10.1016/j.hydromet.2011.09.002

Sun S., Jin C., He W., Li G., Zhu H. and Huang J. (2022). A review on management of waste three-way catalysts and strategies for recovery of platinum group metals from them. *Journal of Environmental Management*, **305**, 114383, https://doi.org/10.1016/j.jenvman.2021.114383

Suoranta T., Zugazua O., Niemelä M. and Perämäki P. (2015). Recovery of palladium, platinum, rhodium and ruthenium from catalyst materials using microwave-assisted leaching and cloud point extraction. *Hydrometallurgy*, **154**, 56–62, https://doi.org/10.1016/j.hydromet.2015.03.014

Tang H., Peng Z., Tian R., Ye L., Zhang J., Rao M. and Li G. (2022). Recycling of platinum-group metals from spent automotive catalysts by smelting. *Journal of Environmental Chemical Engineering*, **10**, 108709, https://doi.org/10.1016/j.jece.2022.108709

Tang H., Peng Z., Tian R., Ye L., Zhang J., Rao M. and Li G. (2023). Platinum-group metals: demand, supply, applications and their recycling from spent automotive catalysts. *Journal of Environmental Chemical Engineering*, **11**, 110237, https://doi.org/10.1016/j.jece.2023.110237

Taninouchi Y. K. and Okabe T. H. (2023). Trends of technological development of platinum group metal recycling: solubilization and physical concentration processes. *Materials Transactions*, **64**, 627–637, https://doi.org/10.2320/matertrans.MT-M2022150

Thethwayo B. M. (2018). Extraction of platinum group metals. In: M. S. Seehra and A. D. Bristow (eds.), Noble and Precious Metals – Properties, Nanoscale Effects and Applications, IntechOpen, London, W1W 5PF, United Kingdom, pp. 393–403.

Trinh H. B., Lee J. C., Srivastava R. R. and Kim S. (2019). Total recycling of all the components from spent auto-catalyst by NaOH roasting-assisted hydrometallurgical route. *Journal of Hazardous Materials*, **379**, 120772, https://doi.org/10.1016/j.jhazmat.2019.120772

Trinh H. B., Lee J. C., Suh Y. J. and Lee J. (2020). A review on the recycling processes of spent auto-catalysts: towards the development of sustainable metallurgy. *Waste Management*, **114**, 148–165, https://doi.org/10.1016/j.wasman.2020.06.030

Trivedi H., Meshram A. and Gupta R. (2023a). Recycling of photovoltaic modules for recovery and repurposing of materials. *Journal of Environmental Chemical Engineering*, **11**, 109501, https://doi.org/10.1016/j.jece.2023.109501

Trivedi H., Verma K. D. and Kar K. K. (2023b). Recycling of supercapacitor materials. In: Springer Series in Materials Science. Springer, Cham, pp. 393–411.

Van Schalkwyk R. F., Eksteen J. J. and Akdogan G. (2013). Leaching of Ni–Cu–Fe–S converter matte at varying iron endpoints; mineralogical changes and behaviour of Ir, Rh and Ru. *Hydrometallurgy*, **136**, 36–45, https://doi.org/10.1016/j.hydromet.2013.02.008

Volesky B. and Holan Z. R. (1995). Biosorption of heavy metals. *Biotechnology Progress*, **11**, 235–250, https://doi.org/10.1021/bp00033a001

Wilburn D. R. and Bleiwas D. I. (2004). Platinum-Group Metals: World Supply and Demand. US Geological Survey Open-File Rep 1224:152.

Williams C. T., Chen E. K. Y., Takoudis C. G. and Weaver M. J. (1998). Reduction kinetics of surface rhodium oxide by hydrogen and carbon monoxide at ambient gas pressures as probed by transient surface-enhanced Raman spectroscopy. *The Journal of Physical Chemistry B*, **102**, 4785–4794, https://doi.org/10.1021/jp981287d

Willner J., Fornalczyk A., Cebulski J. and Janiszewski K. (2014). Preliminary studies on simultaneous recovery of precious metals from different waste materials by pyrometallurgical method. *Archives of Metallurgy and Materials*, **59**, 801–804, https://doi.org/10.2478/amm-2014-0136

Xia J. and Ghahreman A. (2023). Platinum group metals recycling from spent automotive catalysts: Metallurgical extraction and recovery technologies. *Separation and Purification Technology*, **311**, 123357, https://doi.org/10.1016/j.seppur.2023.123357

Xu S. Q. and Xu L. (1998). Some recovery process for precious metals from spent automobile catalyst. *China National Resources Recycling*, **10**, 7–11.

Yakoumis I., Moschovi A., Panou M. and Panias D. (2020). Single-step hydrometallurgical method for the platinum group metals leaching from commercial spent automotive catalysts. *Journal of Sustainable Metallurgy*, **6**, 259–268, https://doi.org/10.1007/s40831-020-00272-9

Yakoumis I., Panou M., Moschovi A. M. and Panias D. (2021). Recovery of platinum group metals from spent automotive catalysts: A review. *Cleaner Engineering and Technology*, **3**, 100112, https://doi.org/10.1016/j.clet.2021.100112

Yong P., Rowson N. A., Farr J. P. G., Harris L. R. and Macaskie L. E. (2003). A novel electrobiotechnology for the recovery of precious metals from spent automotive catalysts. *Environmental Technology*, **24**, 289–297, https://doi.org/10.1080/09593330309385561

Yoo J. S. (1998). Metal recovery and rejuvenation of metal-loaded spent catalysts. *Catalysis Today*, **44**, 27–46, https://doi.org/10.1016/S0920-5861(98)00171-0

You G., Fang W., Li Q., Ma Y., Yang X. T. and Yang H. (2016). Study on enrichment method of platinum, palladium and rhodium in spent auto-catalysts. *Yejin Fenxi/Metallurgical Anal*, **36**, 7–11, https://doi.org/10.13228/j.boyuan.issn1000-7571.009680

Zheng H., Ding Y., Wen Q., Zhao S., He X., Zhang S. and Dong C. (2022). Slag design and iron capture mechanism for recovering low-grade Pt, Pd, and Rh from leaching residue of spent auto-exhaust catalysts. *Science of the Total Environment*, **802**, 149830, https://doi.org/10.1016/j.scitotenv.2021.149830

Part V

Manufacturing of platinum group element materials from waste and wastewater

doi: 10.2166/9781789065084_0245

Chapter 10

Platinum group metal nanoparticles: bacteriogenic synthesis and applications

*T. J. Mou, M. Capeness and L. E. Horsfall**

Institute of Quantitative Biology, Biochemistry and Biotechnology, School of Biological Sciences, The University of Edinburgh, Edinburgh EH9 3FF, UK
**Corresponding author: Louise.Horsfall@ed.ac.uk*

ABSTRACT

Platinum group nanoparticles (NPs) have gained notable attention owing to their exceptional catalytic, electronic, and optical properties, with a wide range of applications. Bacteriogenic synthesis of platinum group metallic NPs provides a sustainable and environmental-friendly alternative to traditional chemical and physical synthesis methods. Utilizing the metabolic aptitude of certain bacterial species, this approach promotes the synthesis of NPs with tuneable size, shape, and functionality. Bacterial reduction of metal ions results in the formation of NPs, often stabilized by biogenic capping agents, which also enhance their stability and dispersion. This chapter focuses on the mechanistic aspects of bacterial interaction with platinum group metals especially platinum and palladium, highlighting the role of microbial proteins and reducing agents in the formation of NPs. This chapter also addresses the feasibility and scalability of the bacterial approach as a greener route to the synthesis of high-value platinum group NPs paving the way for novel applications in a variety of fields.

Keywords: Bacteria, green synthesis, metal reduction, nanoparticles, platinum group, sustainable nanotechnology.

10.1 INTRODUCTION

Nanotechnology has emerged as a cornerstone of the current scientific world, and it studies the synthesis, characterization, and application of nanoscale materials (Vijayaram *et al.*, 2024). This interdisciplinary branch of science has transmuted various fields, including medicine, electronics, and environmental science, by providing innovative solutions through the manipulation of matter at the nanoscale (Malik *et al.*, 2023; Vijayaram *et al.*, 2024). The term 'nano' is derived from the Greek word 'nanos' which means 'dwarf' or 'very small' and refers to the measurement on a scale of 'one billionth' or 10^{-9} of a unit. Richard Feynman, a physics Nobel winner, proposed the concept of nanotechnology in his 1959 presentation titled 'There's Plenty of Room at the Bottom' at the American Physical Society meeting (Feynman, 1960). However, the term 'nanotechnology' was coined by Professor Norio Taniguchi in 1974 (Taniguchi *et al.*, 1974). Any tiny particle with at least one dimension in the range of 1–100 nm is defined as a 'nanoparticle (NP)' (Bayda *et al.*, 2019). Owing to the small size and enormous surface area-to-volume ratio, NPs exhibit unique physical, chemical, and biological properties that are different from their bulk counterparts (Bayda *et al.*, 2019; Jeevanandam *et al.*, 2018). A range of materials such as metals and non-metals including polymers and carbon-based substances such as graphene can be utilized as a precursor for the synthesis of NPs (Altammar, 2023). Currently, researchers worldwide are especially fascinated with metallic NPs due to their unique optical, electrical, magnetic, and catalytic properties. Furthermore, metallic NPs can bind a wide range of substances such as targeted agents and ligands such as pharmaceuticals, biomolecules, and antibodies in a straightforward manner which promotes their application in biotechnology, biomedical science, and green chemistry (Altammar, 2023; Bayda *et al.*, 2019).

Platinum group metals (PGMs), as bulk materials, have some common physical and chemical features such as higher thermal stability, resistance to oxidation and corrosion, electrical conductivity, and excellent catalytic properties. These properties are of strong industrial interest, particularly in the chemicals manufacturing, medical, and electronic sectors. In spite of being in high demand, PGMs are limited in terms of their abundance, supply, and annual production, which make them expensive (Jeyaraj *et al.*, 2019). To maximize the utilization of PGMs, using them in the NP form is an effectual alternative, as they elicit the desired outcome, often with increased efficiency, but require lower amounts of metallic input, due to their increased surface-to-volume ratio and thus are widely applied in the field of heterogenous catalysis (Jeyaraj *et al.*, 2019; Mikhailova, 2022).

10.2 SYNTHESIS APPROACHES FOR PLATINUM GROUP METALLIC NPs

Metallic NPs can be synthesized employing two basic approaches – top-down and bottom-up (Figure 10.1). Top-down approaches comprise mainly of physical methods to transform bulk materials into nanoscale particles (Isaacoff & Brown, 2017). For PGM NPs, the most employed physical methods are flame pyrolysis, vapour deposition, mechanical milling, laser ablation, sputter discharge, lithography, and arc discharge (Habibullah *et al.*, 2021). Bottom-up approaches are also referred to as building-up approaches, where NPs are generated from individual atoms or molecules. Chemical and biological methods are classified under this bottom-up approach.

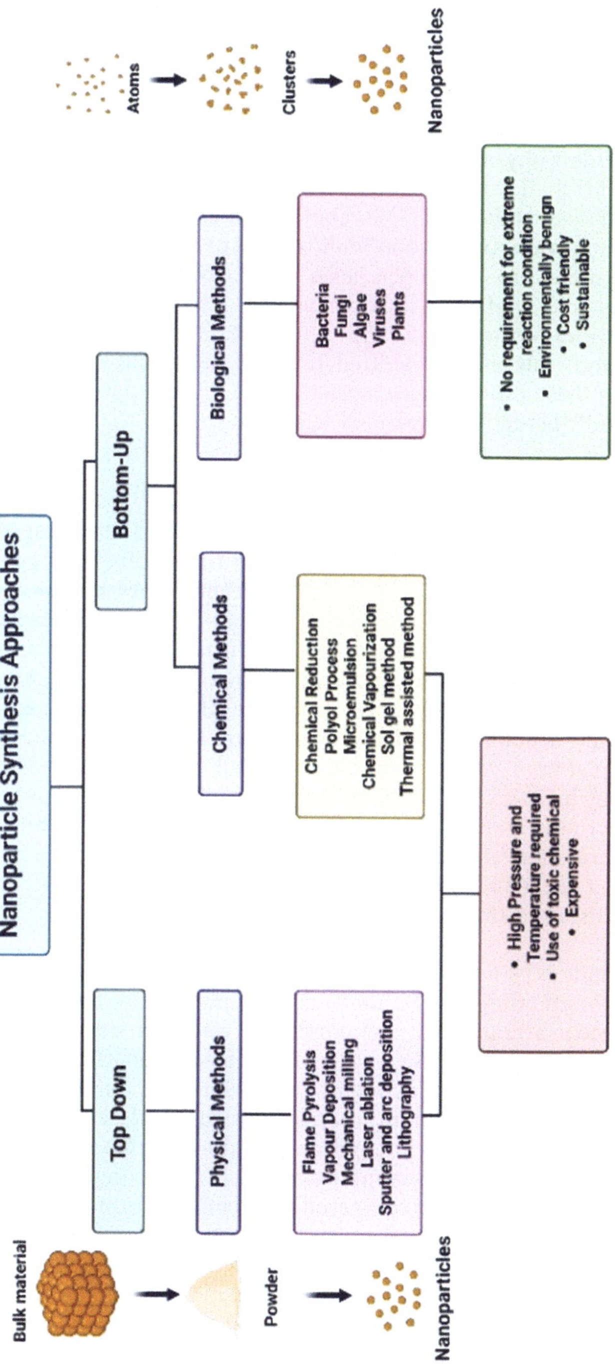

Figure 10.1 Schematic illustration of different NP synthesis approaches with comparative features.

For PGM NPs, commonly used bottom-up approaches are chemical reduction, polyol process, microemulsion, plasma-enhanced chemical vapour deposition, sol–gel method, template-assisted, and green biological synthesis methods (Rzelewska-Piekut *et al.*, 2022). However, the requirement of high energy, expensive and high-purity precursors, and use of hazardous chemicals renders physical and chemical methods incompatible for continuous production (Mikhailova, 2022; Rzelewska-Piekut *et al.*, 2022). For instance, certain reactions specific for PGMs require temperatures of 100–180°C and the use of pure hydrogen and carbon monoxide. Apart from the safety risk, even at ambient temperature, hydrogen has a broad explosive potential and carbon monoxide becomes highly toxic at concentrations as minimal as 10 parts per million (Martinez-Mora *et al.*, 2023). Hence to address these challenges among others, research interest has shifted towards biological synthesis of metal-based NPs as these processes are cost-effective, biocompatible, non-toxic, and environmentally benign (Campaña *et al.*, 2023).

10.3 GREEN SYNTHESIS OF PLATINUM GROUP NPs

10.3.1 Bacterial synthesis of platinum group NPs and advantages

Green synthesis of NPs is an emergent era of nanobiotechnology which has been sought to create eco-friendly sustainable pathways for the synthesis of diverse nanostructures with novel shapes and sizes using biological agents. Diverse ranges of bacteria, fungi, algae, viruses, and plants have been investigated to explore their potential to produce metallic NPs. Among these, bacteriogenic synthesis of NPs is preferred over other biological methods owing to its efficiency, scalability, and cost-effectiveness. Bacterial species reported to produce platinum group NPs are presented in Table 10.1. Metallic NPs synthesized by bacteria have a higher specific surface area, stability, enhanced catalytic reactivity, and better interaction between the enzyme and metal salt due to the presence of bacterial carrier matrices (Cueva & Horsfall, 2017; Edmundson *et al.*, 2014). Moreover, these highly stable and monodispersed NPs could be synthesized through controlled and up-scalable processes (Campaña *et al.*, 2023).

Bacteriogenic NPs provide distinct benefits compared to those synthesized chemically. Firstly, they can be derived from impure sources such as metal-containing waste streams and metal-polluted water, broadening the range of feasible starting materials. Secondly, their integration into bioremediation procedures facilitates the extraction of valuable metals, thereby enhancing the overall value of the remediation process (Vaseghi *et al.*, 2018). Thirdly, their synthesis occurs under ambient conditions and mild temperatures (20–30°C), resulting in reduced operational and manufacturing expenses. Bacteriogenic NPs have also higher stability compared to chemically synthesized NPs owing to the presence of capping agents, which prevent the agglomeration/aggregation and protects them from oxidation (Durán & Seabra, 2012). Additionally, faster reproduction of bacteria and the use of genetic manipulation allow large-scale production of NPs in a shorter time frame and at lower costs, as well as with more tuned properties of the resulting nanomaterials (Capeness *et al.*, 2019a).

Table 10.1 Bacteria reported for the synthesis of platinum and palladium NPs.

NP	Bacteria	Size (nm)	Shape	Localization	Reference
Palladium	*Shewanella oneidensis*	2–25	Spherical	Extracellular/supernatant	Zhang *et al.* (2022)
	Desulfovibrio alaskensis	–	Spherical	Intracellular/periplasmic	Capeness *et al.* (2015)
	Desulfovibrio desulfuricans	1.1–13	Cubo-octahedron	Intracellular	Omajali *et al.* (2015)
	Geobacter sulfurreducens	14–36	–	Extracellular	Yates *et al.* (2013)
	Escherichia coli	2–3	Spherical	Intracellular	Bachar *et al.* (2020)
	Escherichia coli	1–30	Anisotropic	Intracellular/membrane	Gomez-Bolivar *et al.* (2019)
	Bacillus thuringiensis	3–16		Intracellular/extracellular/ membrane/periplasmic space	Chen *et al.* (2022a, 2022b)
	Bacillus benzeovorans	1.7–5.8	Anisotropic	Intracellular	Omajali *et al.* (2015)
	Bacillus wiedmannii	10–36	–	Membrane	Chen*et al.* (2018)
	Bacillus megaterium	10–40	Spherical	Intracellular/extracellular	Chen *et al.* (2022a, 2022b)
	Bacillus sphaericus	1–10	Square/circular	Extracellular	Wahl *et al.* (2001)
	Cupriavidus metallidurans	20–40	Dendrite- shaped	Intracellular	Tan *et al.* (2020)
	Citrobacter sp.	11.3–25.8		Anisotropic	Matsena *et al.* (2020)
	Shewanella loihica	4–10	Spherical	Intracellular/extracellular	Lv *et al.* (2018)

(*Continued*)

Table 10.1 Bacteria reported for the synthesis of platinum and palladium NPs. (*Continued*)

NP	Bacteria	Size (nm)	Shape	Localization	Reference
Platinum	*Acetobacter xylinum*	6.3–9.3	Granular	Extracellular and intracellular	Aritonang *et al.* (2014)
	Acinetobacter calcoaceticus	2–3.5	Cuboidal	Intracellular/membrane	Gaidhani *et al.* (2014)
	Desulfovibrio alaskensis	–	Amorphous	Extracellular	Capeness *et al.* (2015)
	Desulfovibrio vulgaris	–	–	Extracellular	Martins *et al.* (2017)
	Escherichia coli	2.3–4.5	Spherical	–	Attard *et al.* (2012)
	Streptomyces sp.	20–50	Spherical	Extracellular	Baskaran *et al.* (2017)
	Shewanella algae	5	–	Periplasmic	Konishi *et al.* (2007)
	Shewanella oneidensis	61.03	Snowflake	Extracellular	Tuo *et al.* (2013)
	Pseudomonas kunmingensis	3.95	Spherical	Supernatant	Eramabadi *et al.* (2020)
	Psychrobacter faecalis	2.49	Spherical	Supernatant	Eramabadi *et al.* (2020)
	Vibrio fischeri	3.84	Spherical	Supernatant	Eramabadi *et al.* (2020)
	Leptolyngbya foveolarum			Intracellular	Brayner *et al.* (2007)
	Jeotgalicoccus coquinae	5.74	Spherical	Supernatant	Eramabadi *et al.* (2020)
	Sporosarcina psychrophila	4.24	Spherical	Supernatant	Eramabadi *et al.* (2020)
	Kocuria rosea	5.85	Spherical	Supernatant	Eramabadi *et al.* (2020)
	Pseudomonas putida	8.06	Spherical	Supernatant	Eramabadi *et al.* (2020)
	Plectonema boryanum	30–300	Spherical	Intracellular and extracellular	Lengke *et al.* (2006)
	Anabaena flos-aquae	–	–	Intracellular	Brayner *et al.* (2007)

Platinum group metallic NPs have received considerable attention because of their wide range of uses in the catalysis, electronics, pharmaceutical, and biomedical sectors (Fahmy *et al.*, 2020). With regards to bacteriogenic synthesis, most of the studies focused on platinum and palladium, while ruthenium, osmium, and iridium have been less explored as they are less frequently used, albeit there are reports of bimetallic NP biosynthesis that includes ruthenium (Omajali *et al.*, 2019). Hence, the following sections detail bacteriogenic synthesis of platinum and palladium, their characterization and applications.

10.3.2 Bacterial mechanism of metallic NP synthesis

Bacteria are promising candidates for the synthesis of NPs as they possess remarkable capacity to reduce metal ions into metallic forms. Some metals are essential nutrients for bacteria, whereas some are toxic to them even at relatively low concentrations. Bacteria have different mechanisms specific for metal transport, detoxification, and oxidation/reduction (Campaña *et al.*, 2023). Upon exposure to toxic metal ions, some bacteria produce NPs as part of their defence mechanism transforming soluble ions to less toxic insoluble forms as a by-product of adenosine triphosphate generation. During the bacterial reduction of metal ions into NPs, cations are transformed into lower valent ions, except in cases for some precious metals such as palladium, platinum, silver, and gold which are reduced to elemental forms (Bloch *et al.*, 2021; Campaña *et al.*, 2023; Capeness *et al.*, 2019b; Tsekhmistrenko *et al.*, 2020). Bacterial synthesis of metallic NPs can be summarized in the following steps:

(1) *Binding and/or uptake of metal ions*: The positively charged metal ions and the negatively charged groups on microbial cells (e.g. amino groups, sulfhydryl groups, and carboxyl groups) interact electrostatically. Metal ions are stabilized by the adhesive substances on the bacterial cells which are complex and mesh-like. This is made possible by the secreted extracellular polymeric compounds or by electrostatic contact, which attach the ions to the bacterial cells. Bacterial cells can take up metals actively through trans-membrane transport, passively through ion channels, or directly by penetration through the lipid membrane.

(2) *Reduction of the metal ions*: To combat the toxic effects of metal ions, bacterial cells produce proteins involved in electron shuttling, such as hydrogenases, cytochromes, and nicotinamide adenine dinucleotide (NADH) dependent reductases, which in turn reduce the metal ions to zero-valent NPs.

(3) *Nucleation and growth*: Reduced metal ions then form nuclei within the bacterial cells by aggregation. The nuclei further draw in metal ions from their surroundings, promoting the growth of metallic NPs.

(4) *Stabilization and surface modification*: Some bacteria release biomolecules that bind to the surface of the NPs, such as proteins, polysaccharides, or other organic substances, either during or after the production of the NPs. Variables such as metal ion concentration, temperature, pH, and the presence of stabilizing agents or biomolecules within the bacterial cells influence the size and shape of the synthesized NPs.

Depending on the localization, NP synthesis in bacteria can be categorized as intracellular and extracellular (Hulkoti & Taranath, 2014). According to the intracellular mechanism, positively charged metal ions are trapped within the negatively charged cell wall (Figure 10.2). This is followed by an enzymatic bioreduction of the metal ions within the cell wall, which results in the formation of nanoclusters. Finally, the NPs are widely distributed around the cell wall.

Extracellular synthesis comprises the events of enzyme secretion, bioreduction followed by particle capping (Figure 10.3). Both intracellular and extracellular syntheses are reported for the bacteriogenic synthesis of platinum and palladium NPs. For example, *Cupriavidus metallidurans*, *Bacillus benzeovorans*, and *Leptolyngbya foveolarum* are reported to synthesize palladium and platinum NPs intracellularly, whereas *Geobacter sulfurreducens*, *Desulfovibrio vulgaris*, and *Streptomyces* sp. are known for extracellular synthesis (Capeness *et al.*, 2015; Gaidhani *et al.*, 2014; Omajali

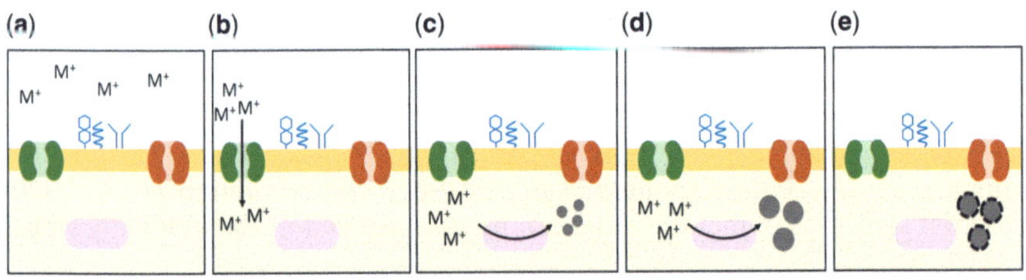

Figure 10.2 Bacterial intra-cellular production of metallic NPs. (a) Metal ions and the cell, (b) metal ions enter the cell through transportation machinery, (c) metal ions are reduced by free or membrane-bound enzymes such as cytochromes and hydrogenases to form elemental 'seed' NPs, (d) NPs bound to the cell continue to grow through the incorporation of other NPs or the self-reduction of metal ions, or (e) NPs stop growing, either through lack of new metal ions or are capped with bio-molecules.

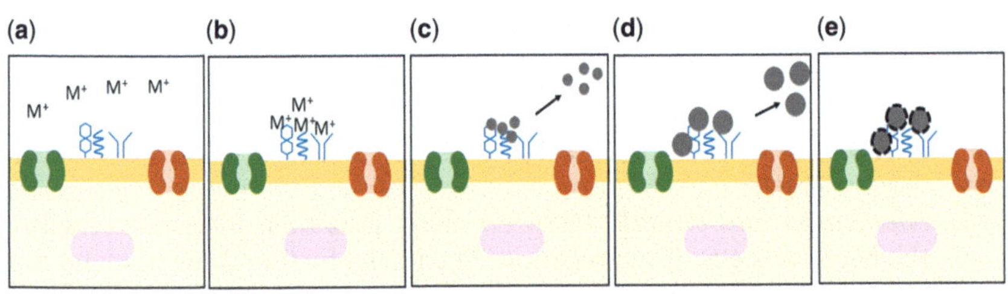

Figure 10.3 Extracellular synthesis of metallic NPs in bacteria. (a) Metal ions and the cell, (b) metal ions binding to bio-molecules, such as lipopolysaccharides, peptides, and proteins, displayed on the outside of the cell, and (c) metal ions are reduced to elemental forms to form 'seed' NPs. These can also be from the cells surface, (d) more metal ions are reduced, leading to the growth of the NPs. The NPs can also be released or (e) NPs stop growing, either through lack of new metal ions, or are capped with other bio-molecules.

et al., 2015; Tan *et al.*, 2020; Tuo *et al.*, 2013; Yates *et al.*, 2013). In addition, *Shewanella algae* were found to synthesize NPs in their periplasmic space (Chen *et al.*, 2022a, 2022b; Konishi *et al.*, 2007) and some bacteria are reported to have a combination of the above-mentioned synthesis mechanisms. For instance, both *Desulfovibrio alaskensis* and *Shewanella oneidensis* can synthesize NPs intracellularly in the periplasm and extracellularly (Aritonang *et al.*, 2014; Chen *et al.*, 2022a, 2022b).

10.3.3 Palladium NP synthesis by bacteria

The study of Pd nanoparticle (Pd NP) synthesis by bacteria has advanced quickly due to their versatile application as catalysts and precursors for the construction of diverse pharmaceuticals, agrochemicals, and polymers. Many different bacterial species and strains have been investigated for their ability to produce palladium NPs effectively and a more detailed mechanism has been elucidated (Law *et al.*, 2022). Bacterial synthesis of Pd NPs often occurs in three steps:

(1) biological adsorption (biosorption) on the cell surface;
(2) bioreduction of Pd^{2+} to Pd^0 inside and outside the cell wall; and
(3) autocatalytic reduction of Pd^{2+} to Pd^0 (non-enzymatic process, where Pd^{2+} ions cluster onto the preformed Pd NPs on the cell wall promoting further NP growth).

Production of Pd NPs by bacteria is mainly dependent on periplasmic and membrane bound enzymes such as c-type cytochromes which transfer electrons from cellular respiration to the metal ions, reducing them in a respiratory process. Redox proteins, which are crucial for reducing metals and subsequently aid in the creation of metal particles, are also present in the form of hydrogenases. Thus, Pd^{2+} is transformed to Pd^0 and NP synthesis can be extracellular or intracellular and sometimes both (De Corte *et al.*, 2012; Deplanche *et al.*, 2014; Law *et al.*, 2022). Palladium NP synthesis by some of the studied bacteria are discussed in the subsequent sections.

10.3.3.1 *Shewanella oneidensis*

S. oneidensis is a well-known electroactive bacterium in regards to metallic NP synthesis (Figure 10.4). This bacterium was reported for extracellular synthesis of spherical Pd NPs with the size ranging from 2 to 12 nm. It can reduce ~97% of palladium after 24 h with an initial concentration of 250 μM palladium and 500 μM formate, producing membrane-associated Pd NPs (Zhang *et al.*, 2022). Pd NPs can be synthesized by this bacterium by different mechanisms. Biosorption of palladium ions to the cell wall is facilitated by ion exchange, electrostatic interactions, complexation, and physical adsorption. Pd^{2+} is transformed into Pd^0 by chemical or autocatalytic conversion, regardless of the kind of biosorption mechanism. However, physical absorption can also occur in which the functional groups do not impede the Pd^{2+} to bind to the cell wall. In addition to biosorption, the hydroxyl group present on the cell wall can contribute to the reduction of palladium (Figure 10.5).

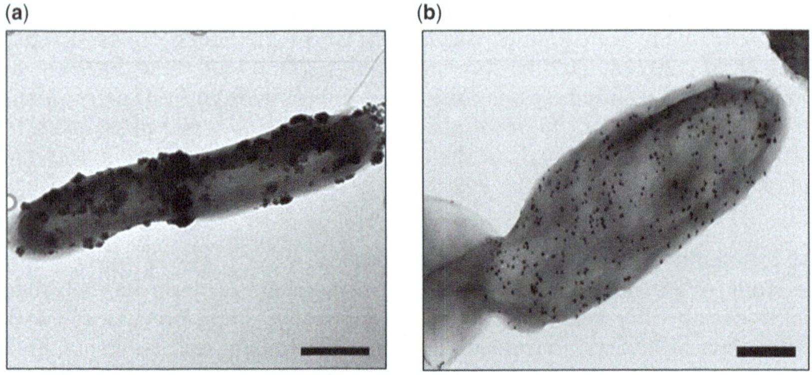

Figure 10.4 Transmission electron microscopy images of palladium NPs produced by (a) *Shewanella oneidensis* MR-1 (Jensen, unpublished) and (b) *Desulfovibrio alaskensis* G20 (Capeness *et al.*, 2019a). Scale bars = 200 nm.

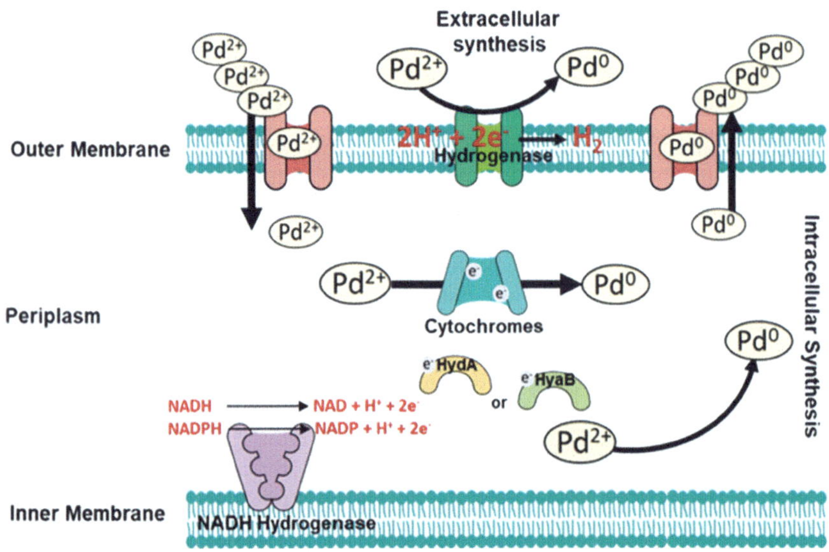

Figure 10.5 Proposed mechanism of Pd NP production in *S. oneidensis* MR-1. (*source*: adapted from Law *et al.*, 2022)

According to genomic studies, 42 c-type cytochromes have been found in *S. oneidensis* during Pd reduction (Coursolle & Gralnick, 2010). Bioreduction of Pd^{2+} to Pd^0 is mediated through the c-type cytochrome (Figure 10.5). Reduction of the metal ions both inside and outside of the cell, the protein complexes of NADH dehydrogenase, [FeFe]-hydrogenase (*Hyd*A), and [NiFe] hydrogenase (*Hya*B) are essential for Pd^{2+} reduction and NP production. NADH dehydrogenase plays a crucial role in the initial reduction step by transferring electrons from NADH. The electrons released by NADH dehydrogenase are carried by [FeFe]-hydrogenase (HydA) and [NiFe] hydrogenase (HyaB), then converting Pd^{2+} to Pd^0. The reduced palladium atoms (Pd^0) aggregate and nucleate to form palladium NPs. The presence of hydrogenases helps to control the growth and stabilization of these NPs by providing a continuous supply of electrons (Choi *et al.*, 2018; Omajali *et al.*, 2015).

10.3.3.2 *Desulfovibrio sp.*

The capacity of various *Desulfovibrio* species to catalyse the reduction of Pd^{2+} to Pd^0 is reported with examples including *D. alaskensis*, *D. desulfuricans*, *D. vulgaris*, and *D. fructosivorans*. In these bacteria, Pd ions undergo a bioreduction process in the periplasm of the cell and are converted to metallic NPs, which may then spread to the outer membrane to generate substantial extracellular deposits. In this reduction process, cytochromes act as redox partners in addition to formate dehydrogenases (FDHs) or hydrogenases which oxidize formate or hydrogen, respectively. The metallic NPs that are produced subsequently catalyse the abiotic reduction of Pd^{2+} after the initial enzymatic bioreduction of Pd^{2+} to Pd^0. The breakdown products CO_2 and H_2 are formed as a result of the delocalized pairs of electrons in the metal weakening the carbon–hydrogen bonds in formate. The residual Pd^{2+} in solution is continuously reduced to Pd^0 by the subsequent oxidation of H_2 on the surface of the Pd^0 particles, which provides electrons for this process (Capeness *et al.*, 2015, 2019b; Mikheenko *et al.*, 2008; Omajali *et al.*, 2015).

Palladium NP synthesis in *D. alaskensis* G20 is illustrated in Figure 10.6. It is believed that *D. alaskensis* has the capacity to reduce metals as a defence

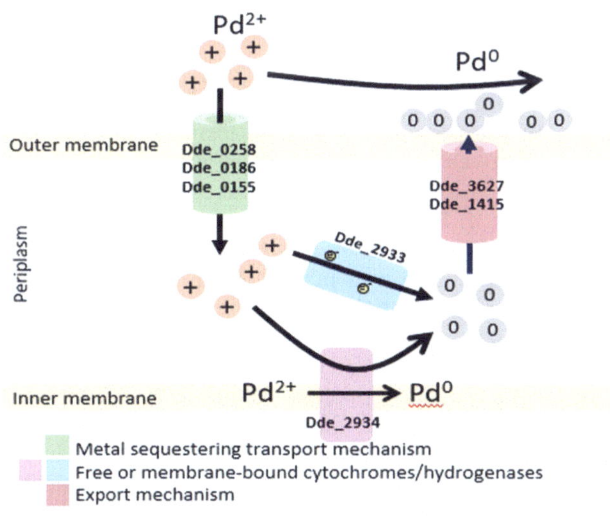

Figure 10.6 Proposed mechanism for synthesis of palladium NP production by *D. alaskensis* (Capeness *et al.*, 2019b). Dde_#### represent proteins predicted to play a role in the process of Pd ion import, its reduction, and the export of reduced Pd.

mechanism against excessive metal ion concentrations. These metals are thought to be incorporated by the organism through their energy pathways, where they serve as the terminal electron acceptors during respiration. Based on proteomic data, *D. alaskensis* absorbs metal ions through the outer membrane into the periplasm via transport proteins, where they are converted to zero-valent NPs by hydrogenases and/or cytochromes. Following production, the NPs are exported from the cell by an export mechanism and deposited on the cell surface, where they catalyse additional metal reduction (Capeness *et al.*, 2015, 2019b).

10.3.3.3 Bacillus sp.

Recently, the Gram-positive bacterium *Bacillus thuringiensis* Y9 has been reported to produce palladium NPs without a supply of external electrons under both aerobic and anaerobic conditions (Chen *et al.*, 2022a, 2022b). Different pathways mediate the Pd NP production collaboratively, with a high correlation between palladium reduction and the genes encoding the quinone cycle, *ribE*, cytochrome c reductase, and NADH-quinone oxidoreductase. The quinone pool and *rib*E can act as exogenous electron mediators which accelerate the cellular electron transfer. These electrons could be eventually utilized by the hydrogenase enzyme to generate hydrogen, which acts as a reducing agent facilitating Pd^{2+} reduction. Additionally, a pyruvate formate lyase-activating protein gene (*pflA*) and formate transporter gene (FNT) were identified from the transcriptome data and they may be involved in hydrogen generation and subsequent palladium reduction. The synthesized NPs were widely distributed on an extracellular polymer, the cell membrane, periplasmic space, and inside the cells (Chen *et al.*, 2022a, 2022b). The extracellular particles tended to form aggregates, whereas the intracellular particles were dispersed with diameters ranging from 3 to 16 nm. The highest removal efficiency was obtained with dry cell weight of the cells (878.83 mg/L) in their log phase with a precursor Pd^{2+} concentration of 117.67 mg/L with incubation conditions of pH 6.3 and a temperature below 30.9°C (Chen *et al.*, 2022a, 2022b).

Bacillus megaterium is reported to reduce Pd^{2+} in the presence of different electron donors. In the presence of sodium formate, this bacterium reduces palladium most effectively when in log phase growth. The highest removal capacity (1658.3 mg/g) was observed using the bacterial dry cell (30 mg/L) in the presence of 50 mg/L Pd and 5 mM sodium formate (Chen *et al.*, 2022a, 2022b). Palladium reduction in *B. megaterium* is mediated by the synergistic action of several proteins involved in different cellular pathways. The genes encoding oxidoreductase, metabolic enzymes, and cell-surface proteins were mainly identified to be responsible for the reduction of Pd^{2+}. Proteins such as NirB, Tdh, FabG, GudB, and Zwf generate NADPH, which serves as an electron donor for the reduction process. Additionally, reductases encoded by genes such as *ssuE*, *trx-1*, and the flavin reductase family contribute to Pd^{2+} reduction. The efficiency of this process is further enhanced by menaquinone produced by the *qoxB* protein and FMN (Flavin Mononucleotide) synthesized by *ssuE* and *ssuD*. The expression levels of *fabG*, *tdh*, *gudB*, and *rocG* are further increased in the presence of sodium formate as an exogenous electron donor. Additionally, endogenous electron mediators such as quinones and flavins in *B. megaterium* can further enhance Pd NP production (Chen *et al.*, 2018, 2024).

In *Bacillus sphaericus*, palladium NP synthesis is linked to an S-layer protein (Wahl *et al.*, 2001). Because of the substantial phosphorylation of its S-layer protein, cells of this strain bind strongly to phosphate groups as well as to carboxylic groups, such as glutamic acid, via cationic metals (Fahmy *et al.*, 2020). Similarly, two other bacterial species, that is, *Bacillus wiedmannii* and *B. benzeovorans*, have been reported to produce membrane-associated palladium NPs with size ranges of 10–36 nm and 1.7–5.8 nm, respectively (Chen *et al.*, 2018; Omajali *et al.*, 2019).

10.3.3.4 *Escherichia coli*

Escherichia coli, a facultatively anaerobic bacterium, is also reported to produce Pd NPs (Bachar *et al.*, 2020; Gomez-Bolivar *et al.*, 2019) and has well-defined molecular tools to clarify reaction pathways. Under anaerobic conditions, the nickel-dependent hydrogenase enzymes Hyd-1, Hyd-2, and Hyd-3 as well as the formate dehydrogenase molybdoenzymes FDH-N and FDH-H are involved in the bioreduction of palladium. Among these, membrane bound proteins Hyd-1, Hyd-2, FDH-O, and FDH-N are periplasmically oriented, whereas Hyd-3 and FDH-H face into the cytoplasm. Hyd-3 and FDH-H are subunits of the intracellular enzyme complex formate hydrogenlyase (FHL). Upon absorption of palladium ions, hydrogenase enzymes play the key role in producing Pd NPs from the reduced palladium atoms. *E. coli* was found to produce spherical Pd NPs with the size range from 2 to 3 nm (Gomez-Bolivar *et al.*, 2019).

10.3.3.5 *Geobacter sulfurreducens*

G. sulfurreducens can reduce soluble Pd^{2+} to Pd^0 NPs mostly outside the cell, lowering the toxicity of metal ions and permitting NP formation without cell damage. The bacterium reduced 50 (±3) mg/L Pd^{2+} with 1% hydrogen gas (v/v headspace) from 100 mg/L Pd^{2+} as the precursor. Palladium NP formation in *G. sulfurreducens* is facilitated by a set of genes. These include the genes, *pilA*, *omcB*, and *mtrABC* encoding c-type cytochromes that enable the bacterium to transfer electrons from within the cell to palladium ions (Pd^{2+}). Two other genes, *pilR* and *pilS*, which are involved in bacterial pili formation, can also play a central role in the Pd NP formation by this bacterium. The process of reduction occurs when electrons are transmitted to palladium ions, resulting in the transformation of Pd^{2+} ions into metallic palladium atoms. As more reduced palladium atoms are deposited onto their surfaces, the initial tiny clusters of reduced palladium atoms grow. The extracellular polymeric materials and proteins that *G. sulfurreducens* produces interact with the developing palladium NPs to stabilize them and modify their size and structure (Tuo *et al.*, 2013).

10.3.4 Platinum NP synthesis by bacteria

Platinum nanoparticles (Pt NPs) are produced both intra- and extracellularly by a range of different bacteria through the reduction of metal ions (Bloch *et al.*, 2021). Platinum metal ions are absorbed by the bacteria and are reduced through the action of different enzymes such as cytochromes, reductase, and metallothionein (Figure 10.7). Pt NP synthesis by some of the reported bacterial strains is discussed in the sections below.

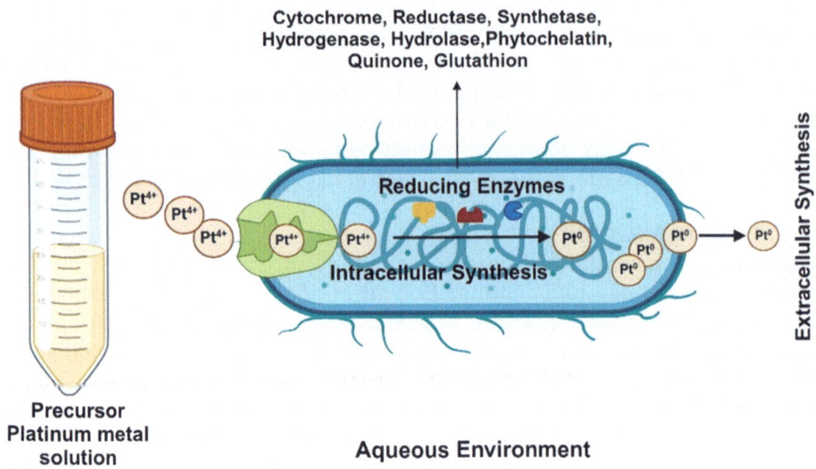

Figure 10.7 Representative summary of the bacterial mechanism for platinum NP synthesis.

10.3.4.1 *Acetobacter xylinum*

Pt NPs were reported to be synthesized from the bacterial cellulose matrix of *Acetobacter xylinum* using hydrogen gas as the reducing agent. K_2PtCl_4 is produced when $PtCl_2(H_2O_2)_2$ in solution passes through solvolysis, and as a result, hydrogen bonds are formed between the oxygen atoms of the bacterial cellulose and the hydrogen atoms of $PtCl_2(H_2O_2)_2$. The resulting extracellular production of Pt NPs was observed on the surface and inner side of the bacterial cellulose membrane. The NPs produced were granular shaped in the range of 6.3–9.3 nm. The platinum concentration used for this bacterium was 3–30 mM and the size of Pt NPs increased with the precursor metal concentration. In this study, the yield of the Pt NPs was measured by two methods: energy-dispersive X-ray spectroscopy (EDS) and thermogravimetric analysis (TGA). With the precursor platinum concentration of 30 mM, NP yield was 0.64 mg/cm^2 measured by EDS and 6.61 mg/cm^2 by TGA. TGA, however, is more representative for NP yield as it quantifies mass loss and distinguishes components based on thermal stability, giving clearer insights into purity and yield. Conversely EDS is mostly qualitative, focused on elemental composition without providing mass or thermal behaviour information (Aritonang *et al.*, 2014).

10.3.4.2 *Acinetobacter calcoaceticus*

The synthesis of Pt NPs is carried out by four proteins in *Acinetobacter calcoaceticus*, with respective molecular masses of 97, 66, 43, and 29 kDa (Gaidhani *et al.*, 2014). Therefore, it was suggested that a multistep process is involved in the intracellular reduction of Pt^{4+} to Pt NPs, which were subsequently released as a result of cell death caused by rupture of the cell membrane. The stability of the biogenic NPs was also due to coating of the NPs by bacterial proteins. This bacterium synthesizes cuboidal Pt NPs with the size range from 2 to 3.5 nm. The yield of Pt NPs from this bacterium was 1.07×10^{-3} mol/L with the precursor platinum concentration of 1 mM and incubation conditions of 30°C for 72 h at pH 7 (Gaidhani *et al.*, 2014).

10.3.4.3 Cyanobacteria

The cyanobacterium *Plectonema boryanum* produces spherical Pt NPs both inside and outside of the cells. Metal attachment to the bacterial cell surface is facilitated by carboxyl groups found in polysaccharides in the cell envelope with a high uronic acid concentration. Furthermore, metallothioneins (polypeptides that include abundant sulfur amino acids) can bind to generate metal–thiolate interactions. Dispersed Pt NPs are formed due to the entry of platinum into the cells as complexed $PtCl_4$ that is successively reduced to Pt^{2+} and then to Pt^0 by the cyanobacteria. In some cyanobacteria, the platinum reduction occurs inside the heterocyst and vegetative cells. Here, two classes of reducing enzymes, hydrogenases and nitrogenases, play crucial roles in the synthesis of Pt NPs (Lengke *et al.*, 2006). Cyanobacterial strains such as *Anabaena flos-aquae*, *L. Foveolarum*, and *Calothrix pulvinata* were observed with intracellular production of Pt NPs of size 3.2 (±0.3) nm (Brayner *et al.*, 2007).

10.3.4.4 Sulfate-reducing bacteria

The anaerobic sulfate-reducing bacterium *D. alaskensis* synthesizes Pt NPs by the use of cytochrome and/or hydrogenase enzymes (Capeness *et al.*, 2019b). The incubation conditions used were 30°C in the absence of oxygen for 2 h with a platinum concentration of 1–2 mM. Extracellular Pt NPs are created when seed particles are transferred to the cell's outer surface following production (Figure 10.8) (Capeness *et al.*, 2015).

D. vulgaris Hildenborough is another sulfate-reducing bacterium which is reported to synthesize Pt NPs. The bacterium was grown in Postgate's media C under anaerobic conditions (pH 7.2, 37°C) with a precursor platinum concentration of 100 mg/L. Seven putative hydrogenase enzymes are encoded by *D. vulgaris* Hildenborough of which four are found in the periplasm and three in the cytoplasm. These enzymes are considered essential for the manufacture of Pt NPs.

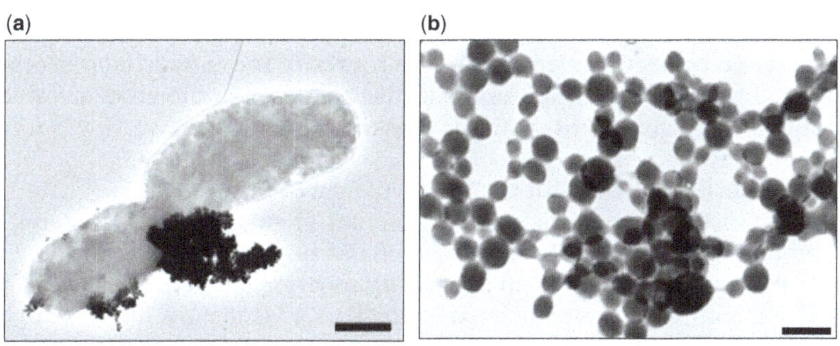

Figure 10.8 Transmission electron micrograph of platinum NPs produced by (a) *D. alaskensis* G20 (Capeness *et al.*, 2019a) and (b) spherical Pt NPs produced by cyanobacteria (Bloch et al., 2021). Scale bars = 200 nm.

The cell-free, cell-soluble protein extract from a consortium of sulfate-reducing bacteria, produced Pt NPs involving two distinct hydrogenase enzymes according to Riddin *et al.* (2010). Platinum(IV) was first reduced to platinum(II) by a two-electron bioreduction employing a new cytoplasmic hydrogenase that was protected from oxygen. Second, the platinum(II) was reduced to platinum(0) NPs with the aid of another two-electron bioreduction involving an oxygen-sensitive periplasmic hydrogenase (Riddin *et al.*, 2010).

10.3.4.5 *Escherichia coli*

Many hydrogenases are present in *E. coli* MC4100 cells, which aid in the catalysis of hydrogen into protons and electrons, potentially facilitating the biogenic production of Pt NPs. This bacterium was grown anaerobically in nutrient broth at pH 7.3 and 37°C with 2 mM platinum as the precursor (Attard *et al.*, 2012). It produced spherical Pt NPs intracellularly with size 2.3 (±0.7) and 4.5 (±0.7) nm in diameter for 1 and 20% initial metal loading, respectively (Attard *et al.*, 2012).

10.3.4.6 *Shewanella* sp.

Platinum NPs can be deposited in an eco-friendly way by employing the metal ion reducing bacterium *S. algae*. This bacterium was able to convert aqueous $PtCl_6^{2-}$ ions (1 mM) into elemental platinum at room temperature and neutral pH within 1 h when lactate was supplied as the electron donor. Biogenic platinum NPs of ~5 nm were found in the periplasm (Konishi *et al.*, 2007).

S. oneidensis was also reported to produce Pt NPs under anaerobic reaction conditions of 30°C, neutral pH, and 1 mM platinum precursor (Tuo *et al.*, 2013). This bacterium produced clustered snowflake-shaped Pt NPs on the surface of the cell with a size of 61.03 nm. It produced platinum NPs primarily through enzymatic reduction by c-type cytochromes and hydrogenases present on its cell surface. These enzymes play a crucial role in transferring electrons from the bacterial cell to platinum ions, leading to the formation of platinum NPs (Tuo *et al.*, 2013).

10.3.4.7 *Streptomyces* sp.

Pt NPs with a size of 20–50 nm were formed by *Streptomyces* sp. (Baskaran *et al.*, 2017). The bacterial chloride reductase enzyme, which is normally involved in the nitric acid cycle and plays a significant role in the reduction of chloride to chlorine, was identified as the source of the extracellular synthesis of Pt NPs by *Streptomyces* sp. Amino acids in this bacterium were revealed as the capping agent for Pt NP synthesis (Baskaran *et al.*, 2017).

10.3.5 Bacterial synthesis of bimetallic platinum group NPs

Although most studies focus on the reduction of single metals, bacteria can also produce bimetallic NPs. Combinations of metals can have a synergistic impact that produces NPs with unique or superior catalytic, electrical, and optical capabilities (Campaña *et al.*, 2023). Consequently, bimetallic NPs frequently show enhanced performance, which may be intriguing for a variety of applications (Campaña *et al.*, 2023; Xu *et al.*, 2019). Research into the ability to form NPs in the presence of other metals is crucial for the use of a bacterial chassis for the capture and reduction of metals, as these technologies are rolled out to use different waste streams containing many different metals.

Bacteria may synthesize bimetallic NPs when exposed to high concentrations of mixed metals (Sasireka & Lalitha, 2021). Bimetallic NPs can have a variety of configurations depending on the metal combination and their concentration used. Therefore, optimizing the composition of metals present in the incubation medium can enhance the bimetallic NPs with more tuned properties and performance.

Bimetallic NPs are categorized into two types: mixed and segregated. Depending on the configuration of atoms, they can be further classified into intermetallic, alloy, subclusters, and core–shell types. Mixed bimetallic NPs are synthesized as alloyed or intermetallic from two different metals where atoms are arranged as random and ordered, respectively. In regards to the segregated structure, the initial structure is formed by one metal followed by the addition of the second metal (Arora *et al.*, 2020). There are a number of studies which have reported the bacterial synthesis of bimetallic platinum group metallic NPs, which are discussed below. Table 10.2 provides examples of bacteriogenic PGM-based bimetallic NPs.

S. oneidensis can produce spherical bimetallic Pd–Pt NPs when cultured in a solution containing both Pd^{2+} and Pt^{4+}. When compared to pure Pd or Pt NPs, the synthesized polycrystalline Pd–Pt NPs have better catalytic performance

Table 10.2 PGM-based bimetallic NP synthesis by bacteria.

Bacterium	Bimetallic NPs	Size (nm)	Shape	Localization	Reference
B. benzeovorans	Pd–Ru	1–8	Core shell	Intracellular	Omajali *et al.* (2019)
Cupriavidus necator	Pd–Au	10–50	Spherical	Membrane	Hosseinkhani *et al.* (2012)
D. alaskensis	Pd–Pt	–	Core shell	Extracellular	Capeness *et al.* (2019a)
E. coli	Au–Pd	16	Spherical	Intracellular/ membrane	Deplanche (2008)
	Pd–Ru	1–3	Core shell	Intracellular/ membrane	Gomez-Bolivar *et al.* (2019)
S. oneidensis	Pd–Fe_3O_4	5.5	–	–	Tuo *et al.* (2013)
	Pd–Pt	3–40	Spherical	Intracellular/ extracellular	Xu *et al.* (2019)
	Pd–Au	1–50	–	Intracellular/ extracellular	De Corte *et al.* (2012)
	Pd–Ag	5–60	Core shell	Membrane	Kimber *et al.* (2021)
	Pd–Au	4–15	Core shell	Extracellular	Kimber *et al.* (2021)
Shewanella putrefaciens	Pd–Pt	4–60	Flower shaped	Membrane	Tuo *et al.* (2017)

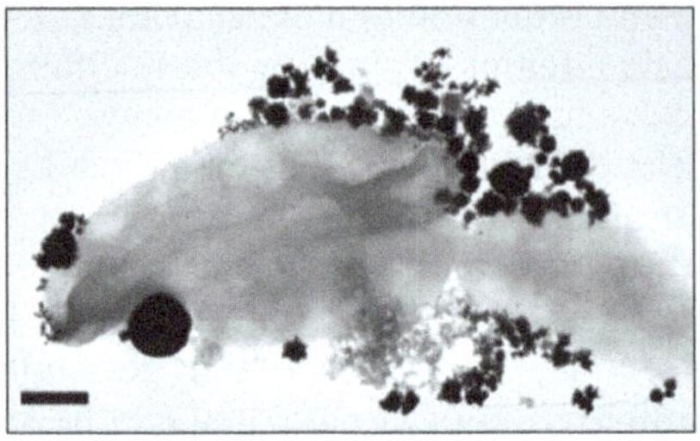

Figure 10.9 Transmission electron microscopy image of bimetallic palladium–platinum nanoparticles (Pd–Pt NPs) synthesized on the cell surface of *D. alaskensis*. Scale bar = 200 nm (Capeness *et al.*, 2019a).

(Xu *et al.*, 2019). *D. alaskensis* produces elemental Pd–Pt NPs extracellularly (Figure 10.9), where a core of platinum is surrounded by palladium (Capeness *et al.*, 2019a).

In order to produce core shell Pd–Ru bimetallic NPs, Gomez-Bolivar *et al.* (2019) cultured *E. coli* bacteria first in a Pd^{2+} solution to reduce Pd on the surface of the cells under H_2. Next, the biomass containing Pd^0 was again suspended in Ru^{3+} for the production of Pd–Ru NPs. Small metallic NPs were found intracellularly. However Pd–Ru NPs were primarily synthesized at the cell surface (Gomez-Bolivar *et al.*, 2019). *B. benzeovorans* was also reported to synthesize bimetallic Pd/Ru NPs (Omajali *et al.*, 2019).

10.4 FACTORS TO BE OPTIMIZED FOR INDUSTRIAL PRODUCTION OF PGM NPs USING BACTERIA

The bacteriogenic synthesis of PGM NPs can be made more efficient, scalable, and controllable by understanding and optimizing certain factors which are summarized below (Campaña *et al.*, 2023; Capeness & Horsfall, 2020; Carmona *et al.*, 2023; Koul *et al.*, 2021).

10.4.1 Bacterial species

The NP synthesis potential of bacteria varies according to their species or genes and their intrinsic properties (Tsekhmistrenko *et al.*, 2020). These include growth rate, biochemical pathways and enzymatic activities, and resistance to the metallic ions. Even different strains of the same bacterium may differ in their capacity to synthesize NPs. This is due to their genetic variations or environmental adaptation. Environmental factors that can affect the bacterial ability to produce NPs include exposure to heavy metals, oxidative stress, and variations in osmotic pressure.

Synthetic biology can improve the bacterial chassis for NP production using genetic engineering strategies to improve the bacteriogenic synthesis of NPs. By modifying bacterial genomes, improving metabolic pathways, and adjusting NP properties, synthetic biology allows for control over NP synthesis. It can also increase biomass production and creates designer microbiomes for better NP synthesis. Furthermore, synthetic biology allows for real-time monitoring and control of synthesis parameters for NPs (Capeness & Horsfall, 2020).

10.4.2 Metal source

The precursor or starting metal source plays a crucial role in the bacteriogenic synthesis of metallic NPs (Campaña *et al.*, 2023). They directly interact with bacterial cells and affect the reduction process. Parameters of the precursor metals such as biocompatibility, solubility, and catalytic activity have profound impact on the interaction with the bacterial cells. Additionally, it affects the nucleation and growth kinetics of NPs, ultimately shaping their size, shape, and dispersity. Therefore, the precursor metal should be carefully selected for the efficient synthesis of NPs with desired properties.

The concentration of precursor metal ions impacts NP synthesis in several ways. Usually, higher levels accelerate the bacterial uptake of metal ions which can result in the production of more NPs. Conversely, lower concentrations reduce the availability of metal ions for the bacteria which results in the slower synthesis of NPs with reduced sizes. A further consideration is that the metal concentration should be within the threshold range of the targeted bacteria, so that its growth is not inhibited upon the toxic exposure. Hence, to control the structural features and yield of NPs, the metal ion concentration should be considered as a crucial factor.

Longer exposure times enhance the chance of bacteria–metal interactions as well as the uptake of ions which subsequently leads to the reduction of metal ions to form NPs with higher yields. Conversely, shorter incubation times may result in incomplete reduction of metal ions and lower NP production.

10.4.3 Growth and culture conditions

The production of NPs can be significantly impacted by bacterial growth parameters such as pH, temperature, oxygen content, and availability of nutrients. Depending on the type of bacteria and the required characteristics of the NPs, growth conditions should be optimized. Appropriate design of a bioreactor can improve the efficiency of bacterial synthesis of NPs and it is a vital factor in the transition from lab scale to industrial scale operations for the production of biogenic NPs.

For the bacterial synthesis of NPs, pH can play an important role as it changes the chemical environment of the cells. Optimum pH conditions are essential for bacteria to effectively synthesize NPs because more extreme pH levels can interfere with bacterial metabolism and enzyme activity and what form the metal ions are found in. This in turn can affect the structural properties of the synthesized NPs and the production efficiency.

Temperature influences NP synthesis by bacteria by affecting metabolic activities, enzyme secretion, cell growth, and gene expression. Optimal temperatures can enhance enzyme efficiency, promote cell growth, and facilitate the reduction of metal ions and even their shape, leading to increased NP production. Conversely, high temperatures can cause high expense for the process and also interrupt the cellular processes, denature proteins, and lower the chance of NP synthesis.

10.5 PURIFICATION, CHARACTERIZATION, AND APPLICATION OF NPs

NP purification and characterization are pivotal to understand their physical, chemical, and structural properties, which ultimately determine their possible application (Mourdikoudis *et al.*, 2018).

10.5.1 Purification of NPs

The purification of platinum group NPs is achieved through several steps possibly including cell lysis, centrifugation, filtration, size exclusion chromatography, or density gradient centrifugation. Sometimes, additional steps such as chemical treatments are required to remove stabilizing agents or surface-modifying ligands. There could be some challenges during purification of bacteriogenic NPs which includes contamination, agglomeration, size distribution variations, surface modifications interfering with analysis, losses during purification, chemical transformations, and limitations of detection by certain techniques (Jeyaraj *et al.*, 2019).

With regards to the intracellular and extracellular bacteriogenic synthesis of NPs, extracellular, cell lysate, or cell-free synthesis is preferable due to its simplicity and the resulting NPs, requiring less complex downstream processing. In contrast, intracellularly synthesized NPs often require additional purification steps to separate the NPs from the cytosol and ultrasound treatment to reduce agglomeration. Bacteriogenic-synthesized NPs use bacterial-derived materials for nucleation and capping which can affect the properties of NPs, either positively or negatively. Depending on the application, those materials may be required to be removed from the NPs. Purification procedures include dialysis, filtering, and sucrose density centrifugation (Campaña *et al.*, 2023; Carmona *et al.*, 2023). A third option with cell-surface display of NPs is also possible, the cells together with the NPs can be taken and used. The microbial cells act as the support for the NPs, making NP retention and recovery easier, and preventing agglomeration of the NPs as they are fixed to the cells.

10.5.2 Characterization of NPs

NP characterization is crucial because their size, shape, surface properties, and composition directly influence their performance, stability, and interactions in specific applications. Smaller NPs have larger surface area-to-volume ratios, increasing their reactivity and cell absorption. The shape of the NPs also impacts their biodistribution and cellular absorption. Therefore, characterizing platinum group NPs especially in terms of their size, shape, and surface chemistry is crucial for understanding their practical applications. The optimization of purification protocols and careful selection of characterization methods are essential to overcome these challenges and ensure accurate analysis of PGM NPs. Some commonly used techniques for characterizing PGM NPs are summarized in Table 10.3 (Jeyaraj *et al.*, 2019).

10.5.3 Applications of bacterially synthesized PGM NPs

PGM-based NPs are promising candidates for applications in different fields owing to their remarkable catalytic, electrical, optical, and magnetic capabilities.

Table 10.3 Characterization techniques for platinum group metallic NPs.

Feature	Technique
Size	Transmission electron microscopy (TEM), atomic force microscopy (AFM), X-ray diffraction (XRD)
Size distribution	Scanning electron microscopy (SEM), inductively coupled plasma optical emission spectroscopy (ICP-OES), dynamic light scattering (DLS)
Shape	Scanning electron microscopy (SEM) and/or transmission electron microscopy (TEM), dynamic light scattering (DLS)
Elemental composition	Energy-dispersive X-ray spectroscopy (EDS), inductively coupled plasma mass spectrometry (ICP-MS), X-ray diffraction (XRD)
Surface chemistry	UV–visible spectroscopy, Fourier-transform infrared (FTIR) spectroscopy, zeta potential analysis
Plasmonic activity	UV–visible spectroscopy
Electrochemical and catalytic properties	Cyclic voltammetry, chronoamperometry, Brunauer–Emmett–Teller (BET) analysis

Some of the key areas where bacteriogenic platinum and palladium NPs could potentially be applied are discussed below.

10.5.3.1 Environmental application

Owing to the catalytic potential, platinum group NPs could be employed in environmental pollution remediation. They can be used for reductive removal of heavy metals and other organic contaminants such as chromium, nitrophenol, trichloromethane, and azo-based dyes in water (Xu *et al.*, 2019). They aid in the transformation of dangerous pollutants into less toxic or inactive forms, which helps to purify water supplies (Bhavya *et al.*, 2021). Moreover, biogenic platinum and palladium NPs have also been reported to remove pharmaceutical products from wastewaters such as diclofenac, antibiotics, and endocrine-disrupting compounds, which are serious concern for the water treatment. Additionally, by aiding in the transformation of hazardous gases such as carbon monoxide, nitrogen oxides, and volatile organic compounds into less hazardous forms, these NPs can be used in catalytic converters to reduce air pollution. Palladium NPs are also used in soil remediation for the breakdown of persistent organic pollutants (Kapoor *et al.*, 2021).

10.5.3.2 Catalytic activities

Bacteriogenic platinum and palladium NPs demonstrated catalytic activities in a number of organic transformations, including oxidation, hydrogenation, and C–C bond formation reactions. For instance, Suzuki–Miyaura cross-coupling as well as Sonogashira and Heck couplings enable the synthesis of complex organic molecules with high selectivity and efficiency (Era *et al.*, 2021, 2022). They offer a greener and more efficient process for production of pharmaceuticals, fine and bulk chemicals as well as agrochemicals (Capeness *et al.*, 2019a; Era *et al.*, 2021, 2022).

10.5.3.3 Sustainable and clean energy sources

In fuel cell technology, bacterial NPs operate as catalysts, assisting the electrochemical reactions that transform chemical energy into electrical energy (Jiang *et al.*, 2014). They improve the efficiency and functionality of fuel cells by catalysing the reduction of oxygen at the cathode and the oxidation of fuel (such as hydrogen) at the anode (Stephen *et al.*, 2019). These NPs also reduce the energy barrier and play a role in the hydrogen evolution reaction through water electrolysis or photoelectrochemical water splitting. Thus, platinum group NPs could be potent for application in the field of sustainable clean energy (Law *et al.*, 2022).

10.5.3.4 Sensor technology

PGM NPs are widely used in sensor technology due to their sensitivity, stability, and selectivity towards target analytes (Segev-Bar & Haick, 2013). In industrial and environmental monitoring applications, they are used in gas sensors to detect harmful gases such as hydrogen sulfide, carbon monoxide, and nitrogen dioxide. These NPs can also detect biomolecules including glucose, DNA, and proteins, providing quick and precise diagnostic capabilities in the fields of biotechnology (Chen *et al.*, 2022a, 2022b; Law *et al.*, 2022; Vijayaram *et al.*, 2024).

10.5.3.5 Biomedical application

Pharmaceutical chemicals and drug intermediates can be synthesized using bacteriogenic platinum group NPs (Bloch *et al.*, 2021). They are advantageous for pharmaceutical applications because of their strong catalytic activity, gentle reaction conditions, and biocompatibility. These attributes of the NPs facilitate the efficient encapsulation and controlled release of therapeutic agents, offering potential in drug delivery, tissue engineering, gene therapy, and regenerative medicine (Jeyaraj *et al.*, 2019). Moreover, having superior imaging qualities these NPs can be used as contrast agents in medical imaging procedures including computed tomography scans and magnetic resonance imaging. Platinum-based NPs can also have therapeutic applications, such as in chemotherapy for treatment of ovarian and testicular cancer (Mikhailova, 2022).

10.6 FUTURE PERSPECTIVES

There are still many features of the bacteriogenic synthesis of platinum group metallic NPs yet to be explored, in particular the mechanistic aspects have not been fully understood. Besides platinum and palladium, the other four metals of the group rhodium, ruthenium, osmium, and iridium have also potential catalytic application which should be explored for green bacterial synthesis. The main challenges for bacteriogenic synthesis of platinum group NPs are: (i) optimizing bacterial growth conditions for efficient NP production, (ii) controlling NP size and shape, (iii) minimizing contamination, ensuring biocompatibility and safety, and (iv) scaling up production methods for industrial applications. Therefore, more thorough research is required to depict the precise mechanisms as well as the proteins and enzymes involved in the manufacturing of NPs. Using bacteria in the large-scale production of NPs is attractive because it eliminates the need for costly toxic chemicals in the stabilization and synthesis processes. This needs to be undertaken in concert with thorough life cycle sustainability assessment to ensure developments are beneficial to society and the environment. Interdisciplinary research and collaboration might unlock the full potential of the bacterial synthesis of PGM NPs and their applications in a wide array of different fields.

10.7 CONCLUSION

The bacteriogenic synthesis of NPs of PGMs offers a transformative and environmentally sustainable approach with immense potential across nanotechnology and various industrial sectors. Investigating different bacterial species provides a deeper understanding of their diverse mechanisms for NP synthesis, ranging from enzymatic reduction to the secretion of bioactive molecules that mediate NP formation. In addition to providing an economical and environmentally friendly alternative for conventional physicochemical techniques, these bio-based technologies make it possible to produce NPs with precise and application-specific characteristics. The efficiency and scalability of NP synthesis of PGMs can be greatly increased by combining synthetic biology and nanotechnology, opening the door to increased yields and better control over the characteristics of NPs. This synergy further underscores the role of bacteriogenic synthesis in promoting sustainable practices and reducing the environmental footprint of nanomaterial production. In summary, bacterial synthesis of PGM NPs not only advances the field of sustainable nanotechnology but also aligns with global efforts towards achieving environmental sustainability, offering a promising pathway for both innovation and ecological balance.

REFERENCES

Altammar K. A. (2023). A review on nanoparticles: characteristics, synthesis, applications, and challenges. *Frontiers in Microbiology*, **14**, 1155622, https://doi.org/10.3389/fmicb.2023.1155622

Aritonang H. F., Onggo D., Ciptati C. and Radiman C. L. (2014). Synthesis of platinum nanoparticles from K_2PtCl_4 solution using bacterial cellulose matrix. *Nanoparticles*, **285954**, 1–6.

Arora N., Thangavelu K. and Karanikolos G. N. (2020). Bimetallic nanoparticles for antimicrobial applications. *Frontiers in Chemistry*, **8**, 412, https://doi.org/10.3389/fchem.2020.00412

Attard G., Casadesús M., Macaskie L. E. and Deplanche K. (2012). Biosynthesis of platinum nanoparticles by *Escherichia coli* MC4100: can such nanoparticles exhibit intrinsic surface enantioselectivity? *Langmuir: The ACS Journal of Surfaces and Colloids*, **28**, 5267–5274, https://doi.org/10.1021/la204495z

Bachar O., Meirovich M. M., Kurzion R. and Yehezkeli O. (2020). In vivo and in vitro protein mediated synthesis of palladium nanoparticles for hydrogenation reactions. *Chemical Communications*, **56**, 11211–11214, https://doi.org/10.1039/D0CC04812G

Baskaran B., Muthukumarasamy A., Chidambaram S., Sugumaran A., Ramachandran K. and Rasu Manimuthu T. (2017). Cytotoxic potentials of biologically fabricated platinum nanoparticles from *Streptomyces* sp. on MCF-7 breast cancer cells. *IET Nanobiotechnology*, **11**, 241–246, https://doi.org/10.1049/iet-nbt.2016.0040

Bayda S., Adeel M., Tuccinardi T., Cordani M. and Rizzolio F. (2019). The history of nanoscience and nanotechnology: From chemical–physical applications to nanomedicine. *Molecules*, **25**, 112, https://doi.org/10.3390/molecules25010112

Bhavya G., Belorkar S. A., Mythili R., Geetha N., Shetty H. S., Udikeri S. S. and Jogaiah S. (2021). Remediation of emerging environmental pollutants: A review based on advances in the uses of eco-friendly biofabricated nanomaterials. *Chemosphere*, **275**, 129975, https://doi.org/10.1016/j.chemosphere.2021.129975

Bloch K., Pardesi K., Satriano C. and Ghosh S. (2021). Bacteriogenic platinum nanoparticles for application in nanomedicine. *Frontiers in Chemistry*, **9**, 624344, https://doi.org/10.3389/fchem.2021.624344

Brayner R., Barberousse H., Hemadi M., Djedjat C., Yéprémian C., Coradin T., Livage J., Fiévet F. and Couté A. (2007). Cyanobacteria as bioreactors for the synthesis of Au, Ag, Pd, and Pt nanoparticles via an enzyme-mediated route. *Journal of Nanoscience and Nanotechnology*, **7**, 2696–2708, https://doi.org/10.1166/jnn.2007.600

Campaña A. L., Saragliadis A., Mikheenko P. and Linke D. (2023). Insights into the bacterial synthesis of metal nanoparticles. *Frontiers in Nanotechnology*, **5**, https://doi.org/10.3389/fnano.2023.1216921

Capeness M. J. and Horsfall L. E. (2020). Synthetic biology approaches towards the recycling of metals from the environment. *Biochemical Society Transactions*, **48**, 1367–1378, https://doi.org/10.1042/BST20190837

Capeness M. J., Edmundson M. C. and Horsfall L. E. (2015). Nickel and platinum group metal nanoparticle production by *Desulfovibrio alaskensis* G20. *New Biotechnology*, **32**, 727–731, https://doi.org/10.1016/j.nbt.2015.02.002

Capeness M. J., Echavarri-Bravo V. and Horsfall L. E. (2019a). Production of biogenic nanoparticles for the reduction of 4-nitrophenol and oxidative laccase-like reactions. *Frontiers in Microbiology*, **10**, 997, https://doi.org/10.3389/fmicb.2019.00997

Capeness M. J., Imrie L., Muhlbauer L. F., Le Bihan T. and Horsfall L. E. (2019b). Shotgun proteomic analysis of nanoparticle-synthesizing *Desulfovibrio alaskensis* in response to platinum and palladium. *Microbiology*, **165**, 1282–1294, https://doi.org/10.1099/mic.0.000840

Carmona M., Poblete-Castro I., Rai M. and Turner R. J. (2023). Opportunities and obstacles in microbial synthesis of metal nanoparticles. *Microbial Biotechnology*, **16**, 871–876, https://doi.org/10.1111/1751-7915.14254

Chen Y., Chen Y., Wu J. and Zhang J. (2018). The effect of biotic and abiotic environmental factors on Pd(II) adsorption and reduction by *Bacillus wiedmannii* MSM. *Ecotoxicology and Environmental Safety*, **162**, 546–553, https://doi.org/10.1016/j.ecoenv.2018.07.043

Chen Y., Chen Y., Wang B., Mao C., Liu C., Liu R. and Yan B. (2022a). From trash to treasure: microbial conversion of palladium contaminants into valuable Pd nanoparticles by *Bacillus thuringiensis* Y9. *Journal of Cleaner Production*, **366**.

Chen Y., Chen Z., Yang D., Zhu L., Liang Z., Pang Y. and Zhou L. (2022b). Novel microbial palladium nanoparticles with a high photothermal effect for antibacterial applications. *ACS Omega*, **8**, 1534–1541, https://doi.org/10.1021/acsomega.2c07037

Choi Y., Park T. J., Lee D. C. and Lee S. Y. (2018). Recombinant *Escherichia coli* as a biofactory for various single- and multi-element nanomaterials. *Proceedings of the National Academy of Sciences of the United States of America*, **115**, 5944–5949, https://doi.org/10.1073/pnas.1804543115

Coursolle D. and Gralnick J. A. (2010). Modularity of the Mtr respiratory pathway of *Shewanella oneidensis* strain MR-1. *Molecular Microbiology*, **77**, 995–1008, https://doi.org/10.1111/j.1365-2958.2010.07266.x

Cueva M. E. and Horsfall L. E. (2017). The contribution of microbially produced nanoparticles to sustainable development goals. *Microbial Biotechnology*, **10**, 1212–1215, https://doi.org/10.1111/1751-7915.12788

De Corte S., Hennebel T., De Gusseme B., Verstraete W. and Boon N. (2012). Bio-palladium: From metal recovery to catalytic applications. *Microbial Biotechnology*, **5**, 5–17, https://doi.org/10.1111/j.1751-7915.2011.00265.x

Deplanche K. (2008). New Nanocatalysts Made by Bacteria from Metal Solutions and Recycling of Metal Wastes. *PhD Thesis*. Birmingham: University of Birmingham. http://ethos.bl.uk/OrderDetails.do?uin=uk.bl.ethos.532277

Deplanche K., Bennett J. A., Mikheenko I. P., Omajali J., Wells A. S., Meadows R. E., Wood J. and Macaskie L. E. (2014). Catalytic activity of biomass-supported Pd nanoparticles: Influence of the biological component in catalytic efficacy and potential application in 'green' synthesis of fine chemicals and pharmaceuticals. *Applied Catalysis B: Environmental*, **147**, 651–665, https://doi.org/10.1016/j.apcatb.2013.09.045

Durán N. and Seabra A. B. (2012). Metallic oxide nanoparticles: State of the art in biogenic syntheses and their mechanisms. *Applied Microbiology and Biotechnology*, **95**, 275–288, https://doi.org/10.1007/s00253-012-4118-9

Edmundson M. C., Capeness M. and Horsfall L. E. (2014). Exploring the potential of metallic nanoparticles within synthetic biology. *New Biotechnology*, **31**, 572–578, https://doi.org/10.1016/j.nbt.2014.03.004

Era Y., Dennis J. A., Wallace S. and Horsfall L. E. (2021). Micellar catalysis of the Suzuki Miyaura reaction using biogenic Pd nanoparticles from *Desulfovibrio alaskensis*. *Green Chemistry*, **23**, 8886–8890, https://doi.org/10.1039/D1GC02392F

Era Y., Dennis J. A., Horsfall L. E. and Wallace S. (2022). Palladium nanoparticles from *Desulfovibrio alaskensis* G20 catalyze biocompatible Sonogashira and biohydrogenation cascades. *JACS Au*, **2**, 2446–2452, https://doi.org/10.1021/jacsau.2c00366

Eramabadi P., Masoudi M., Makhdoumi A., Mashreghi M. (2020). Microbial cell lysate supernatant (CLS) alteration impact on platinum nanoparticles fabrication, characterization, antioxidant and antibacterial activity. *Materials science & engineering. C, Materials for biological applications*, **117**, 111292, https://doi.org/10.1016/j.msec.2020.111292

Fahmy S. A., Preis E., Bakowsky U. and Azzazy H. M. E. (2020). Platinum nanoparticles: green synthesis and biomedical applications. *Molecules*, **25**, 4981.

Feynman R. (1960). There's plenty of room at the bottom. *Engineering Science*, **23**, 22–36.

Gaidhani S. V., Yeshvekar R. K., Shedbalkar U. U., Bellare J. H. and Chopade B. A. (2014). Bio-reduction of hexachloroplatinic acid to platinum nanoparticles employing *Acinetobacter calcoaceticus*. *Process Biochemistry*, **49**, 2313–2319, https://doi.org/10.1016/j.procbio.2014.10.002

Gomez-Bolivar J., Mikheenko I. P., Orozco R. L., Sharma S., Banerjee D., Walker M., Hand R. A., Merroun M. L. and Macaskie L. E. (2019). Synthesis of Pd/Ru bimetallic nanoparticles by *Escherichia coli* and potential as a catalyst for upgrading 5-hydroxymethyl furfural into liquid fuel precursors. *Frontiers in Microbiology*, **10**, 1276, https://doi.org/10.3389/fmicb.2019.01276

Habibullah G., Viktorova J. and Ruml T. (2021). Current strategies for noble metal nanoparticle synthesis. *Nanoscale Research Letters*, **16**, 47, https://doi.org/10.1186/s11671-021-03480-8

Hosseinkhani B., Søbjerg L. S., Rotaru A. E., Emtiazi G., Skrydstrup T. and Meyer R. L. (2012). Microbially supported synthesis of catalytically active bimetallic Pd-Au nanoparticles. *Biotechnology and Bioengineering*, **109**, 45–52, https://doi.org/10.1002/bit.23293

Hulkoti N. I. and Taranath T. C. (2014). Biosynthesis of nanoparticles using microbes – A review. *Colloids and Surfaces B: Biointerfaces*, **121**, 474–483, https://doi.org/10.1016/j.colsurfb.2014.05.027

Isaacoff B. P. and Brown K. A. (2017). Progress in top-down control of bottom-up assembly. *Nano Letters*, **17**, 6508–6510, https://doi.org/10.1021/acs.nanolett.7b04479

Jeevanandam J., Barhoum A., Chan Y. S., Dufresne A. and Danquah M. K. (2018). Review on nanoparticles and nanostructured materials: history, sources, toxicity and regulations. *Beilstein Journal of Nanotechnology*, **9**, 1050–1074, https://doi.org/10.3762/bjnano.9.98

Jeyaraj M., Gurunathan S., Qasim M., Kang M. H. and Kim J. H. (2019). A comprehensive review on the synthesis, characterization, and biomedical application of platinum nanoparticles. *Nanomaterials (Basel)*, **9**, 1719, https://doi.org/10.3390/nano9121719

Jiang X., Hu J., Lieber A. M., Jackan C. S., Biffinger J. C., Fitzgerald L. A., Ringeisen B. R. and Lieber C. M. (2014). Nanoparticle facilitated extracellular electron transfer in microbial fuel cells. *Nano Letters*, **14**, 6737–6742, https://doi.org/10.1021/nl503668q

Kapoor R. T., Salvadori M. R., Rafatullah M., Siddiqui M. R., Khan M. A. and Alshareef S. A. (2021). Exploration of microbial factories for synthesis of nanoparticles – A sustainable approach for bioremediation of environmental contaminants. *Frontiers in Microbiology*, **12**, 658294, https://doi.org/10.3389/fmicb.2021.658294

Kimber R. L., Parmeggiani, F., Neill T. S., Merroun M. L., Goodlet G., Powell N. A., Turner N. J. and Lloyd J. R. (2021). Biotechnological synthesis of Pd/Ag and Pd/Au nanoparticles for enhanced Suzuki-Miyaura cross-coupling activity. *Microbial Biotechnology*, **14**, 2435–2447, https://doi.org/10.1111/1751-7915.13762

Konishi Y., Ohno K., Saitoh N., Nomura T., Nagamine S., Hishida H., Takahashi Y. and Uruga T. (2007). Bioreductive deposition of platinum nanoparticles on the bacterium *Shewanella algae*. *Journal of Biotechnology*, **128**, 648–653, https://doi.org/10.1016/j.jbiotec.2006.11.014

Koul B., Poonia A. K., Yadav D. and Jin J-O. (2021). Microbe-mediated biosynthesis of nanoparticles: applications and future prospects. *Biomolecules*, **11**, 886, https://doi.org/10.3390/biom11060886

Law C. K. Y., Bonin L., Gusseme B. D., Boon N. and Kundu K. (2022). Biogenic synthesis of palladium nanoparticles: New production methods and applications. *Nanotechnology Reviews*, **11**, 3101–3124. https://doi.org/10.1515/ntrev-2022-0482

Lengke M. F., Fleet M. E. and Southam G. (2006). Synthesis of platinum nanoparticles by reaction of filamentous cyanobacteria with platinum(IV)–chloride complex. *Langmuir: The ACS Journal of Surfaces and Colloids*, **22**, 7318–7323, https://doi.org/10.1021/la060873s

Lv Q., Zhang B., Xing X., Zhao Y., Cai R., Wang W. and Gu Q. (2018). Biosynthesis of copper nanoparticles using *Shewanella loihica* PV-4 with antibacterial activity: Novel approach and mechanisms investigation. *Journal of Hazardous Material*, **347**, 141–149.

Malik S., Muhammad K. and Waheed Y. (2023). Nanotechnology: A revolution in modern industry. *Molecules*, **28**, 661.

Martinez-Mora O., Pozo G., Leon-Fernandez L. F., Fransaerb J. and Dominguez-Benetton X. (2023). Synthesis of platinum group metal nanoparticles assisted by CO_2 reduction and H_2 cogeneration at gas-diffusion electrodes. *RSC Sustainability*, **1**, 454–458, https://doi.org/10.1039/D3SU00046J

Martins M., Mourato C., Sanches S., Noronha J. P., Crespo M. T. B. and Pereira I. A. C. (2017). Biogenic platinum and palladium nanoparticles as new catalysts for the removal of pharmaceutical compounds. *Water Research*, **108**, 60–168, https://doi.org/10.1016/j.watres.2016.10.071

Matsena M. T., Tichapondwa S. M. and Chirwa E. M. N. (2020). Synthesis of biogenic palladium nanoparticles using Citrobacter sp. for application as anode electrocatalyst in a microbial fuel cell. *Catalysts*, **10**, 838, https://doi.org/10.3390/catal10080838

Mikhailova E. O. (2022). Green synthesis of platinum nanoparticles for biomedical applications. *Journal of Functional Biomaterials*, **13**, 260, https://doi.org/10.3390/jfb13040260

Mikheenko I. P., Rousset M., Dementin S. and Macaskie L.E. (2008). Bioaccumulation of palladium by *Desulfovibrio fructosivorans* wild-type and hydrogenase-deficient strains. *Applied and Environmental Microbiology*, **74**, 6144–6146, https://doi.org/10.1128/AEM.02538-07

Mourdikoudis S., Pallares R. M. and Thanh N. T. K. (2018). Characterization techniques for nanoparticles: Comparison and complementarity upon studying nanoparticle properties. *Nanoscale*, **10**, 12871–12934, https://doi.org/10.1039/C8NR02278J

Omajali J. B., Mikheenko I. P., Merroun M. L., Wood J. and Macaskie L. E. (2015). Characterization of intracellular palladium nanoparticles synthesized by *Desulfovibrio desulfuricans* and *Bacillus benzeovorans*. *Journal of Nanoparticle Research*, **17**, 264, https://doi.org/10.1007/s11051-015-3067-5

Omajali J. B., Gomez-Bolivar J., Mikheenko I. P., Sharma S., Kayode B., Al-Duri B., Banerjee D., Walker M., Merroun M. L. and Macaskie L. E. (2019). Novel catalytically active Pd/Ru bimetallic nanoparticles synthesized by *Bacillus benzeovorans*. *Scientific Reports*, **9**, 4715, https://doi.org/10.1038/s41598-019-40312-3

Riddin T., Gericke M. and Whiteley C. G. (2010). Biological synthesis of platinum nanoparticles: Effect of initial metal concentration. *Enzyme and Microbial Technology*, **46**, 501–505, https://doi.org/10.1016/j.enzmictec.2010.02.006

Rzelewska-Piekut M., Wiecka M. and Regel-Rosocka M. (2022). Studies on the formation of catalytically active PGM nanoparticles from model solutions as a basis for the recycling of spent catalysts. *Molecules*, **27**, 390, https://doi.org/10.3390/molecules27020390

Sasireka K. S. and Lalitha P. (2021). Biogenic synthesis of bimetallic nanoparticles and their applications. *Reviews in Inorganic Chemistry*, **41**, 223–244, https://doi.org/10.1515/revic-2020-0024

Segev-Bar M. and Haick H. (2013). Flexible sensors based on nanoparticles. *ACS Nano*, **7**, 8366–8378, https://doi.org/10.1021/nn402728g

Stephen A. J., Rees N. V., Mikheenko I. and Macaskie L. E. (2019). Platinum and palladium bio-synthesized nanoparticles as sustainable fuel cell catalysts. *Frontiers in Energy Research*, **7**, https://doi.org/10.3389/fenrg.2019.00066

Tan L., Ray Jones T., Poitras J., Xie J., Liu X. and Southam G. (2020). Biochemical synthesis of palladium nanoparticles: The influence of chemical fixatives used in electron microscopy on nanoparticle formation and catalytic performance. *Journal of Hazardous Materials*, **398**, 122945, https://doi.org/10.1016/j.jhazmat.2020.122945

Taniguchi N., Arakawa C. and Kobayashi T. (1974). On the basic concept of nano-technology. International Conference on Production Engineering, Tokyo, Japan.

Tsekhmistrenko S. I., Bityutskyy V. S., Tsekhmistrenko O. S., Horalskyi L. P., Tymoshok N. O. and Spivak M. Y. (2020). Bacterial synthesis of nanoparticles: A green approach. *Biosystems Diversity*, **28**, 9–17, https://doi.org/10.15421/012002

Tuo Y., Liu G., Zhou J., Wang A., Wang J., Jin R. and Lv H. (2013). Microbial formation of palladium nanoparticles by *Geobacter sulfurreducens* for chromate reduction. *Bioresource Technology*, **133**, 606–611, https://doi.org/10.1016/j.biortech.2013.02.016

Tuo Y., Liu G., Dong B., Yu H., Zhou J., Wang J. and Jin R. (2017). Microbial synthesis of bimetallic PdPt nanoparticles for catalytic reduction of 4-nitrophenol. *Environmental Science Pollution Research*, **24**, 5249–5258, https://doi.org/10.1007/s11356-016-8276-7

Vaseghi Z., Nematollahzadeh A. and Tavakoli O. (2018). Green methods for the synthesis of metal nanoparticles using biogenic reducing agents: A review. *Reviews in Chemical Engineering*, **34**, 529–559, https://doi.org/10.1515/revce-2017-0005

Vijayaram S., Razafindralambo H., Sun Y. Z., Vasantharaj S., Ghafarifarsani H., Hoseinifar S. H. and Raeeszadeh M. (2024). Applications of green synthesized metal nanoparticles – A review. *Biological Trace Element Research*, **202**, 360–386, https://doi.org/10.1007/s12011-023-03645-9

Wahl R., Mertig M., Raff J., Selenska-Pobell S. and Pompe W. (2001). Electron-beam induced formation of highly ordered palladium and platinum nanoparticle arrays on the S layer of *Bacillus sphaericus* NCTC 9602. *Advanced Materials*, **13**, 736–740, https://doi.org/10.1002/1521-4095(200105)13:10<736::AID-ADMA736>3.0.CO;2-N

Xu H., Xiao Y., Xu M., Cui H., Tan L., Feng N., Liu X., Qiu G., Dong H. and Xie J. (2019). Microbial synthesis of Pd–Pt alloy nanoparticles using *Shewanella oneidensis* MR-1 with enhanced catalytic activity for nitrophenol and azo dyes reduction. *Nanotechnology*, **30**, 065607, https://doi.org/10.1088/1361-6528/aaf2a6

Yates M. D., Cusick R. D. and Logan B. E. (2013). Extracellular palladium nanoparticle production using *Geobacter sulfurreducens*. *ACS Sustainable Chemistry & Engineering*, **1**, 1165–1171, https://doi.org/10.1021/sc4000785

Zhang S., Zhou H., Liao H., Tan P., Tian W. and Pan J. (2022). Microbial synthesis of efficient palladium electrocatalyst with high loadings for oxygen reduction reaction in acidic medium. *Journal of Colloid and Interface Science*, **611**, 161–171, https://doi.org/10.1016/j.jcis.2021.12.080

Index

J

K

L

M

N

O

P

IWA Publishing's authorised EU representative for General Product Safety Regulations is Diane D'Arras, 15 rue Duret, 75116 Paris, France, e-mail: safety@iwap.co.uk.

Printed and bound by CPI Group (UK) Ltd, Croydon, CR0 4YY
12/05/2026
02108869-0005